D1532843

ORACLE 10*g*: SQL

ORACLE 10*g* : SQL

COURSE TECHNOLOGY
CENGAGE Learning™

Australia • Brazil • Japan • Korea • Mexico • Singapore • Spain • United Kingdom • United States

Oracle 10g: SQL
Joan Casteel

Senior Acquisitions Editor: Maureen Martin

Senior Product Manager: Alyssa Pratt

Developmental Editor: Betsey Henkels

Production Editor: Danielle Chouhan

Associate Product Manager: Jennifer Smith

Senior Marketing Manager: Karen Seitz

Editorial Assistant: Allison Murphy

Composition House: GEX Publishing Services

Cover Designer: Laura Rickenbach

Senior Manufacturing Coordinator: Justin
  Palmeiro

Quality Assurance Coordinators: Serge
  Palladino, Chris Scriver

Copyeditor: Mary Kemper

Proofreader: Karen Annett

Indexer: Joan Green

For product information and technology assistance, contact us at
Cengage Learning Customer & Sales Support, 1-800-354-9706

For permission to use material from this text or product, submit all requests online at **cengage.com/permissions**
Further permission questions can be emailed to
permissionrequest@cengage.com

ISBN-13: 978-1-4188-3629-0
ISBN-10: 1-4188-3629-X

Course Technology Cengage Learning
25 Thomson Place
Boston, Massachusetts, 02210
USA

Cengage Learning is a leading provider of customized learning solutions with office locations around the glode, including Singapore, the United Kingdom, Australia, Mexico, Brazil and Japan. Locate your office at:
international.cengage.com/region

Oracle is a registered trademark, and Oracle 10g, PL/SQL, and SQL*Plus are trademarks or registered trademarks of Oracle Corporation and/or its affiliates.

Boson and BOSONSOFTWARE are trademarks or registered trademarks of Boson Software, Inc. in the United States and certain other countries. All other trademarks are trademarks of their respective owners.

Cengage Learning products are represented in Canada by Nelson - Education, Ltd.

For your lifelong learning solutions, visit **course.cengage.com**

Visit our corporate website at **cengage.com**

Printed in Canada
5 6 7 8 9 10 09 08

TABLE OF CONTENTS

The last few decades have seen a proliferation of organizations that rely heavily on information technology. These organizations store their data in databases, and many choose Oracle® database management systems to access their data. The current Oracle database version, Oracle 10g, is an object-oriented database management system that allows users to create, manipulate, and retrieve data. In addition, Oracle 10g has increased database security and includes new features that ease the administrative duties associated with databases.

To allow user interaction, most database management systems support some aspect of the industry standard ANSI-SQL. Oracle 10g allows the JOIN keyword to support the ANSI approach to linking tables. The purpose of this textbook is to introduce the student to basic SQL commands used to interact with an Oracle 10g database in a business environment. In addition, concepts relating specifically to the objectives of the Oracle9i and Oracle 10g SQL certification exams have been incorporated in the text for those individuals wishing to pursue certification.

The Intended Audience

This textbook has been designed for students in technical two-year or four-year programs who need to learn how to interact with Oracle 10g databases. Although it is preferred that students have some understanding of database design, an introductory chapter has been included to review the basic concepts of E-R Modeling and the normalization process.

Oracle Certification Program (OCP)

This textbook covers the objectives of *Exam 1Z0-007, Introduction to Oracle9i: SQL* and the SQL portions of *1Z0-042 Oracle Database 10g: Administration*. Exam 1Z0-007 is optionally administered on-line. Information about registering for these exams, along with other reference material, can be found at **www.oracle.com/education/certification**.

The Approach

The concepts introduced in this textbook are presented in the context of a hypothetical "real world" business—an online book retailer named JustLee Books. First, the business operation and the database structure are introduced and analyzed. Then, as commands are introduced throughout the text, examples using the JustLee Books' database model the commands. This allows students to not only learn the syntax of a command, but also how it can be used in a real-world environment. In addition, a script file that generates the database is available to allow students hands-on practice in re-creating the examples and practicing variations of SQL commands to enhance their understanding.

To expose students to what a database is and how it is created, the initial focus of this textbook is on creating tables and learning how to perform data manipulation operations. Once the student is familiar with the database structure, the focus then turns to querying a database in the second half of the text. In chapters 8 through 14, students learn how to retrieve data from the database, using SELECT statements, functions, and subqueries.

To reinforce the material presented, each chapter includes a chapter summary and, when appropriate, a syntax guide of the commands covered in the chapter. In addition, at the end of each chapter, groups of activities are presented that test students' knowledge and challenge them to apply that knowledge to solving business problems. A continuous case study introduces a second database involving a city jail system to provide a second database scenario.

Overview of This Book

The examples, assignments, and cases in this book will help students to achieve the following objectives:

- Issue SQL commands that will retrieve data based on criteria specified by the user.

- Use SQL commands to join tables and retrieve data from the joined tables.

- Perform calculations based on data contained within the database.

- Use subqueries to retrieve data based on unknown conditions.

- Create, modify, and drop database tables.

- Manipulate data stored in database tables.

- Enforce business rules through the use of table constraints.

- Create users and assign the privileges necessary for a user to complete various tasks.

- Create printable reports through various SQL*Plus commands.

- Understand the basic concepts regarding SQL statement tuning

- Identify SQL differences amongst various databases

The contents of the chapters build in complexity while reinforcing previous ideas. **Chapter 1** introduces basic database management concepts, including database design. **Chapter 2** shows how to retrieve data from a table. **Chapter 3** presents how to create new database tables **Chapter 4** addresses the use of constraints to enforce business rules and to ensure the integrity of table data **Chapter 5**. explains adding data to a table, modifying existing data, and deleting data. **Chapter 6** shows how to use a sequence to generate numbers, create indexes to speed up data retrieval, and create synonyms to provide aliases for tables and views. **Chapter 7** steps the reader through creating user accounts and roles and demonstrates how to grant (and revoke) privileges to those accounts and roles. **Chapter 8** demonstrates how to restrict rows retrieved from a table, based on a given condition. **Chapter 9** presents how to link tables with common columns. **Chapter 10** details the various single-row functions supported by Oracle 10g. **Chapter 11** covers multiple-row functions used to derive a single value for a group of rows—and how to restrict groups of rows. **Chapter 12** covers the use of subqueries to retrieve rows based on an unknown condition already contained within the database. **Chapter 13** covers the use of views to restrict access to data and reduce the complexity of certain types of queries. **Chapter 14** reveals how to create printable reports, including grouping and subtotaling within the report.

Chapter 15 explores SQL in application development including tuning concepts and how SQL statements differ amongst databases.

The Appendices provide support and reinforcement for the chapter materials. **Appendix A** provides a printed version of the tables and data in the JustLee Books' database. This database serves as a sustained example from chapter to chapter. **Appendix B** introduces the operation of both the client-based and Internet-based versions of SQL*Plus software and identifies some of the execution differences that you may encounter when using *i*SQL*Plus versus client SQL*Plus. **Appendix C** provides a syntax guide of the commands presented within each chapter **Appendix D introduces you to a useful database utility tool,** TOAD (Tool for Oracle Application Developers).. **Appendix E** contains five practice exams that can be used to prepare for certification exams or to assess comprehension of groups of chapters. **Appendix F** provides a list of Oracle resources. Appendix G demonstrates two examples of file loads to introduce you to the SQL*Loader utility.

Features

To enhance students' learning experience, each chapter in this book includes the following elements:

- **Chapter Objectives:** Each chapter begins with a list of the concepts to be mastered by the chapter's conclusion. This list provides a quick overview of chapter contents as well serving as a useful study aid.

- **Running Case:** A sustained example, the business operation of JustLee Books, serves as the basis for introducing new commands and for practicing the material introduced in each chapter.

- **Methodology:** As new commands are presented in each chapter, the syntax of the command is presented and then an example, using the JustLee Books' database, illustrates the command in the context of a business operation. This methodology shows the student not only *how* the command is used but also *when* and *why* it is used. The script file used to create the database is available so students can work through the examples in this textbook, engendering a hands-on environment in which students can reinforce their knowledge of chapter material.

- **Tip:** This feature, designated by the *Tip* icon, provides students with practical advice. In some instances, Tips explain how a concept applies in the workplace.

- **Note:** These explanations, designated by the *Note* icon, provide further information about loading files and operations with the databases.

- **Caution:** This warning, designated by the *Caution* icon, call out database operations that, if misused, could have devastating results.

- **Chapter Summaries:** Each chapter's text is followed by a summary of chapter concepts. These summaries are a helpful recap of chapter contents.

- **Syntax Summaries:** Beginning with Chapter 2, a Syntax Guide table is given after each Chapter Summary. It recaps the command syntax presented in the chapter.

- **Review Questions:** End-of-chapter assessment begins with a set of review questions that reinforce the main ideas introduced in each chapter. These questions ensure that students have mastered the concepts and understand the information presented.

- **Multiple Choice Questions:** Each chapter contains multiple choice questions that cover the material presented within the chapter. Oracle certification-type questions are included to prepare students for the type of questions that can be expected on a certification exam, as well as to measure the students' level of understanding.

- **Hands-on Assignments:** Along with conceptual explanations and examples, each chapter provides hands-on assignments related to the chapter's contents. The purpose of these assignments is to provide students with practical experience. In most cases, the assignments are based on the JustLee Books' database and serve as a continuation of the examples given within the chapter.

- **Advanced Challenge:** This section provides another problem regarding the JustLee Book's database which is larger in scope than the Hands-on assignments.

- **Case Study:** One major case is presented at the end of each chapter. These cases are designed to help students apply what they have learned to real-world situations. The cases give students the opportunity to independently synthesize and evaluate information, examine potential solutions, and make recommendations, much as students will do in an actual business situation. The case uses a database based on a city jail system.

The Oracle Database 10g (Version 10.1.0) CD, which is in the envelope adhered to this book, enables users to install this software on their own computers at home. Users can then connect to either an Oracle 10g Enterprise Edition, Standard Edition, or Personal Edition database. You can use the Database 10g software with Microsoft Windows NT, Windows 2000 Professional or Server, Windows 2003 Server, and Windows XP Professional operating systems. The installation and configuration instructions for Database 10g are available at *www.course.com/cdkit*. Look for this book's title and front cover, and click the link to access the information specific to this book.

Before proceeding to use the software, **you must** register the software and agree to the Oracle Technology Network Developer License Terms in order to receive the key code to unlock the software. Please go to *http://otn.oracle.com/books/*. Upon registering the software, you agree that Oracle may contact you for marketing purposes. You also agree that any information you provide Oracle may be used for marketing purposes.

Supplemental Materials

The following supplemental materials are available when this book is used in a classroom setting. All teaching tools available with this book are provided to the instructor on a single CD-ROM as well as on the Thomson Course Technology Web site at www.course.com.

- **Electronic Instructor's Manual:** The Instructor's Manual that accompanies this textbook includes the following elements:
  - Additional instructional material to assist in class preparation, including suggestions for lecture topics.
  - A sample syllabus.
  - When appropriate, information about potential problems that can occur in networked environments is identified.

- **ExamView®:** This objective-based test generator lets the instructor create paper, LAN, or Web-based tests from testbanks designed specifically for this Thomson Course Technology text. Instructors can use the QuickTest Wizard to create tests in fewer than five minutes by taking advantage of Thomson Course Technology's question banks—or create customized exams.

- **PowerPoint Presentations:** Microsoft PowerPoint slides are included for each chapter. Instructors might use the slides in three ways: As teaching aids during classroom presentations, as printed handouts for classroom distribution, or as network-accessible resources for chapter review. Instructors can add their own slides for additional topics introduced to the class.

- **Data Files:** The script files necessary to create the JustLee Books' and City Jail databases are provided through the Thomson Course Technology Web site at **www.course.com**, and is also available on the Teaching Tools CD-ROM. Additional script files needed for chapters 15 are also available through the Web site and the Instructor's Resource Kit.

- **Solution Files:** Solutions to the chapter examples, end-of-chapter review questions and multiple-choice questions, Hands-On Assignments, and the Case are provided on the Teaching Tools CD-ROM. Solutions may also be found on the Thomson Course Technology Web site at **www.course.com**. The solutions are password protected.

- **Figure Files:** Figure files allow instructors to create their own presentations using figures taken directly from the text. These are found on the Teaching Tools CD-ROM only.

ACKNOWLEDGMENTS

I feel fortunate that Thomson Course Technology pursued my authorship of this text and supported my efforts. I am one lucky person – I have two angels in heaven, my mother and grandmother, and one angel here on earth, Scott. Without them watching over me, I would not be able to tackle such challenges. I also want to thank my father who was more excited than me that I actually finished this book.

However, this text is the result of an incredible effort by many people whom I wish I had to opportunity to thank personally. In particular, I would like to thank Betsey Henkels, Development Editor, Danielle Chouhan, Production Editor, and Serge Palladino and Chris Scriver, Quality Assurance Testers. Hats off to all these folks who had to turn my writing into a successful learning tool and had to test all the coding examples. There is no way to express adequate gratitude for their "magic" performed, as these tasks play a most critical role in the development of an effective text.

I would like to express my appreciation to Maureen Martin, Senior Acquisitions Editor, Alyssa Pratt, Senior Product Manager, Mirella Misiaszek, Associate Product Manager, and Jennifer Smith, Associate Product Manager. This does not cover all the important people who made this text a reality, but I truly appreciate each and every effort made on behalf of this text.

In addition, I need to recognize the enormous contribution of colleagues and reviewers who provided helpful suggestions and insight into the development of this textbook. Reviewers include Dr. Douglas D. Bickerstaff of Eastern Washington University; Dr. Muhammad A. Razi of Haworth College of Business, Western Michigan University, Kalamazoo; and Professor Eli J. Weissman of DeVry Institute of Technology. And last, but not least, thanks to all my students who put up with all my ramblings and, in turn, teach me as well.

READ THIS BEFORE YOU BEGIN

TO THE USER

Data Files

To work through the examples and complete the projects in this book, you will need to load the data files created for this book. Your instructor will provide you with those data files, or you can obtain them electronically from the Thomson Course Technology Web site by accessing **www.course.com** and then searching for this book's title. The data files are designed to provide you with the same data shown in the chapter examples, so you can have hands-on practice re-creating the example queries and their output. The tables in the database can be reset if you encounter problems, such as accidentally deleting data. It is highly recommended that you work through all the examples to re-enforce your learning.

The database used throughout this book is created with the **Bookscript.sql** file. The file is located in the JustLee Database folder on your Data Disk. When you begin Chapter 2, you should run this file in Oracle 10g. (A printed version of the tables and data generated by that file are provided in Appendix A.). Many chapters will require a script to be executed and these files are found in their respective chapter folders (Chapter05, Chapter10, etc.) on your Data Disk and have the file names **prech#.sql**, where the number sign (#) is replaced by the chapter number e.g., **prech10.sql**. A note is included at the beginning of each of these chapters indicates the script file that should be executed. If the computer in your school lab—or your own computer—has Oracle 10g database software installed, you can work through the chapter examples and complete the Hands-on Assignments and Case projects. At a minimum, you will need the Oracle 9i Release 2 Personal Edition of the software to complete the examples and assignments in this textbook.

Using Your Own Computer

To use your own computer to work through the chapter examples and to complete the Hands-on Assignments and Case projects, you will need the following:

- Hardware: A computer capable of using the Microsoft Windows NT, 2000 Professional, or XP Professional operating system. You should have at least 500MB of RAM and between 2.75GB and 4.75GB of hard disk space available before installing the software.

- Software: Oracle 10g Personal Edition or Oracle9i Release 2 Personal Edition. The database software necessary to perform all the tasks shown in this textbook is included with the book. Detailed installation, configuration, and logon information for the software in this kit are provided at **www.course.com/cdkit** on the Web page for this title.

Before proceeding to use the software, **you must** register the software and agree to the Oracle Technology Network Developer License Terms in order to receive the key code to unlock the software. Please go to *http://otn.oracle.com/books/*. Upon registering the software, you agree that Oracle may contact you for marketing purposes. You also agree that any information you provide Oracle may be used for marketing purposes.

- When you install the Oracle software, you will be prompted to change the password for certain default administrative user accounts. Make certain that you record the names and passwords of the accounts because you may need to log in to the database with one of these administrative accounts in later chapters. After you install Oracle, you will be required to enter a user name and password to access the software. One default user name created during the installation process is "scott". The default password for the user name is "tiger". If you have installed the Personal Edition of Oracle, you will not need to enter a Connect String during the log in process.

- Data files: You will not be able to use your own computer to work through the chapter examples and complete the projects in this book unless you have the data files. You can get the data files from your instructor, or you can obtain the data files electronically by accessing the Thomson Course Technology Web site at **www.course.com** and then searching for this book's title.

- When you download the data files, they should be stored in a directory separate from any other files on your hard drive or diskette. You will need to remember the path or folder containing the files, because the file name of each script must be prefixed with its location when it is executed in SQL*Plus. (SQL*Plus is the interface tool you will use to interact with the database.)

Visit Our World Wide Web Site

Additional materials designed especially for you might be available on the World Wide Web. Go to **www.course.com** periodically and search this site for more details.

TO THE INSTRUCTOR

To complete the chapters in this book, your users must have access to a set of data files. These files are included in the Instructor's Resource Kit. They may also be obtained electronically through the Thomson Course Technology Web site at **www.course.com**.

The set of data files consists of the JustLee Database folder and a folder for each chapter. Many chapters will require a script to be executed and these files are found in their respective chapter folders (Chapter05, Chapter10, etc.) on your Data Disk and have the file names **prech#.sql**, where the number sign (#) is replaced by the chapter number e.g., **prech10.sql**. A note is included at the beginning of each of these chapters indicates the script file that should be executed. Initially, the **Bookscript.sql** file will be executed at the beginning of Chapter 2 to create the JustLee Books database Each user should execute the scripts as instructed to have a copy of the tables stored in his or her schema. You should instruct your users in how to access and copy the data files to their own computers or workstations.

The chapters and projects in this book were tested using the Microsoft Windows Server 2003 operating system with Oracle 10.1.0.2 Enterprise Edition.

Oracle9i and Oracle 10g include a browser-based version of SQL*Plus, which operates much like the client SQL*Plus software product and is named iSQL*Plus. However, there are a few differences in commands and settings. iSQL*Plus is used in the examples within chapters of this textbook.

Thomson Course Technology Data Files

You are granted a license to copy the data files to any computer or computer network used by individuals who have purchased this book.

CHAPTER 1

OVERVIEW OF DATABASE CONCEPTS

LEARNING OBJECTIVES

After completing this chapter, you should be able to do the following:

- Define database terms
- Identify the purpose of a database management system (DBMS)
- Explain database design using entity-relationship models and normalization
- Explain the purpose of a Structured Query Language (SQL)
- Understand how this textbook's topics are sequenced and how the two sample databases are used

INTRODUCTION

Imagine that it is 20 years ago, and you work in the Billing Department of a local telephone company. One day, your supervisor requests a list of all customers. The list must include each customer's name, telephone number, and current account balance. For most of today's businesses, this would be a simple request. However, this telephone company's investment in technology has been spent on faster and clearer telephone connection services for its customers, not on information management for its Billing Department. Thus, customer information for all of the company's 55,000 customers—including names, addresses, balance due, and payments—is stored on index cards. To create the list for your supervisor, you must take the names and telephone numbers from the index cards and type the list on a typewriter.

Two weeks later, after the list is completed, your supervisor returns and requests a list of all telephone numbers. The list must contain the names of all customers and their telephone numbers; however, the data must be sorted and presented in order of the telephone numbers. Because all the index cards are organized alphabetically by customers' last names, you can expect at least three more months of job security. Of course, by the time your list is completed, it will be out of date because new customers won't have been added and updates won't have been made for individuals who are no longer customers or who have changed their telephone numbers. This scenario describes an ideal setting for a database. A **database** is a storage structure that allows data to be entered, manipulated, and retrieved in a variety of formats—all without having to retype the data each time it is needed.

The database used throughout this textbook is based on the activities of a hypothetical business, an online bookseller named JustLee Books. The company sells books via the Internet to customers throughout the United States. When a new customer places an order, a customer service representative collects data regarding the customer's name, billing and shipping addresses, and the items ordered. The company also uses a database for the books in inventory.

To access the data required for the operation of JustLee Books, management relies upon a DBMS. A **database management system (DBMS)** is used to create and maintain the structure of a database, and then to enter, manipulate, and retrieve the data it stores. Creating an efficient database design is the key to effectively using a database to support an organization's business operations.

This chapter introduces basic database terminology and discusses the process of designing JustLee Books' database.

DATABASE TERMINOLOGY

Whenever a customer opens an account with a company, certain data must be collected. In many cases, the customer completes an online form that asks for the customer's name, address, and so on, as shown in Figure 1-1.

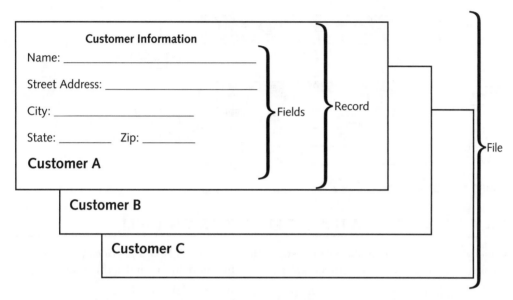

FIGURE 1-1 A customer's information creating data fields and a record and then becoming part of a file

    While completing the form, the customer or a service representative fills in each blank with a series of characters. A **character** is the basic unit of data, and it can be a letter, number, or special symbol. A group of related characters (for example, the characters that make up a customer's name) is called a **field**. A field represents one attribute or characteristic (for example, the name) of the customer. A collection of fields (for example, name, address, city, state, and zip code) about one customer is called a **record**. A group of records about the same type of entity (for example, customers, inventory items) is stored in a **file**. A collection of interrelated files, such as those relating to customers, their purchases, and their payments, is stored in a **database**. Although these terms relate to the logical database design, in many cases, they are used interchangeably with the terminology for the physical database design. When creating the physical database, a field is commonly referred to as a **column**, a record is called a **row**, and a file is known as a **table**. A table is quite similar to a spreadsheet in that it contains columns and rows. Figure 1-2 shows a representation of these terms.

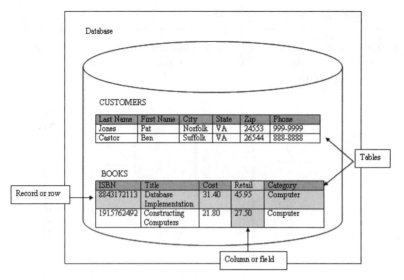

FIGURE 1-2 Database terminology

DATABASE MANAGEMENT SYSTEM

As mentioned earlier, a database is housed within a DBMS. A DBMS provides the functionality to create and work with a database. This functionality includes the following:

- *Data storage*: manage the physical structure of the database
- *Security*: control user access and privileges
- *Multiuser access*: manage concurrent data access
- *Backup*: enable recovery options for database failures
- *Data access language*: provide a language that allows database access
- *Data integrity*: enable constraints or checks on data
- *Data dictionary*: maintain information about database structure

DATABASE DESIGN

To determine the most appropriate structure of fields, records, and files in a database, developers go through a design process. The design and development of a system is accomplished through a series of steps. The process is formally called the **Systems Development Life Cycle (SDLC)** and consists of the following steps:

1. *Systems investigation*: understanding the problem
2. *Systems analysis*: understanding the solution to the previously identified problem
3. *Systems design*: defining the logical and physical components
4. *Systems implementation*: creating the system
5. *Systems integration and testing:* placing the system into operation for testing
6. *Systems deployment:* placing the system into production
7. *Systems maintenance and review*: evaluating the implemented system

Although the SDLC is a methodology designed for any type of system needed by an organization, this chapter specifically addresses the development of a DBMS. For the purposes of this discussion, assume that the problem identified was the need to collect and maintain data about customers and their orders. The identified solution was to use a database to store all needed data. The discussion that follows presents the steps necessary to design the database.

NOTE

A variety of SDLC models have been developed to address different development environments. The steps presented here represent a traditional waterfall model. Other models, such as fountain and rapid prototyping, involve a different series of steps.

To design a database, the requirements of the database—the inputs, processes, and outputs—must first be identified. Usually, the first question asked is, "What information, or output, must come from this database?" or "What questions should this database be able to answer?" By understanding the necessary output, the designer can then determine what information should be stored in the database. For example, if the organization wants to send out birthday cards to its customers, the database must store the birth date of each customer. After the requirements of a database have been identified, an Entity-Relationship (E-R) Model is usually drafted to obtain a better understanding of the data to be stored in the database. In an E-R Model, an **entity** is any person, place, or thing with characteristics or attributes that will be included in the system. An **E-R Model** is a diagram that identifies the entities (customers, books, orders, and such) in the database, and it shows how the entities are related to one another. It serves as the logical representation of the physical system to be built.

The next two sections demonstrate the construction of an E-R Model and the normalization process used to determine the appropriate entities for a database.

Entity-Relationship (E-R) Model

In an E-R Model, an entity is usually represented as a square or rectangle. As shown in Figure 1-3, a line depicts how an entity's data relates to another entity. If the line connecting two entities is solid, the relationship between the entities is mandatory. However, if the relationship between two entities is optional, a dashed line is used.

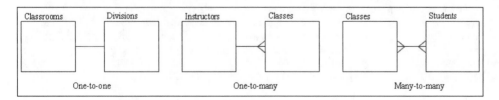

FIGURE 1-3   E-R Model notation examples

As shown in Figure 1-3, the following types of relationships can exist between two entities:

1. *One-to-one:* In a one-to-one relationship, each occurrence of data in one entity is represented by only one occurrence of data in the other entity. For example, if each classroom is assigned to only one academic division, this creates a one-to-one relationship between the classroom and division entities. This type of relationship is depicted in an E-R Model as a simple straight line.

2. *One-to-many:* In a one-to-many relationship, each occurrence of data in one entity can be represented by many occurrences of the data in the other entity. For example, a class has only one instructor, but an instructor may teach many classes. A one-to-many relationship is represented by a straight line with a "crowfoot" at the "many" end to indicate "many."

3. *Many-to-many:* In a many-to-many relationship, data can have multiple occurrences in both entities. For example, a class can consist of more than one student, and a student can take more than one class. A straight line with a "crowfoot" at each end indicates a many-to-many relationship.

Figure 1-4 shows a simplified E-R Model for the JustLee Books database used throughout this textbook. A more thorough E-R Model would include a list of attributes for each entity.

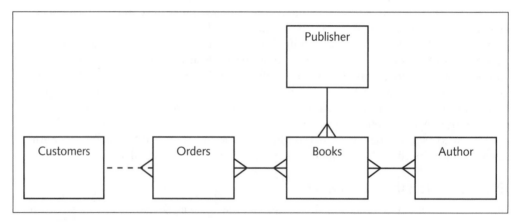

FIGURE 1-4   An E-R Model for JustLee Books

NOTE

The notations in the sample E-R Models in this chapter reflect only one example of diagramming entity relationships. If you are using a modeling software tool, you may encounter different notations to represent relationships. For example, Microsoft products typically represent the many side of a relationship with the infinity symbol (∞). In addition, modeling tools may automatically add the common fields or foreign key columns needed as relationships are defined.

The following relationships are defined in the E-R Model in Figure 1-4:

1. Customers can place multiple orders, but each order can be placed only by one customer (one-to-many). The dashed line between Customers and Orders means a customer may exist in the database without having a current order stored in the ORDERS table. This is considered an optional relationship.
2. An order can consist of more than one book, and a book can appear on more than one order (many-to-many).
3. Books can have more than one author, and an author can write more than one book (many-to-many).
4. A book can have only one publisher, but a publisher can publish more than one book (one-to-many).

Although some E-R modeling approaches are much more complex, the simplified notations used in this chapter do point out the important relationships among the entities, and using them helps the designer identify potential problems in table layouts. Upon examination of the E-R Model in Figure 1-4, you should have noticed the two many-to-many relationships. Before creating the database, all many-to-many relationships must be reduced to a set of one-to-many relationships, as you will learn.

The identification of entities and relationships in the database design process is important because the entities in the database are usually represented as a table, and the relationships can reveal whether additional tables are needed in the database. If the problem arising from the many-to-many relationship in the E-R Model is not apparent to the designer at this point, it will become clear during the normalization process.

Database Normalization

Many people new to or unfamiliar with database design principles often ask, "Why not just put all the data in one big table?" This single table approach leads to problems of data redundancy (duplication) and data anomalies (data inconsistencies). For example, review the order data being recorded in Figure 1-5. The customer information is repeated for each order that a customer places (redundancy). Also, the city data in the last row is different from the first two rows. Under these circumstances, it is not clear if the last row actually represents a different customer, if the previous customer had an address change, or if the city information is incorrect (data anomaly).

LAST NAME	FIRST NAME	CITY	STATE	ZIP	ORDER DATE	ORDER #
Jones	Pat	Norfolk	VA	24553	3/22/2005	45720
Jones	Pat	Norfolk	VA	24553	5/28/2005	48243
Jones	Pat	Suffolk	VA	26544	9/05/2005	51932

FIGURE 1-5 Single table approach issues

To avoid these data issues, database **normalization** is used to create a successful design that reduces or eliminates data redundancy and, therefore, avoids data anomalies. In general, normalization helps the database designer to determine which attributes, or fields, belong to each entity. In turn, this helps to determine which fields belong in each table(s). Normalization is a multistage process that allows the designer to take the raw data to be collected about an entity and evolve the data into a structured, normalized form that will reduce the risks associated with data redundancy. Data redundancy presents a special problem in databases because storing the same data in different places can cause problems when updates or changes to the data are required.

It is difficult for most novices to understand the impact of storing **unnormalized** data or data that has not been designed using a normalization process. Here's an example. Suppose you work for a large company and submit a change-of-address form to the Human Resources (HR) Department. If all the data stored by HR is normalized, then a data entry clerk would need to update only the EMPLOYEES master table with your new address. However, if the data used by HR is *not* stored in a normalized format, it is likely the data entry clerk will need to enter the change in each table that contains your address—your EMPLOYEE RECORD table, HEALTH INSURANCE table, CAFETERIA PLANS table, SICK LEAVE table, ANNUAL TAX INFORMATION table, and so on—even though all this data is stored in the same database. Thus, if your mailing address is stored in a variety of tables (or even duplicated in the same table) and the data entry clerk fails to make the change in one table, you might get a paycheck that shows one address and, at the end of the year, have your W-2 form mailed to a different address! Storing the data in a normalized format means that only one update is required to reflect the new address, and it should always be the one that appears whenever your mailing address is needed.

A portion of the database for JustLee Books will be used to step through the normalization process—specifically, the books that are sold to customers. For each book, you need to store the book's International Standard Book Number (ISBN), its title, publication date, wholesale cost, retail price, category (literature, self-help, etc.), publisher name, contact person at the publisher for reordering the book (and telephone number), and name of the book's author or authors.

Figure 1-6 shows a sample of the data that must be maintained. For ease of illustration, the publishers' telephone numbers are eliminated, and the authors' names use just the first initial.

The first step in determining which data should be contained in each table is to identify a primary key. A **primary key** is a field that uniquely identifies each record. You might select the ISBN to identify each book because no two books will ever have the same ISBN.

ISBN	TITLE	PUBLICATION DATE	COST	RETAIL	CATEGORY	PUBLISHER	CONTACT	AUTHOR
8843172113	Database Implementation	04-JUN-99	31.40	55.95	Computer	American Publishing	Davidson	T. Peterson, J. Austin, J. Adams
1915762492	Hand-cranked Computers	21-JAN-01	21.80	25.00	Computer	American Publishing	Davidson	W. White, L. White

FIGURE 1-6 The BOOKS table for JustLee Books

N O T E

When data that already exists, such as a book ISBN, is used to serve as a primary key, it is often referred to as an intelligent or natural key. At times, data that serves as a primary key does not exist and, therefore, a system-generated unique value is used as a primary key. For example, JustLee Books does not have any ID to associate with book authors, so an ID number is generated. This is referred to as a surrogate or artificial key.

However, note that in Figure 1-6, if a book has more than one author, the Author field contains more than one data value. When a record contains repeating groups (that is, multiple entries for a single column), it is considered unnormalized. First-normal form (1NF) indicates that all the values of the columns are atomic—that is, they contain no repeating values. To convert the record to 1NF, remove the repeating values by making each author entry a separate record, as shown in Figure 1-7.

In Figure 1-7, the repeating values of authors' names is eliminated—each record now contains no more than one data value for the Author field. Notice that you can no longer use the book's ISBN as the primary key, because more than one record will have the same value in the ISBN field. *The only combination of fields that will uniquely identify each record is the ISBN and Author fields together*. When more than one field is used as the primary key for a table, the combination of fields is usually referred to as a **composite primary key**. Now that the repeating values have been eliminated and the records can be uniquely identified, the data is in 1NF, but a few design problems remain.

A problem known as partial dependency may occur when the primary key consists of more than one field. **Partial dependency** means that the fields contained within a record (row) depend only upon one portion of the primary key. For example, a book's title, publication date, publisher name, and so on, all depend upon the book itself, not upon who wrote the book (the author). *The simplest way to resolve a partial dependency is to break the composite primary key into two parts—each representing a separate table*. In this case, you can create a table for books and a table for authors. By removing the partial dependency, you have converted the BOOKS table to **second-normal form (2NF)**, as shown in Figure 1-8.

ISBN	TITLE	PUBLICATION DATE	COST	RETAIL	CATEGORY	PUBLISHER	CONTACT	AUTHOR
8843172113	Database Implementation	04-JUN-99	31.40	55.95	Computer	American Publishing	Davidson	T. Peterson
8843172113	Database Implementation	04-JUN-99	31.40	55.95	Computer	American Publishing	Davidson	J. Austin
8843172113	Database Implementation	04-JUN-99	31.40	55.95	Computer	American Publishing	Davidson	J. Adams
1915762492	Handcranked Computers	21-JAN-01	21.80	25.00	Computer	American Publishing	Davidson	W. White
1915762492	Handcranked Computers	21-JAN-01	21.80	25.00	Computer	American Publishing	Davidson	L. White

FIGURE 1-7 BOOKS table from Figure 1-6 after conversion to 1NF with fields containing multiple values eliminated

ISBN	TITLE	PUBLICATION DATE	COST	RETAIL	CATEGORY	PUBLISHER	CONTACT
8843172113	Database Implementation	04-JUN-99	31.40	55.95	Computer	American Publishing	Davidson
1915762492	Handcranked Computers	21-JAN-01	21.80	25.00	Computer	American Publishing	Davidson

FIGURE 1-8 The BOOKS table in 2NF after elimination of partial dependency

Now that the BOOKS records are in 2NF, you must look for any transitive dependencies. A **transitive dependency** means that at least one of the values in the record is not dependent upon the primary key, but upon another field in the record. In this case, the contact person from the publisher's office is actually dependent upon the publisher of the book, not the book itself. To remove the transitive dependency from the BOOKS table, remove the contact information and place it in a separate table. Because the table was in 2NF and has had all transitive dependencies removed, the BOOKS table is now in **third-normal form (3NF)**, as shown in Figure 1-9.

ISBN	TITLE	PUBLICATION DATE	COST	RETAIL	CATEGORY	PUBLISHER
8843172113	Database Implementation	04-JUN-99	31.40	55.95	Computer	American Publishing
1915762492	Handcranked Computers	21-JAN-01	21.80	25.00	Computer	American Publishing

FIGURE 1-9 The BOOKS table in 3NF with transitive dependency of the publisher's contact person removed

There are several levels of normalization beyond 3NF. However, in the "real world," tables are normalized only to 3NF. A summary of the normalization steps presented in this section is as follows:

1. *1NF*: Eliminate all repeating values, and identify a primary key or primary composite key.
2. *2NF*: Make certain the table is in 1NF, and eliminate any partial dependencies.
3. *3NF*: Make certain the table is in 2NF, and remove any transitive dependencies.

Relating Tables Within the Database

After the BOOKS table is in 3NF, you can then normalize each of the remaining tables of the database. After each table has been normalized, make certain all relationships among the tables have been established. For example, you will need a way to determine the author(s) for each book in the BOOKS table. Because the authors' names are stored in a separate table, there must be some way to join data together. In most cases, a connection between two tables is established through a common field. A **common field** is a field that exists in both tables. In many cases, the common field is a primary key for one of the tables. In the second table, it is referred to as a foreign key. The purpose of a **foreign key** is to establish a relationship with another table or tables. The foreign key appears in the "many" side of a one-to-many relationship.

Also, an accepted industry standard is to use an ID code (numbers and/or letters) to represent an entity; this reduces the chances of data entry errors. For example, rather than entering the entire name of each publisher into the BOOKS table, you assign each publisher an ID code in the PUBLISHER table, and then list that ID code in the BOOKS table as a foreign key to retrieve the publisher's name for each book. In this case, the publisher ID code could be the primary key in the PUBLISHER table and a foreign key in the BOOKS table.

During the normalization of JustLee Books' database, the many-to-many relationships prevented data from being normalized to 3NF. The unnormalized version of the data had repeating groups for authors in the BOOKS table and for books in the ORDERS table. As part of the conversion of the data into 3NF, two additional tables were created that did not appear on the original E-R Model: ORDERITEMS and BOOKAUTHOR.

A many-to-many relationship cannot exist in a relational database. The most common approach used to eliminate a many-to-many relationship is to create two one-to-many relationships through the addition of a bridging entity. A **bridging entity** is placed between the original entities and serves as a "filter" for the data. The ORDERITEMS table, a bridging entity, created one-to-many relationships with the ORDERS and BOOKS tables. The BOOKAUTHOR table, another bridging table, created one-to-many relationships with the BOOKS and AUTHOR tables.

After normalization, the final table structures are as shown in Figure 1-10. Notice the following about the table structures:

- The underlined field(s) in each table indicates the primary key for that particular table. As previously mentioned, the primary key is the field that uniquely identifies each record in the table.
- For the bridging entities that were added, note that composite primary keys uniquely identify each record. The composite primary key for the BOOKAUTHOR table was created using the primary key from each table it joins together (BOOKS and AUTHOR).

Figure 1-11 shows a portion of the BOOKS table and the fields it contains after normalization. As mentioned previously, each field represents a characteristic, or attribute, that is being collected for an entity. The group of attributes for a specific occurrence (for example, a customer or a book) is called a record. In Oracle 10g, a list of a table's contents uses *columns to represent fields and rows to represent records.* These terms are used interchangeably throughout this textbook.

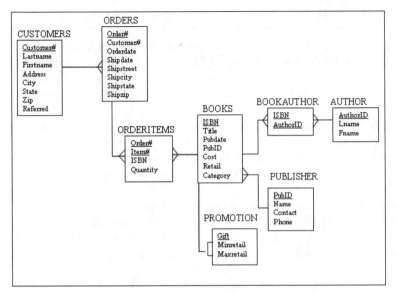

FIGURE 1-10 JustLee Book's table structures after normalization

ISBN	TITLE	PUBDATE	...	CATEGORY
1059831198	BODYBUILD IN 10 MINUTES A DAY	21-JAN-01	...	FITNESS
0401140733	REVENGE OF MICKEY	14-DEC-01	...	FAMILY LIFE
4981341710	BUILDING A CAR WITH TOOTHPICKS	18-MAR-02	...	CHILDREN
8843172113	DATABASE IMPLEMENTATION	04-JUN-99	...	COMPUTER
...	...	...	...	...
0299282519	THE WOK WAY TO COOK	11-SEP-00	...	COOKING
8117949391	BIG BEAR AND LITTLE DOVE	08-NOV-01	...	CHILDREN
0132149871	HOW TO GET FASTER PIZZA	11-NOV-02	...	SELF HELP
9247381001	HOW TO MANAGE THE MANAGER	09-MAY-99	...	BUSINESS
2147428890	SHORTEST POEMS	01-MAY-01	...	LITERATURE

← Record

— Field

FIGURE 1-11 A portion of the BOOKS table after normalization

Overview of Database Concepts

STRUCTURED QUERY LANGUAGE (SQL)

The industry standard for interacting with a relational database is through SQL—pronounced "sequel." **Structured Query Language (SQL)** is not considered a programming language, such as COBOL or Java. It is a data sublanguage, and unlike a programming language, it processes sets of data as groups, and it can navigate the data stored within various tables.

Through the use of SQL statements, users can instruct the DBMS to create and modify tables, enter and maintain data, and retrieve data for a variety of situations. You will be issuing the SQL commands used in this textbook through Oracle 10g's **SQL*Plus®** tool, which is the interface that allows users to issue SQL statements that create, maintain, and search stored data.

Two industry-accepted committees set the industry standards for SQL: the **American National Standards Institute (ANSI)** and the **International Organization for Standardization (ISO)**. The use of industry-established standards allows individuals to use the same skills to work with various relational database management systems and enables various programs to communicate with different databases without major redevelopment efforts. The benefit this creates for users (and students) is that the SQL statements learned with Oracle 10g can be transferred to another DBMS program, such as Informix. To work correctly in another environment, you may need to substitute minor items (such as a square bracket for a parenthesis in a SQL statement), but the basic structure and keywords are usually the same.

DATABASES USED IN THIS TEXTBOOK

Two main databases are referenced throughout this text. The first, JustLee Books, is described in detail in the following paragraphs. This database is used primarily to learn data retrieval statements. The second, City Jail, is used in the case studies. You will design and create this database as you work through the chapter case studies. The Chapter 1 case study presents the information needs of the city jail and challenges you to design the database. Then, the case studies in Chapters 3 through 6 challenge you to construct the various objects in the database.

The initial organization of the database structure for JustLee Books is shown in Figure 1-10. The database is used first to record customers' orders. Customers and JustLee's employees can identify a book by its ISBN, title, or author name(s). Employees can also determine when a particular order was placed and when, or if, the order was shipped. The database also stores the publisher contact information so the bookseller can reorder a book.

Basic Assumptions

Three assumptions made when designing the database are as follows:

- An order is not shipped until all items for that order are available (in other words, there are no back orders or partial order shipments).
- All addresses are within the United States; otherwise, the structure of the Address/Zip Code fields would need to be altered because many countries use different address information such as province names.

- Only orders for the current month or orders from previous months that have not yet shipped are stored in the ORDERS table. At the end of each month, all completed orders are transferred to an annual SALES table. This allows for faster processing of data within the ORDERS table; when necessary, users can still access information pertaining to previous orders through the annual SALES table.

In addition to recording data, management also wants to be able to track the type of books that customers purchase. Although databases were originally developed to record, maintain, and report data for their collected purpose, organizations have realized the importance of having data to support other business functions. These secondary sets of data are used for a purpose other than originally collected. Organizations that deal with thousands (even millions) of sales transactions each month usually store copies of transactions in a separate database for various types of research. Analyzing historical sales data and other information stored in an organization's database is generally referred to as **data mining**. Thus, the bookseller's database also includes data to be used by the Marketing Department to determine which categories of books customers frequently purchase. By knowing a buyer's purchasing habits, new items in inventory can be promoted to customers who frequently purchase that type of book. For example, if a customer has placed several orders for children's books, then that customer might purchase similar books in the future. The Marketing Department can thus target promotions for other children's books to that customer, knowing there's an increased likelihood of a purchase.

Tables Within JustLee Books' Database

Next, let's consider each of the tables within JustLee Books' database. Refer to the table structures in Figure 1-10.

CUSTOMERS table: Notice that the CUSTOMERS table is the first table in Figure 1–10. It serves as a master table for storing the basic data relating to any customer who has placed an order with JustLee Books. It stores the customer's name and mailing address, plus the name of the person who referred that customer to the company. As a promotion to obtain new customers, the bookstore sends a 10% discount coupon to any customer who refers a friend who makes a purchase.

You should also notice in Figure 1-10 that there is a Customer# field in the CUSTOMERS table. Why? Because you might have two customers with the same name, and by assigning each customer a number, you can uniquely identify each person. Using account numbers, or codes can also decrease the likelihood of data entry errors due to incorrect spelling or abbreviations.

BOOKS table: The BOOKS table stores the ISBN, title, publication date, publisher ID, wholesale cost, and retail price of each book. The table also stores a category name for each book (for example, Fitness, Children, Cooking) to track customers' purchasing patterns, as previously mentioned. At the present time, the actual name of the category is entered into the database. Because someone might spell the name of one of the categories incorrectly, a CATEGORY table will be created in Chapter 3, and only the code for each category will be stored in the BOOKS table.

AUTHOR and BOOKAUTHOR tables: As shown in Figure 1-10, the AUTHOR table maintains a list of the authors' names. Because a many-to-many relationship originally existed between the books entity and the authors entity, the BOOKAUTHOR table was created as a bridging table between the two entities. The BOOKAUTHOR table stores the ISBN and the author's ID for each book. If you need to know who wrote a particular book, you would first have the DBMS look up the ISBN of the book in the BOOKS table, then look up each entry of the ISBN in the BOOKAUTHOR table, and finally trace the name of the author(s) back to the AUTHORS table through the AuthorID field.

ORDERS and ORDERITEMS tables: Data about a customer's order is divided into two tables: ORDERS and ORDERITEMS. The ORDERS table identifies which customer placed each order, the date on which the order was placed, and the date on which it was shipped. Because the shipping address might be different from a customer's billing address, the shipping address is also stored within the ORDERS table. If a customer orders two or more books in one order, the ORDERS table could contain a repeating group. Therefore, the individual items purchased on each order are stored in the ORDERITEMS table.

The ORDERITEMS table records the order number, the ISBN of the book being purchased, and the quantity for each book. To uniquely identify each item purchased in an order for multiple products, the table includes an Item# field that corresponds to the item's position in the sequence of products ordered. For example, if a customer places an order for three different books, the first book listed in the order is assigned an Item# of 1, the second book listed is Item# 2, and so on. A variation of this table could use the combination of the Order# and the book's ISBN to identify each product for a particular order. However, the concept of item# or line# is widely used in the industry to identify various line items on an invoice or in a transaction, and thus it has been included in this table to familiarize you with the concept.

PUBLISHER table: The PUBLISHER table contains the publisher's ID code, the name of the publisher, the publisher's contact person, and the publisher's telephone number. The PUBLISHER table can be joined to the BOOKS table through the PubID field, which is contained in both tables. This linked data from the PUBLISHER and BOOKS table allows you to determine which publisher to contact when you need to reorder books by identifying which books you obtained from each publisher.

PROMOTION table: The last table in Figure 1-10 is the PROMOTION table. JustLee Books has an annual promotion that includes a gift with each book purchased. The gift is based upon the retail price of the book. Customers ordering books that cost less than $12 will receive a certain gift, whereas customers buying books costing between $12.01 and $25 will receive a different gift. The PROMOTION table identifies the gift, the minimum retail value of the range, and the maximum retail value. There is no exact value that matches the Retail field in the BOOKS table; therefore, to determine the appropriate gift, you need to determine whether a retail price falls within a particular range.

An actual online bookseller's database would contain thousands of customers and books. It would, naturally, be much more complex than the database shown in this textbook. For example, this database does not track data such as the quantity on hand for each book, discounted prices, and sales tax. Furthermore, to simplify the display of the data on the screen and in reports, each table only contains a few records.

NOTE

A complete list of the JustLee Books' tables can be found in Appendix A at the end of this textbook.

TOPIC SEQUENCE

The remaining chapters of this textbook introduce the SQL statements and concepts you need to know for the first DBA or Developer Certification exam for Oracle9*i* or Oracle 10*g*. They also prepare you to use Oracle in the workplace. The early chapters are organized in the same sequence you would use to initially create a database. After you learn how to create the database, the focus moves to data retrieval, which covers a vast array of options. The last portion of the text explores other databases (MS SQL Server, MySQL) and how SQL fits into application development such as an e-commerce site. However, before you can build a database you need to understand how to perform basic data queries, which are covered in Chapter 2, "Basic SQL SELECT Statements."

Working through the examples presented in each chapter and completing the assignments will enhance your learning process.

Chapter Summary

- A DBMS is used to create and maintain a database.
- A database is composed of a group of interrelated tables.
- A file is a group of related records. A file is also called a table in the physical database.
- A record is a group of related fields regarding one specific entity. A record is also called a row.
- Before building a database, designers must look at the input, processing, and output requirements of the system. Tables to be included in the database can be identified through the E-R Model. An entity in the E-R Model usually represents a table in the physical system.
- Through the normalization process, designers can determine whether additional tables are needed and which attributes or fields belong in each table.
- A record is considered unnormalized if it contains repeating groups.
- A record is in first-normal form (1NF) if no repeating groups exist and it has a primary key.
- Second-normal form (2NF) is achieved if the record is in 1NF and has no partial dependencies.
- After a record is in 2NF and all transitive dependencies have been removed, then it is in third-normal form (3NF), which is generally sufficient for most databases.
- A primary key is used to uniquely identify each record.
- A common field is used to join data contained in different tables.
- A foreign key is a common field that exists between two tables but is also a primary key in one of the tables.
- A Structured Query Language (SQL) is a data sublanguage that navigates the data stored within a database's tables. Through the use of SQL statements, users can instruct the DBMS to create and modify tables, enter and maintain data, and retrieve data for a variety of situations.

Review Questions

1. What is the purpose of an E-R Model?
2. What is an entity?
3. Give an example of three entities that might exist in a database for a medical office and list some of the attributes that would be stored in a table for each entity.
4. Define a one-to-many relationship.
5. Discuss the problems that can be caused by data redundancy.
6. Explain the role of a primary key.
7. Describe how a foreign key is different from a primary key.
8. List the steps of the normalization process.

9. What type of relationship cannot be stored in a relational database? Why?

10. Identify at least three reasons an organization might analyze historical sales data stored in its database.

Multiple Choice

1. Which of the following represents a row in a table?

   a. an attribute

   b. a characteristic

   c. a field

   d. a record

2. Which of the following defines a relationship in which each occurrence of data in one entity is represented by multiple occurrences of the data in the other entity?

   a. one-to-one

   b. one-to-many

   c. many-to-many

   d. none of the above

3. An entity is represented in an E-R Model as a(n):

   a. arrow

   b. crowfoot

   c. dashed line

   d. none of the above

4. Which of the following is not an E-R Model relationship?

   a. sometimes-to-always

   b. one-to-one

   c. one-to-many

   d. many-to-many

5. Which of the following symbols represents a many-to-many relationship in an E-R Model?

   a. a straight line

   b. a dashed line

   c. a straight line with a crowfoot at both ends

   d. a straight line with a crowfoot at one end

6. Which of the following may contain repeating groups?

   a. unnormalized data

   b. 1NF

   c. 2NF

   d. 3NF

Overview of Database Concepts

7. Which of the following defines a relationship in which each occurrence of data in one entity is represented by only one occurrence of data in the other entity?

   a. one-to-one

   b. one-to-many

   c. many-to-many

   d. none of the above

8. Which of the following has no partial or transitive dependencies?

   a. unnormalized data

   b. 1NF

   c. 2NF

   d. 3NF

9. Which of the following symbols represents a one-to-many relationship in an E-R Model?

   a. a straight line

   b. a dashed line

   c. a straight line with a crowfoot at both ends

   d. a straight line with a crowfoot at one end

10. Which of the following has no partial dependencies but may contain transitive dependencies?

   a. unnormalized data

   b. 1NF

   c. 2NF

   d. 3NF

11. Which of the following has no repeating groups but may contain partial or transitive dependencies?

   a. unnormalized data

   b. 1NF

   c. 2NF

   d. 3NF

12. The unique identifier for a record is called the:

   a. foreign key

   b. primary key

   c. turn key

   d. common field

13. Which of the following is a primary key in another table when two tables are joined together?

    a. foreign key

    b. primary key

    c. turn key

    d. repeating group key

14. A unique identifier that consists of more than one field is commonly called a:

    a. primary plus key

    b. composite key

    c. foreign key

    d. none of the above

15. Which of the following symbols represents an optional relationship in an E-R Model?

    a. a straight line

    b. a dashed line

    c. a straight line with a crowfoot at both ends

    d. a straight line with a crowfoot at one end

16. Which of the following, when it exists in an E-R Model, indicates the need for an additional table?

    a. sometimes-to-always relationship

    b. one-to-one relationship

    c. one-to-many relationship

    d. many-to-many relationship

17. Which of the following represents a field in a table?

    a. a record

    b. a row

    c. a column

    d. an entity

18. Which of the following defines a relationship in which data can have multiple occurrences in each entity?

    a. one-to-one

    b. one-to-many

    c. many-to-many

    d. none of the above

19. When part of the data in a table depends upon a field in the table that is not the table's primary key, this is known as:

   a. transitive dependency

   b. partial dependency

   c. psychological dependency

   d. a foreign key

20. Which of the following is used to join data contained in two or more tables?

   a. primary key

   b. unique identifier

   c. common field

   d. foreign key

Hands-On Assignments

To perform assignments 1-5, refer to the table structures in Figure 1-10 or the tables in Appendix A.

1. Which tables and fields would you access to determine which books have been purchased by a customer in the current month's orders?

2. How would you determine which orders have not yet been shipped to the customer?

3. If management needed to determine which book category generated the most sales for the last month, which tables and fields would they consult to derive this information?

4. How would you determine how much profit was generated from orders placed this month?

5. If a customer inquired about a book written in 1999 by an author named Thompson, which access path (tables and fields) would you need to follow to find the list of books meeting the customer's request?

In assignments 6-10, create a simple E-R Model depicting entities and relationship lines for each of the data scenarios.

6. A college needs to track placement test scores for incoming students. Each student may take a variety of tests including English and Math. Some students are required to take the placement tests due to previous coursework.

7. Every employee in a company is assigned to one department. Every department can contain many employees.

8. A movie megaplex needs to collect movie attendance data. The company maintains sixteen theaters in a single location. Each movie offered may be shown in one or more of the available theaters and is typically scheduled for three to six showings in a day. The movies are rotated through the theaters to ensure each is shown in one of the stadium seating theaters at least once.

9. An online retailer of coffee beans maintains a long list of unique coffee flavors. The company purchases beans from a number of suppliers, however, each specific flavor of coffee is only purchased from a single supplier. Many of the customers are repeat purchasers and typically order at least five flavors of beans in each order.

10. Data for an information technology conference needs to be collected. The conference has a variety of sessions scheduled over a two-day period. All attendees must register for the sessions they plan to attend. Some speakers are only presenting one session, whereas others are handling multiple sessions. Each session has only one speaker.

Advanced Challenge

To perform this activity, refer to the table structures in Figure 1-10 or the tables in Appendix A.

In this chapter, the normalization process was shown just for the BOOKS table. The other tables in JustLee Books' database are shown after normalization. Because the database needs to contain data for each customer's order, perform the steps necessary to normalize the following data elements to 3NF:

* Customer's name and billing address
* Quantity and retail price of each item ordered
* Shipping address for each order
* Date each order was placed and the date it was shipped

Assume that the unnormalized data in the list are all stored in one table. Provide your instructor with a list of the table(s) that have been identified at each step of the normalization process (that is, 1NF, 2NF, 3NF) and the attributes, or fields, contained in each table. Remember that each customer may place more than one order, each order may contain more than one item, and an item may appear on more than one order.

Case Study: *City Jail*

The following memo is received by your company. First, based on the memo, create an initial database design (E-R Model) for the City Jail that indicates entities, attributes (columns), primary keys, and relationship lines. Consider the columns that are needed to build the relationships between the entities. Use only the entities identified in the memo to develop the entity-relationship diagram, or ERD.

Second, create a list of additional entities or attributes not identified in the memo that may be applicable to a crime-tracking database.

Keep in mind that the memo is written from an end-user perspective—not by a database developer!

MEMO

To: Database Consultant

From: City Jail Information Director

Subject: Establishing a Crime-Tracking Database System

It was a pleasure meeting with you last week. I look forward to working with your company to create a much-needed crime-tracking system. As you requested, our project group has outlined the crime-tracking data needs that we anticipate. Our goal is to simplify the process of tracking criminal activity and provide a more efficient mechanism for data analysis and reporting. Please review the data needs outlined below and contact me with any questions.

Criminals: name, address, phone #, violent offender status (yes/no), probation status (yes/no), and aliases

Crimes: classification (felony, misdemeanor, other), date charged, appeal status (closed, can appeal, in appeal), hearing date, appeal cutoff date (always 60 days after the hearing date), arresting officers (can be more than one officer), crime code (such as burglary, forgery, assault; hundreds of codes exist), amount of fine, court fee, amount paid, payment due date, and charge status (pending, guilty, not guilty)

Sentencing: start date, end date, number of violations (such as not reporting to probation officer), and type (jail period, house arrest, probation)

Note: Criminals may have multiple sentences assigned to a particular crime. For example, a criminal may be required to serve a jail sentence followed by a period of probation.

Appeals: appeal filing date, appeal hearing date, status (pending, approved, and disapproved)

Note: Each crime case can be appealed up to three times.

Police officers: name, precinct, badge #, phone contact, status (active/inactive)

BASIC SQL SELECT STATEMENTS

INTRODUCTION

In Chapter 1, you reviewed database structures and were introduced to the concept of using SQL statements to enter, manipulate, and retrieve data through a DBMS. (DBMS is a generic term that applies to software that allows users to interact with a database.) When you are working with relational databases, however, the DBMS software is considered to be a **relational database management system (RDBMS)**. The RDBMS is the software program used to create the database and allows you to enter, manipulate, and retrieve data. Most RDBMSs include capabilities to create forms for user input screens and to create reports to display output. When trying to retrieve data, most RDBMSs provide the user with an option to interact with the database through a graphical user interface (GUI) or through SQL statements. When a GUI is used, the RDBMS actually converts entries made by the user into SQL statements that are subsequently executed to perform the desired operation.

In this textbook, Oracle 10g Database is used to interact with the database for JustLee Books. Oracle 10g is an **object relational database management system (ORDBMS)** because it can be used to reference not only individual data elements but also objects (such as object fields and maps), which can be composed of individual data elements. However, because the data stored in the database for JustLee Books is composed of simple alphanumeric characters, the examples and concepts presented throughout this textbook also apply to traditional RDBMSs. The use of objects is usually addressed in advanced application development courses.

Oracle 10g Database comes in three editions: Enterprise, Standard, and Personal. All the concepts and examples presented in this textbook can be used with any of these editions. SQL*Plus is the tool used to enter and execute SQL statements. SQL*Plus can be used through a client connection or through an Internet connection (browser), and the Internet version is named iSQL*Plus. The screen captures displayed in this book show the iSQL*Plus interface, however either SQL*Plus interface can be used to accomplish all the examples in this book.

Review Appendix B, "SQL*Plus and iSQL*Plus User's Guide," to become familiar with the SQL*Plus interface you will be using. The appendix is separated into two sections: The first introduces the use of the Internet interface (iSQL*Plus), and the second covers the client interface (SQL*Plus).

In this chapter, you will begin learning how to retrieve data from a database by using SELECT statements. You need this basic understanding of querying a database before you can create a database. As you create the database, you will need to submit queries to verify what objects you have created or what data has been entered.

Figure 2-1 displays a list of all the commands covered in this chapter.

Before beginning this chapter, you need to create the JustLee Books database. The tables in this database are used in the examples and Hands-On Assignments. Work through the appropriate set of steps that follow for either the SQL*Plus or iSQL*Plus interface:
SQL*Plus:

1. Determine the location of the JustLee Database folder in your data files (these are the data files provided with this text). Verify that the bookscript.sql script file is in this folder.
2. Open your browser to the iSQL*Plus site provided by your instructor. Start SQL*Plus and log in.
3. To execute the script file, enter **start d:\JustLeeDatabase\bookscript.sql** at the SQL> prompt inside Oracle 10g. The **d:\JustLeeDatabase** should be substituted with the appropriate drive letter and pathname, if applicable. Then press **Enter**.
4. After running the script, you can verify and view the structure of each table by entering **describe** *tablename* (where *tablename* is substituted with the actual name of one of the tables presented in this chapter) at the SQL> prompt.

COMMAND DESCRIPTION	BASIC SYNTAX STRUCTURE	EXAMPLE
Command to view all columns of a table	SELECT * FROM *tablename*;	**SELECT *** **FROM books;**
Command to view one column of a table	SELECT *columnname* FROM *tablename*;	**SELECT title** **FROM books;**
Command to view multiple columns of a table	SELECT *columnname,* *columnname, ...* FROM *tablename*;	**SELECT title, pubdate** **FROM books;**
Command to assign an alias to a column during display	SELECT *columnname* [AS] *alias* FROM *tablename*;	**SELECT title AS titles** **FROM books;** *or* **SELECT title titles** **FROM books;**
Command to perform arithmetic operations during retrieval	SELECT *arithmetic* *expression* FROM *tablename*;	**SELECT retail - cost** **FROM books;**
Command to eliminate duplication in output	SELECT DISTINCT *columnname* FROM *tablename*; *or* SELECT UNIQUE *columnname* FROM *tablename*;	**SELECT DISTINCT state** **FROM customers;** *or* **SELECT UNIQUE state** **FROM customers;**
Command to perform concatenation of column contents during display	SELECT *columnname* || *columnname* FROM *tablename*;	**SELECT firstname || lastname** **FROM customers;**
Command to view the structure of a table	DESCRIBE *tablename*	**DESCRIBE books**

FIGURE 2-1   List of commands used in this chapter

CAUTION

The first time this script is executed, you will receive errors for the first eight statements of the script (all of these are DROP TABLE commands). If you need to rebuild the database for any reason, these statements will delete the existing tables so that they can be rebuilt. If you receive privilege errors, your user account has not been assigned the appropriate privileges to create tables.

iSQL*Plus:

1. Determine the location of the JustLee Database folder in your data files (these are the data files provided with this text). Verify that the bookscript.sql script file is in this folder.
2. Open a browser to the iSQL*Plus site and log in.

3. To execute the script file, click the **Load Script** button below the workspace area. Use the Browse button to select the bookscript.sql file. Click **Load**. The contents of the script file now appear in the workspace area.

4. Click **Execute**. Messages confirming script processing appear below the workspace area. After running the script, you can verify and view the structure of each table by entering **describe** *tablename* (where *tablename* is substituted with the actual name of the table presented in this chapter).

CAUTION

The first time this script is executed, you will receive errors for the first eight statements of the script (all of these are DROP TABLE commands). If you need to rebuild the database for any reason, these statements will delete the existing tables so that they can be rebuilt. If you receive privilege errors, your user account has not been assigned the appropriate privileges to create tables.

If you want to see the contents of the script, it can be opened with any text editor or word-processing program.

CAUTION

If you open the bookscript.sql file with a word-processing program and then accidentally click **Save**, your word-processing program may save the file with a different extension than "sql," and Oracle 10*g* will be unable to use the script. The script file must have the extension "sql" to execute properly from a line command.

SELECT STATEMENT SYNTAX

As mentioned in this chapter's introduction, the majority of the SQL operations performed on a database in the average organization are SELECT statements. **SELECT statements** allow the user to retrieve data from tables. The user can view all the fields and records within a table or specify only certain fields and records to be displayed. In essence, the SELECT statement asks the database a question, which is why it is also known as a **query**.

After querying a database, the results that are displayed can be based on certain conditions specified in the SELECT statement. In other words, what is displayed is basically the answer to the question asked by the user. For example, in this chapter, you will learn the basic structure of a SELECT statement and how to display only certain fields from a table. Later in the book, you will learn how to modify the SELECT statement to display only certain rows.

The **syntax** for a SQL statement is the basic structure, or rules, required to execute the statement. The syntax for the SELECT statement is shown in Figure 2-2.

```
SELECT    [DISTINCT | UNIQUE] (*, columnname [ AS alias], …)
          FROM      tablename
          [WHERE    condition]
          [GROUP BY group_by_expression]
          [HAVING   group_condition]
          [ORDER BY columnname];
```

FIGURE 2-2   Syntax for the SELECT statement

The capitalized words (SELECT, FROM, WHERE, etc.) in Figure 2-2 are **keywords** (words that have a predefined meaning in Oracle 10g). Each section of the example that begins with a keyword is referred to as a **clause** (SELECT clause, FROM clause, WHERE clause, and so on). Note these important points about SELECT statements:

- The only clauses required for the SELECT statement are SELECT and FROM. (These are the only clauses in Figure 2-2 to be discussed in this chapter.)
- Square brackets are used to indicate portions of the statement that are optional. (Optional clauses are discussed in subsequent chapters.)
- SQL statements can be entered over several lines (as shown in Figure 2-2) or on one line. Most SQL statements are entered with each clause on a separate line to improve readability and make editing easier. As various SELECT commands are demonstrated in this chapter, you will see variations on spacing, number of lines used, and capitalization. These variations are pointed out as they are encountered within the text.
- To execute a SQL statement in SQL*Plus after it is entered, you have two options. Usually the end of a SQL statement is indicated by a semicolon (;) at the end of the statement (as given in the syntax example). If you forget to enter the semicolon and press the Enter key, you can still execute the statement by entering a slash (/) at the SQL> prompt. The iSQL*Plus interface assumes a semicolon to end the statement, so it is optional.

Selecting All Data in a Table

To have the SELECT statement return *all* data from a specific table, type an asterisk (*) after SELECT, as shown in Figure 2-3.

```
SELECT *
FROM customers;
```

FIGURE 2-3   Command to select all data within a table

The asterisk (*) is a symbol that instructs Oracle 10g to include all columns in the table. The symbol can be used only in the SELECT clause of a SELECT statement. If you need to view or display all columns in a table, it is much simpler to type an asterisk than to type the name of each column. The results of the SELECT statement should look like those shown in Figure 2-4.

CUSTOMER#	LASTNAME	FIRSTNAME	ADDRESS	CITY	STATE	ZIP	REFERRED
1001	MORALES	BONITA	P.O. BOX 651	EASTPOINT	FL	32328	
1002	THOMPSON	RYAN	P.O. BOX 9835	SANTA MONICA	CA	90404	
1003	SMITH	LEILA	P.O. BOX 66	TALLAHASSEE	FL	32306	
1004	PIERSON	THOMAS	69821 SOUTH AVENUE	BOISE	ID	83707	
1005	GIRARD	CINDY	P.O. BOX 851	SEATTLE	WA	98115	
1006	CRUZ	MESHIA	82 DIRT ROAD	ALBANY	NY	12211	
1007	GIANA	TAMMY	9153 MAIN STREET	AUSTIN	TX	78710	1003
1008	JONES	KENNETH	P.O. BOX 137	CHEYENNE	WY	82003	
1009	PEREZ	JORGE	P.O. BOX 8564	BURBANK	CA	91510	1003
1010	LUCAS	JAKE	114 EAST SAVANNAH	ATLANTA	GA	30314	
1011	MCGOVERN	REESE	P.O. BOX 18	CHICAGO	IL	60606	
1012	MCKENZIE	WILLIAM	P.O. BOX 971	BOSTON	MA	02110	
1013	NGUYEN	NICHOLAS	357 WHITE EAGLE AVE.	CLERMONT	FL	34711	1006
1014	LEE	JASMINE	P.O. BOX 2947	CODY	WY	82414	
1015	SCHELL	STEVE	P.O. BOX 677	MIAMI	FL	33111	
1016	DAUM	MICHELL	9851231 LONG ROAD	BURBANK	CA	91508	1010
1017	NELSON	BECCA	P.O. BOX 563	KALMAZOO	MI	49006	
1018	MONTIASA	GREG	1008 GRAND AVENUE	MACON	GA	31206	
1019	SMITH	JENNIFER	P.O. BOX 1151	MORRISTOWN	NJ	07962	1003
1020	FALAH	KENNETH	P.O. BOX 335	TRENTON	NJ	08607	

20 rows selected.

FIGURE 2-4   List of all customers in the CUSTOMERS table

NOTE

As previously mentioned, all SQL*Plus output will be shown in iSQL*Plus format, which displays an HTML style table by default. Your output format will be different if you are using the client SQL*Plus interface because output is displayed in pure text format. Appendix B includes example output from each of these interfaces.

When looking at the results of the SELECT statement, pay attention to the column headings. Depending on the SQL*Plus tool being used and the options set, some column headings (such as for the State field) might be truncated. Keep in mind that when you refer to a column in any SQL statement, you still need to specify the entire column name. Be sure you do not depend on the column headings for column names due to the possibility of the name being truncated.

NOTE

The exact name of each column can be viewed by entering **DESCRIBE *tablename***. To view the exact name of each column in the CUSTOMERS table, simply enter **DESCRIBE customers** and execute.

Selecting One Column from a Table

In the previous example, an asterisk was used to indicate that all columns in the table should be displayed. When displaying a table containing a large number of fields, the results may look cluttered. Or there may be sensitive data you do not want other users to see. In these situations, you can instruct Oracle 10g to return only specific columns in the results. Choosing specific columns in a SELECT statement is called **projection**. You can select one column—or as many as all the columns—contained within the table.

For example, suppose that you want to view the titles of all books in inventory. The data regarding books is stored in the BOOKS table. The name of the column you need is Title. As shown in Figure 2-5, you can list the name of the desired column after the SELECT keyword. Type the statement shown in Figure 2-5.

```
SELECT title
FROM books;
```

FIGURE 2-5 Command to select a single column

NOTE

If you receive an error message rather than the results of the query, there may have been a typing error. The error message should display the line in which the error occurred. An asterisk beneath the line serves as an indicator of the error; however, it is not always the exact cause. If the error message indicates that the error is in the second line of the statement, then BOOKS may have been entered as BOOK. Simply retype the statement with the correction, and it should return the desired results.

Results returned from the query should look like those shown in Figure 2-6.

TITLE
BODYBUILD IN 10 MINUTES A DAY
REVENGE OF MICKEY
BUILDING A CAR WITH TOOTHPICKS
DATABASE IMPLEMENTATION
COOKING WITH MUSHROOMS
HOLY GRAIL OF ORACLE
HANDCRANKED COMPUTERS
E-BUSINESS THE EASY WAY
PAINLESS CHILD-REARING
THE WOK WAY TO COOK
BIG BEAR AND LITTLE DOVE
HOW TO GET FASTER PIZZA
HOW TO MANAGE THE MANAGER
SHORTEST POEMS

14 rows selected.

FIGURE 2-6 List of all book titles

The results display only the field specified, which was Title. You might want to practice some variations of the same SELECT statement. Try entering the examples shown in Figure 2-7 and notice that the results are the same.

```
SQL> SELECT TITLE FROM BOOKS;

SQL> select title from books;

SQL> SELECT title FROM books;

SQL> SELECT TITLE
     FROM BOOKS
     /
```

FIGURE 2-7   The SELECT statement can be entered on one or more lines

As shown in these examples, the statement can be entered on one or more lines. *Keywords, table names, and column names are not case sensitive.* To distinguish between keywords and other parts of the SELECT statement, we capitalize the keywords. Keep in mind that this is *not* a requirement of Oracle 10g; it is simply a convention used to improve readability.

Selecting Multiple Columns from a Table

In most cases, displaying only one column from a table is not sufficient output on which to base decisions. If you want to know the date on which each book was published, you could retrieve all the fields from the BOOKS table and manually extract the needed fields. As an alternative, you could issue one SELECT statement to retrieve the Title field, another to retrieve the Publication Date field, and then match up the two results. It is much more practical to issue a query requesting both the title and the publication date for each book, as shown in Figure 2-8.

```
SELECT title, pubdate
FROM books;
```

FIGURE 2-8   Command to select multiple columns from a table

When specifying more than one column in the SELECT clause of the SELECT statement, commas should separate the columns listed. Although a space has been entered after the comma, it is not required. The space serves to improve the readability of the statement and is not part of the required syntax of the SELECT statement. The example shown in Figure 2-9 returns the same results.

```
SELECT title,pubdate FROM books;
```

FIGURE 2-9 Multiple clauses of the SELECT statement on one line

The data returned from the query should look like that shown in Figure 2-10.

TITLE	PUBDATE
BODYBUILD IN 10 MINUTES A DAY	21-JAN-01
REVENGE OF MICKEY	14-DEC-01
BUILDING A CAR WITH TOOTHPICKS	18-MAR-02
DATABASE IMPLEMENTATION	04-JUN-99
COOKING WITH MUSHROOMS	28-FEB-00
HOLY GRAIL OF ORACLE	31-DEC-01
HANDCRANKED COMPUTERS	21-JAN-01
E-BUSINESS THE EASY WAY	01-MAR-02
PAINLESS CHILD-REARING	17-JUL-00
THE WOK WAY TO COOK	11-SEP-00
BIG BEAR AND LITTLE DOVE	08-NOV-01
HOW TO GET FASTER PIZZA	11-NOV-02
HOW TO MANAGE THE MANAGER	09-MAY-99
SHORTEST POEMS	01-MAY-01

14 rows selected.

FIGURE 2-10 Display of title and publication date for all books

When looking at the results of this query, notice the order in which the columns are listed in the output. In this case, Title is listed first, followed by Pubdate. Oracle 10g sequences the columns in the display in the same order that the user sequences them in the SELECT clause of the SELECT statement. To change the order and display the Pubdate column first, simply reverse the order of the columns listed in the SELECT statement, as shown in Figure 2-11.

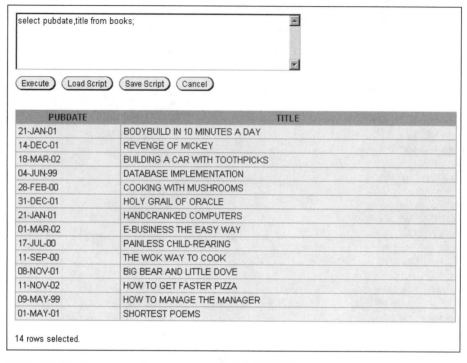

```
select pubdate,title from books;
```

Execute   Load Script   Save Script   Cancel

PUBDATE	TITLE
21-JAN-01	BODYBUILD IN 10 MINUTES A DAY
14-DEC-01	REVENGE OF MICKEY
18-MAR-02	BUILDING A CAR WITH TOOTHPICKS
04-JUN-99	DATABASE IMPLEMENTATION
28-FEB-00	COOKING WITH MUSHROOMS
31-DEC-01	HOLY GRAIL OF ORACLE
21-JAN-01	HANDCRANKED COMPUTERS
01-MAR-02	E-BUSINESS THE EASY WAY
17-JUL-00	PAINLESS CHILD-REARING
11-SEP-00	THE WOK WAY TO COOK
08-NOV-01	BIG BEAR AND LITTLE DOVE
11-NOV-02	HOW TO GET FASTER PIZZA
09-MAY-99	HOW TO MANAGE THE MANAGER
01-MAY-01	SHORTEST POEMS

14 rows selected.

FIGURE 2-11 Reversed column sequence in the SELECT clause

OPERATIONS WITHIN THE SELECT STATEMENT

Now that you've selected columns from tables, let's look at some other operations. In this section, you'll learn how to use column aliases, employ arithmetic operations, eliminate duplicate output, and display rows on multiple lines.

Using Column Aliases

In some cases, a column name can be a vague indicator of the data displayed in a particular column. To better describe the data listed, you can substitute a **column alias** for the column name in the results of a query. For example, if you are presenting a list of all books stored in the database, you might want the column heading to read Title of Books. To instruct the software to use a column alias, simply list the column alias next to the column name in the SELECT clause. Figure 2-12 shows the title and category for each book in the BOOKS table, but it adds a column alias for the title. The **optional keyword** of **AS** has been included in this example to distinguish between the column name and the column alias.

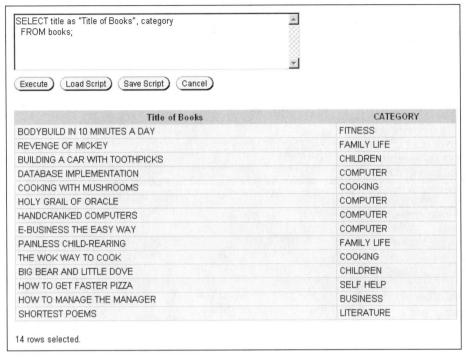

FIGURE 2-12   Using a column alias

You need to keep some guidelines in mind when using a column alias. If the column alias contains spaces, special symbols, or if you do not want it to appear in all capital letters, *it must be enclosed in double quotation marks* (" "). By default, the column headings shown in the results of queries are capitalized. The use of the double quotation marks overrides the default for the column heading. However, notice that the case of the data displayed *within* the column is not altered.

> **NOTE**
>
> As shown in the SELECT statement, *you must separate the list of field names with commas*. If you forget a comma, Oracle 10*g* interprets the subsequent field name as a column alias, and you do not receive the intended results.

If the column alias consists of only one word without special symbols, it does not need to be enclosed in double quotation marks. In Figure 2-13, the Retail field is assigned the column alias of Price. Also note that the optional keyword AS used in Figure 2-12 is not included. Because a comma does not separate the words *retail* and *price*, Oracle 10*g* assumes that Price is the column alias for the Retail column.

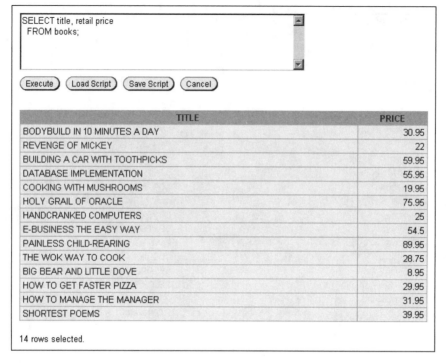

```
SELECT title, retail price
  FROM books;
```

(Execute) (Load Script) (Save Script) (Cancel)

TITLE	PRICE
BODYBUILD IN 10 MINUTES A DAY	30.95
REVENGE OF MICKEY	22
BUILDING A CAR WITH TOOTHPICKS	59.95
DATABASE IMPLEMENTATION	55.95
COOKING WITH MUSHROOMS	19.95
HOLY GRAIL OF ORACLE	75.95
HANDCRANKED COMPUTERS	25
E-BUSINESS THE EASY WAY	54.5
PAINLESS CHILD-REARING	89.95
THE WOK WAY TO COOK	28.75
BIG BEAR AND LITTLE DOVE	8.95
HOW TO GET FASTER PIZZA	29.95
HOW TO MANAGE THE MANAGER	31.95
SHORTEST POEMS	39.95

14 rows selected.

FIGURE 2-13   Using a column alias without the AS keyword

As you look at the results in Figure 2-13, notice the alignment of the column headings:

- By default, the data for text, or character, fields is left-aligned.
- Data for numeric fields is right-aligned.
- Notice that Oracle 10g does not display insignificant zeros (zeros that do not affect the value of the number being displayed). The retail price of the book *Handcranked Computers* is $25.00. Because the zeros in the two decimal positions are insignificant, Oracle 10g does not display them. To force Oracle 10g to display a specific number of decimal positions, formatting codes are required. These are presented later in the text.

Using Arithmetic Operations

If needed, simple arithmetic operations such as multiplication (*), division (/), addition (+), and subtraction (-) can be used in the SELECT clause of a query. Keep in mind that Oracle 10g adheres to the standard order of operations:

1. Moving from left to right within the arithmetic equation, any required multiplication and division operations are solved first.
2. Any addition and subtraction operations are solved after multiplication and division, again moving from left to right within the equation.

To override the order of operations, you can use parentheses to enclose any portion of the equation that should be completed first.

Next, you want to determine the profit generated by the sale of each book. The BOOKS table contains two fields you can use to derive the profit: Cost and Retail. A book's profit is the difference (subtraction) between the amount paid for the book by the bookstore (cost) and the selling price of the book (retail). To clarify the contents of the column, assign the column alias "profit" to the calculated field, as shown in Figure 2-14.

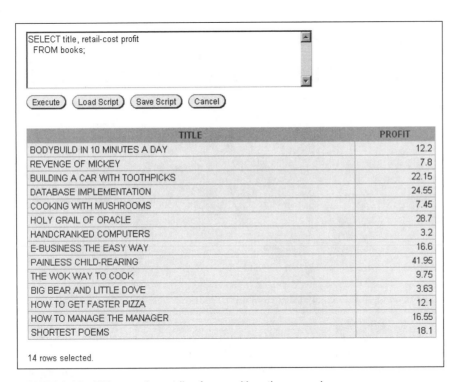

FIGURE 2-14 Using a column alias for an arithmetic expression

Using DISTINCT and UNIQUE

Suppose that you want to know the states in which your customers reside so you can focus a marketing campaign on a particular region of the country. You want a list to identify only the states, not customer names, addresses, and so on. One option is to select the State column from the CUSTOMERS table. You will quickly notice that some states are listed more than once if more than one customer lives in that particular state. If you are working with only 20 records, it is not a problem to simply cross out any duplicate states on a printout. However, if you are dealing with thousands of records, that is a cumbersome task.

To eliminate duplicate listings, you can use the DISTINCT option in your SELECT statement. For example, suppose that you have five customers living in Texas (TX). Without the DISTINCT option, TX appears in your results five times. If you include the DISTINCT option, TX appears only once. To use the DISTINCT option, use the keyword DISTINCT between the SELECT keyword and the first column of the column list, as shown in Figure 2-15.

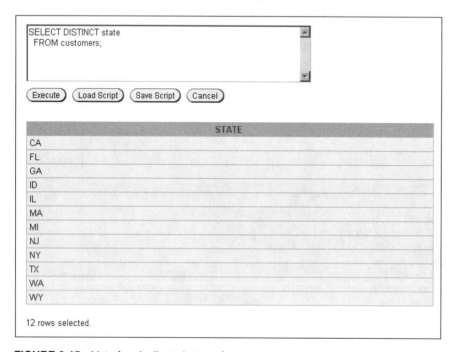

FIGURE 2-15   List of unduplicated states for customers

In Figure 2-15, the database was queried to determine the states in which customers live. Although there are 20 customers in the CUSTOMERS table, they live in only 12 states. You could use this information to determine where you are most likely to attract more customers, or to identify geographical areas that are not responding to a current, nationwide marketing effort.

The DISTINCT keyword is applied to all columns listed in the SELECT clause, even though it is stated directly after the SELECT keyword. In the previous example, if you had also included CITY in the SELECT clause, each different combination of city and state would have been listed only once in the output. In other words, if no two customers in the database live in the same city and state, you still would have had 20 rows of output—one row for each customer.

Using Concatenation

In previous examples, if an output list contained more than one field, each field was placed in a separate column. In some situations, however, you might want the contents of each field to be displayed next to each other, without much blank space. For example, for a list of customer names, you might prefer to have them combined to appear as a single column, rather than as separate first name and last name columns.

Combining the contents of two or more columns is known as **concatenation**. To instruct Oracle 10g to concatenate the output of a query, use two vertical bars or pipes (||). On a keyboard, this symbol is located above the backslash (\ ). In the following example, the goal is to have the last name of the customer listed immediately after the first name, rather than in a separate column.

As you look at the results in Figure 2-16, the first thing you should notice is that the first name and last name of each customer run together, and it is difficult to tell where one name ends and the other begins. To make the results more readable, you need to include a blank space between the Firstname and Lastname fields.

To have Oracle 10g insert a blank space, you must concatenate, or combine, the Firstname and Lastname fields with a **string literal**. A string literal instructs Oracle 10g to interpret the characters you have entered "literally" and not to consider it as a keyword or command. *A string literal must be enclosed within single quotation marks ('').* When you use a string literal, the character or set of characters that you type should appear in the output exactly as you have typed them. In this instance, the string literal is a blank space. Figure 2-17 shows how the customer's list looks after including a blank space in the output.

Although you now have a readable list of all customer names, the display has an unusual column heading. The column heading shown in the results is exactly what you entered for the field list—including the concatenation symbols and the literal. If this list is for management or some individual who is not familiar with Oracle 10g, you might want to give your output a more professional appearance. Rather than have this unusual and unappealing column heading, you can have Oracle 10g substitute a column alias, as you did previously. The query in Figure 2-18 substitutes Customer Name as the column heading in the results.

NOTE

If you get an error message, make sure the blank space is in single quotation marks and the column alias is in double quotation marks.

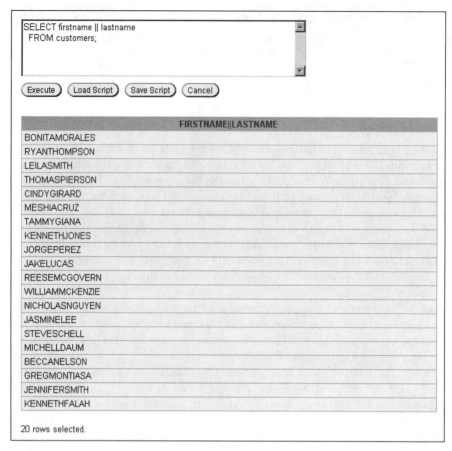

```
SELECT firstname || lastname
  FROM customers;
```

(Execute) (Load Script) (Save Script) (Cancel)

FIRSTNAME||LASTNAME
BONITAMORALES
RYANTHOMPSON
LEILASMITH
THOMASPIERSON
CINDYGIRARD
MESHIACRUZ
TAMMYGIANA
KENNETHJONES
JORGEPEREZ
JAKELUCAS
REESEMCGOVERN
WILLIAMMCKENZIE
NICHOLASNGUYEN
JASMINELEE
STEVESCHELL
MICHELLDAUM
BECCANELSON
GREGMONTIASA
JENNIFERSMITH
KENNETHFALAH

20 rows selected.

FIGURE 2-16   Concatenation of two columns

Inserting a Line Break

Suppose that management needs a list of all customers showing each customer's number and name on separate lines. Also, management requests that the customer names appear as "last name, first name." To display customer names this way, modify the query used in Figure 2-18 to switch the order of the columns and include a comma in the string literal. To have the customer number and name appear on separate lines, include **CHR(10)** after the customer number to instruct Oracle 10g to insert a line break. The CHR(10) code indicates to the computer that a line break should occur at that location. Anything listed after the CHR(10) code is displayed on the next line.

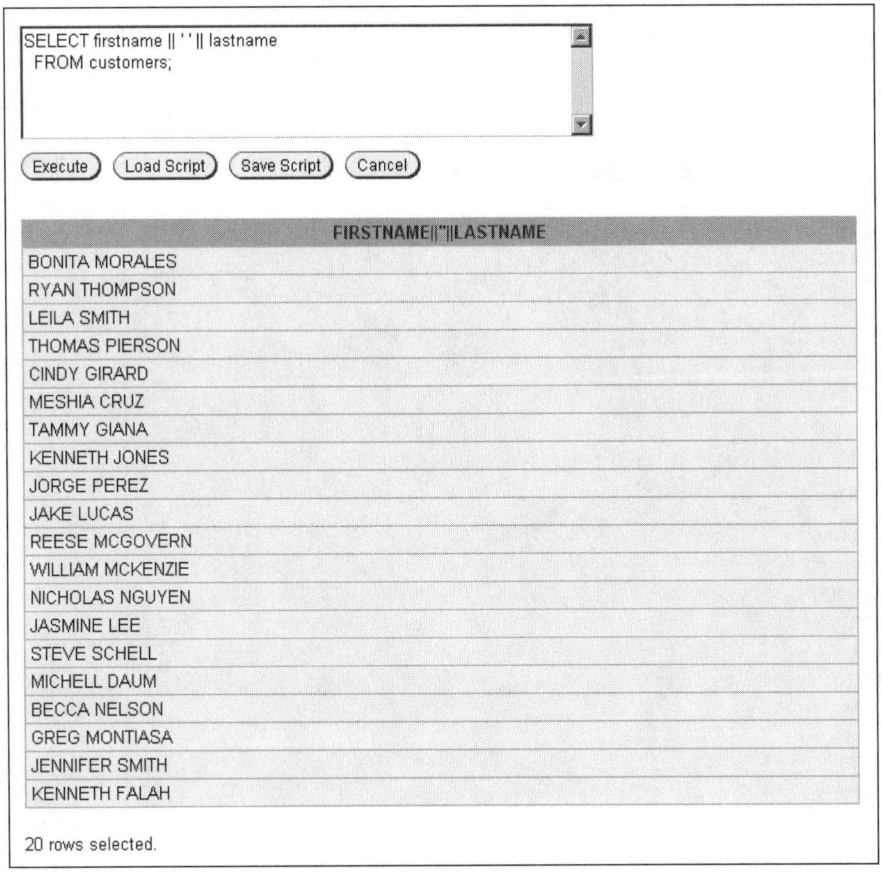

```
SELECT firstname || ' ' || lastname
  FROM customers;
```

(Execute) (Load Script) (Save Script) (Cancel)

FIRSTNAME||"||LASTNAME
BONITA MORALES
RYAN THOMPSON
LEILA SMITH
THOMAS PIERSON
CINDY GIRARD
MESHIA CRUZ
TAMMY GIANA
KENNETH JONES
JORGE PEREZ
JAKE LUCAS
REESE MCGOVERN
WILLIAM MCKENZIE
NICHOLAS NGUYEN
JASMINE LEE
STEVE SCHELL
MICHELL DAUM
BECCA NELSON
GREG MONTIASA
JENNIFER SMITH
KENNETH FALAH

20 rows selected.

FIGURE 2-17   Using a string literal within concatenation

NOTE

The CHR(10) code works properly only with pure text output. The default output format for iSQL*Plus is an HTML table. To change to pure text, click the **Preferences** button and choose the **Script Formatting** link. Set the **Preformatted Output** setting to **On** and click the **Apply** button. Move back to the workspace area and execute your statement. The output from the statement will now be in text format. The client SQL*Plus tool displays output in text format by default. Figure 2-19 displays only this first page of output.

You can use the CHR(10) code whenever you want output results to be displayed over several lines. As shown in Figure 2-19, concatenation (the vertical bars) is used to include the line break in the column. The CHR(10) code must be used in a concatenation operation.

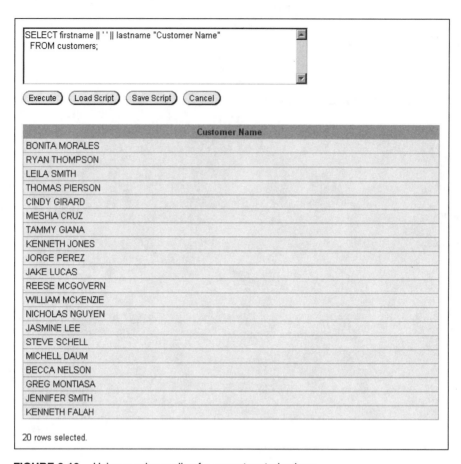

```
SELECT firstname || ' ' || lastname "Customer Name"
  FROM customers;
```

(Execute)   (Load Script)   (Save Script)   (Cancel)

Customer Name
BONITA MORALES
RYAN THOMPSON
LEILA SMITH
THOMAS PIERSON
CINDY GIRARD
MESHIA CRUZ
TAMMY GIANA
KENNETH JONES
JORGE PEREZ
JAKE LUCAS
REESE MCGOVERN
WILLIAM MCKENZIE
NICHOLAS NGUYEN
JASMINE LEE
STEVE SCHELL
MICHELL DAUM
BECCA NELSON
GREG MONTIASA
JENNIFER SMITH
KENNETH FALAH

20 rows selected.

FIGURE 2-18   Using a column alias for concatenated values

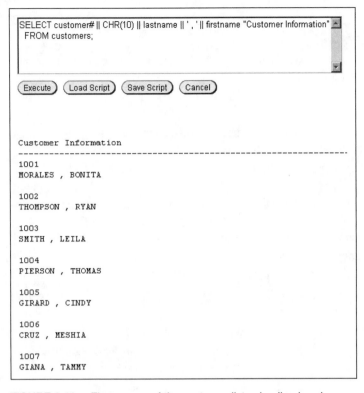

FIGURE 2-19 First screen of the customer list using line breaks

Chapter Summary

- Oracle 10*g* is an ORDBMS.
- A basic query in Oracle 10*g* SQL includes the SELECT and FROM clauses, the only mandatory clauses in a SELECT statement.
- To view all columns in the table, specify an asterisk (*) or list all the column names individually in the SELECT clause.
- To display a specific column or set of columns, list the column names in the SELECT clause (in the order in which you want them to appear).
- When listing column names in the SELECT clause, a comma must separate column names.
- A column alias can be used to clarify the contents of a particular column. If the alias contains spaces or special symbols, or if you want to display the column with any lowercase letters, you must enclose the column alias in double quotation marks (" ").
- Basic arithmetic operations can be performed in the SELECT clause.
- To remove duplicate listings, include either the DISTINCT or UNIQUE keyword.
- To specify which table contains the desired columns, you must list the name of the table after the keyword FROM.
- Use vertical bars (||) to combine, or concatenate, fields, literals, and other data.
- A line break code of CHR(10) can be used to format output on multiple lines. The output must be formatted as text output in SQL*Plus for the line break command to operate properly.

Chapter 2 Syntax Summary

The following table presents a summary of the syntax that you have learned in this chapter. You can use the table as a study guide and reference.

SYNTAX GUIDE		
Element	Description	Example
SELECT clause	Identify the column(s) for retrieval in a SELECT command	`SELECT title`
FROM clause	Identify the table containing selected columns	`FROM books`
SELECT statement	View column(s) in a table	`SELECT title` `FROM books;`

,	Separate column names in a list when retrieving multiple columns from a table	`SELECT title,`
		`pubdate`
		`FROM books;`

SYNTAX GUIDE		
Element	Description	Example
*	Return all data in a table when used in a SELECT clause	`SELECT *FROM books;`

AS	Indicate a column alias to change the heading of a column in output	`SELECT title AS titles,`
		`pubdate`
		`FROM books;`

	Create a column alias to change the heading of a column in output *without* using AS	`SELECT title titles,`
		`pubdate`
		`FROM books;`

" "	Preserve spaces, symbols, or case in an output column heading alias	`SELECT title AS "Book`
		`Name"`
		`FROM books;`

* multiplication
/ division
+ addition
- subtraction
	Solve arithmetic operations (Oracle 10g first solves * and /, then solves + and -, unless parentheses are used)	`SELECT title, retail-cost`
		`profit`
		`FROM books;`

DISTINCT	Eliminate duplicate lists	`SELECT DISTINCT state` `FROM customers;`
UNIQUE	Eliminate duplicate lists	`SELECT UNIQUE state` `FROM customers;`
|| (concatenation)	Combine display of content from multiple columns into a single column	`SELECT city || state FROM` `customers;`
' '(string literal)	Indicate the exact set of characters, including spaces, to be displayed	`SELECT city || ' ' || state` `FROM customers;`

CHR(10)	Insert a line break	`SELECT customer#`
		`|| CHR(10) || city || ' '`
		`|| state`
		`FROM customers;`

DESCRIBE	Display structure of a table	`DESCRIBE books`

; *or* /	Execute a SQL statement	`SELECT Zip`
		`FROM customers;`
		or
		`FROM customers/`

Review Questions

1. What is an RDBMS? An ORDBMS?
2. What are the two required clauses for a SELECT statement?
3. What is the purpose of the SELECT statement?
4. What does an asterisk (*) in the SELECT clause of a SELECT statement represent?
5. What is the purpose of a column alias?
6. How do you indicate that a column alias should be used?
7. When is it appropriate to use a column alias?
8. What are the guidelines to keep in mind when using a column alias?
9. How can you concatenate columns in a query?
10. How do you indicate that a line break should occur in the output of a query?

Multiple Choice

*To determine the exact name of the fields used in the tables for these questions, refer either to Appendix A or use the **DESCRIBE** tablename command to view the structure of the appropriate table.*

1. Which of the following SELECT statements will provide a list of customer names from the CUSTOMERS table?
   a. SELECT customer names FROM customers;
   b. SELECT "Names" FROM customers;
   c. SELECT firstname, lastname FROM customers;
   d. SELECT firstname, lastname, FROM customers;
   e. SELECT firstname, lastname, "Customer Names" FROM customers;

2. Which clause is required in a SELECT statement?
   a. WHERE
   b. ORDER BY
   c. GROUP BY
   d. FROM
   e. all of the above

3. Which of the following is *not* a valid SELECT statement?
   a. SELECT lastname, firstname FROM customers;
   b. SELECT * FROM orders;
   c. Select FirstName NAME from CUSTOMERS;
   d. SELECT lastname Last Name FROM customers;

4. Which of the following symbols is used to represent concatenation?

    a.   *

    b. ||

    c.   [ ]

    d.   ' '

5. Which of the following SELECT statements will return all the fields in the ORDERS table?

    a.   SELECT customer#, order#, orderdate, shipped, address FROM orders;

    b.   SELECT * FROM orders;

    c.   SELECT ? FROM orders;

    d.   SELECT ALL FROM orders;

6. Which of the following symbols is used for a column alias that contains spaces?

    a.   ' '

    b. ||

    c.   " "

    d.   / /

7. Which of the following is a valid SELECT statement?

    a.   SELECT TITLES *TITLE! FROM BOOKS;

    b.   SELECT "customer#" FROM books;

    c.   SELECT title AS "Book Title" from books;

    d.   all of the above

8. Which of the following symbols is used in a SELECT clause to display all columns from a table?

    a.   /

    b.   &

    c.   *

    d.   "

9. Which of the following is *not* a valid SELECT statement?

    a.   SELECT Cost-Retail FROM books;

    b.   SELECT Retail+Cost FROM books;

    c.   SELECT retail*retail*retail FROM books;

    d.   SELECT retail^3 from books;

10. When must a comma be used in the SELECT clause of a query?

    a.   when a field name is followed by a column alias

    b.   to separate the SELECT clause and the FROM clause when only one field is selected

    c.   It is never used in the SELECT clause.

    d.   when listing more than one field name and the fields are not concatenated

    e.   when an arithmetic expression is being included in the SELECT clause

11. Which of the following commands will display a listing of the category for each book in the BOOKS table?

    a. SELECT title books, category FROM books;

    b. SELECT title, books, and category FROM books;

    c. SELECT title, cat FROM books;

    d. SELECT books, ||category "Categories" FROM books;

12. Which clause is *not* required in a SELECT statement?

    a. SELECT

    b. FROM

    c. WHERE

    d. All of the above clauses are required.

13. Which of the following lines of the SELECT statement contains an error?

    1. SELECT title, isbn,

    2. Pubdate "Date of Publication"

    3. FROM books;

    a. line 1

    b. line 2

    c. line 3

    d. There are no errors.

14. Which of the following lines of the SELECT statement contains an error?

    1. SELECT ISBN,

    2. retail-cost

    3. FROM books;

    a. line 1

    b. line 2

    c. line 3

    d. There are no errors.

15. Which of the following lines of the SELECT statement contains an error?

    1. SELECT title, cost,

    2. cost*2

    3. 'With 200% MarkUp'

    4. FROM books;

    a. line 1

    b. line 2

    c. line 3

    d. line 4

    e. There are no errors.

16. Which of the following lines of the SELECT statement contains an error?

    1. SELECT name, contact,
    2. "Person to Call", phone
    3. FROM publisher;

    a. line 1
    b. line 2
    c. line 3
    d. There are no errors.

17. Which of the following lines of the SELECT statement contains an error?

    1. SELECT ISBN, || ' is the ISBN for the book named ' ||
    2. title
    3. FROM books;

    a. line 1
    b. line 2
    c. line 3
    d. There are no errors.

18. Which of the following lines of the SELECT statement contains an error?

    1. SELECT title, category
    2. FORM books;

    a. line 1
    b. line 2
    c. There are no errors.

19. Which of the following lines of the SELECT statement contains an error?

    1. SELECT name, contact, CHR(10),
    2. "Person to Call", phone
    3. FROM publisher;

    a. line 1
    b. line 2
    c. line 3
    d. There are no errors.

20. Which of the following lines of the SELECT statement contains an error?

    1. SELECT *
    2. FROM publishers;

    a. line 1
    b. line 2
    c. There are no errors.

Hands-On Assignments

To determine the exact name of the fields used in the tables for these exercises, refer to Appendix A or use the **DESCRIBE** *tablename command to view the structure of the appropriate table.*

1. Display a list of all data contained within the BOOKS table.

2. List the title only of all the books available in inventory, using the BOOKS table.

3. List the title and publication date for each book in the BOOKS table. Use the column heading of Publication Date for the Pubdate field.

4. List the customer number for each customer in the CUSTOMERS table, along with the city and state in which they reside.

5. Create a list containing the name of each publisher, the person usually contacted, and the telephone number of the publisher. Rename the contact column Contact Person in the displayed results. (*Hint:* Use the PUBLISHER table.)

6. Determine which categories are represented by the current book inventory. List each category only once.

7. List the customer number from the ORDERS table for each customer who has placed an order with the bookstore. List each customer number only once.

8. Create a list of each book title stored in the BOOKS table and the category in which each book belongs. Reverse the sequence of the columns so the category of each book is listed first.

9. List the first and last name of each author in the AUTHORS table.

10. Create a list of authors that displays the last name followed by the first name for each author. The last names and first names should be separated by a comma and a blank space.

Advanced Challenge

The management of JustLee Books has submitted two requests. The first is for a mailing list of all customers stored in the CUSTOMERS table. The second is for a list of the percentage of profit generated by each book in the BOOKS table. The requests are as follows:

1. Create a mailing list from the CUSTOMERS table. The mailing list should display the name, address, city, state, and zip code for each customer. The name of each customer should be listed in order of first name followed by last name. The name of the customer should appear on the first line, the address on the second line, and the city, state, and zip code on the third line.

2. To determine the percentage of profit for a particular item, subtract the cost for the item from the retail price to obtain the dollar amount of profit, and then divide the profit by the cost of the item. The solution is then multiplied by 100 to determine the profit percentage for each book. Use a SELECT statement to display the title of each book and its percentage of profit. For the column displaying the percentage markup, use Profit % as the column heading.

Required: Determine the SQL statements needed to perform the two required tasks. Each statement should be tested to ensure its validity. Submit the appropriate documentation of the commands and their results using the format specified by your instructor.

Case Study: *City Jail*

The case study resumes in Chapter 3.

TABLE CREATION AND MANAGEMENT

LEARNING OBJECTIVES

After completing this chapter, you should be able to do the following:

- Identify the table name and structure
- Create a new table using the CREATE TABLE command
- Use a subquery to create a new table
- Add a column to an existing table
- Modify the definition of a column in an existing table
- Delete a column from an existing table
- Mark a column as unused and then delete it at a later time
- Rename a table
- Truncate a table
- Drop a table

INTRODUCTION

Upon joining JustLee Books as a database specialist, you were able to query the existing database; however, new data-tracking features have been requested. The management of JustLee Books has decided to implement a sales commission program for all account managers who have been employed by the company for more than six months. Beginning with orders placed after March 31, 2005, account managers will receive a commission for each order received from the geographical region they supervise. The commission rate will be based on the retail price of books.

To implement this new policy, several changes must be made to the database. It does not contain any information about the account managers, nor does it identify geographical regions in which customers live. Because the account managers represent a new entity, at least one new table must be added to the database.

Chapters 3–5 demonstrate SQL commands to create and modify tables, add data to tables, and edit existing data. This chapter addresses methods for creating tables and modifying existing tables. Commands that are used to create or modify database tables are called **data definition language (DDL)** commands. These commands are basically SQL commands that are used specifically to create or modify database objects. A **database object** is a defined, self-contained structure in Oracle 10*g*. In this chapter, you will create database tables, which are considered database objects. Later, in Chapter 6, you will learn how to create and modify other types of database objects. Figure 3-1 provides an overview of this chapter's topics.

NOTE

This chapter assumes you have created the initial JustLee Books database as instructed in Chapter 2.

OVERVIEW OF CHAPTER CONTENTS	
Commands and Clauses	Description
Creating Tables	
CREATE TABLE	Creates a new table in the database. The user names the columns and identifies the type of data to be stored. To view a table, use the SQL*PLUS command DESCRIBE.
CREATE TABLE...AS	Creates a table from existing database tables, using the AS clause and subqueries.
Modifying Tables	
ALTER TABLE... ADD	Adds a column to a table.
ALTER TABLE... MODIFY	Changes a column size, datatype, or default value.
ALTER TABLE... DROP COLUMN	Deletes one column from a table.
ALTER TABLE... SET UNUSED *or* SET UNUSED COLUMN	Marks a column for deletion at a later time.
DROP UNUSED COLUMNS	Completes the deletion of a column previously marked with SET UNUSED.
RENAME...TO	Changes a table name.
TRUNCATE TABLE	Deletes all table rows, but table name and column structure remain.
Deleting Tables	
DROP TABLE	Removes an entire table from the Oracle 10g database.
PURGE TABLE	Permanently deletes a table in the recyclebin.
Recovering Tables	
FLASHBACK TABLE . . . TO BEFORE DROP	Recovers a dropped table if PURGE option not used when table dropped.

FIGURE 3-1 Overview of chapter contents

TABLE DESIGN

Before issuing a SQL command to create a table, you must first complete the entity design as discussed in Chapter 1. For each entity, you must choose the table's name and determine its structure—that is, you must determine what columns will be included in the table. In addition, you will need to determine the necessary width of any character or numeric columns.

Let's look at these requirements in more depth. Oracle 10g has the following rules for naming both tables and columns:

- The names of tables and columns can be up to 30 characters in length and must begin with a letter. These limitations apply only to the name of a table or column, not to the data in a column.
- The names of tables and columns cannot contain any blank spaces.
- Numbers, the underscore symbol (_), and the number sign (#) are allowed in table and column names.
- Each table owned by a user should have a unique table name, and the column names within each table should be unique.
- Oracle 10g "reserved words," such as SELECT, DISTINCT, CHAR, and NUMBER, cannot be used.

Because the new table will contain data about account managers, the name of the table will be ACCTMANAGER. The table will need to contain each account manager's name, employment date, and assigned region. In addition, the ACCTMANAGER table should contain an ID code to act as the table's primary key and to uniquely identify each account manager. Although it is unlikely that JustLee Books will ever have two account managers with the same name, the ID code can reduce data entry errors because users will need only to type a short code instead of a manager's entire name in SQL commands.

NOTE

The ID column will be included in the ACCTMANAGER table to serve as the primary key for the table. However, the primary key constraint is not defined in this chapter. The next chapter expands the coverage of table creation capabilities by introducing the various constraints that can be defined for a table.

Now that the contents of the table have been determined, the columns can be designed. When you create a table in Oracle 10g, you must define each of the columns. Before you can create the columns, you must do the following:

1. Choose a name for each column.
2. Determine the type of data each column will store.
3. Determine (in some cases) the maximum width of the column.

Before choosing column names, it is useful to look at some datatypes and their default values. Valid datatypes are shown in Figure 3-2.

ORACLE 10*g* DATATYPES	
Datatype	Description
VARCHAR2(*n*)	**Variable**-length **char**acter data, where *n* represents the maximum length of the column. Maximum size is 4000 characters. There is no default size for this datatype; a minimum value must be specified. *Example:* VAR-CHAR2(9) can contain up to nine letters, numbers, or symbols.
CHAR(*n*)	Fixed-length character column, where *n* represents the length of the column. Default size is 1. Maximum size is 2000 characters. *Example:* CHAR(9) can contain nine letters, numbers, or symbols. However, if fewer than nine are entered, spaces are added to the right to force the data to reach a length of nine.
NUMBER(*p,s*)	Numeric column, where *p* indicates **precision**, or the total number of digits to the left and right of the decimal position, to a maximum of 38 digits; and *s*, or **scale**, indicates the number of positions to the right of the decimal. *Example:* NUMBER(7, 2) can store a numeric value up to 99999.99. If precision or scale is not specified, the column defaults to a precision of 38 digits.
DATE	Stores date and time between January 1, 4712 B.C. and December 31, 9999 A.D. Seven bytes are allocated to the column to store the century, year, month, day, hour, minute, and second of a date. Oracle 10*g* displays the date in the format DD-MON-YY. Other aspects of a date can be displayed by using the TO_CHAR format. The width of the field is predefined by Oracle 10*g* as seven bytes.

FIGURE 3-2   Oracle 10*g* datatypes

N O T E

You will discover that Oracle currently has two variable-length character datatypes: VARCHAR and VARCHAR2. Oracle recommends the use of VARCHAR2 rather than VARCHAR, and, therefore, VARCHAR2 is used throughout this text. Oracle's SQL reference states "The VARCHAR datatype is currently synonymous with the VARCHAR2 datatype. Oracle recommends that you use VARCHAR2 rather than VARCHAR. In the future, VARCHAR might be defined as a separate datatype used for variable-length character strings compared with different comparison semantics."

A **datatype** identifies the type of data Oracle 10g is expected to store in a column. Identifying the type of data not only assists in verifying that you input appropriate data, but it also allows you to manipulate data in ways that are specific to that datatype. For example, imagine you need to calculate the difference in terms of days between two date values such as the Orderdate and Shipdate columns from the ORDERS table. To accomplish this task, the system would need to be able to associate the date values to a calendar. If the columns have a DATE datatype, Oracle 10g automatically associates the values to calendar days.

Let's return to the new ACCTMANAGER table, to which you need to add five columns. The first column will serve to uniquely identify each account manager and will be named AmID. The ID code assigned to each account manager will consist of the first letter of a manager's last name, followed by a three-digit number. Because the column will consist of both letters and numbers, the column will be defined to store the datatype of CHAR and have a width of four. The CHAR datatype should be used only when the length of values to be stored in the column are consistent. In this case, every AmID will have a length of four. Oracle 10g pads a CHAR column to the specified length if an entry does not fill the entire width of the column. This can result in wasted storage space if the majority of the values stored by Oracle 10g must include blank spaces to force the contents of a column to a specific length. If character data to be stored in a column will not be a consistent length, the VARCHAR2 datatype should be used. The VARCHAR2 datatype uses only the physical space required to hold the value input and storage space is conserved because no padding effect takes place.

Why not just always use the VARCHAR2 datatype and avoid the use of CHAR? There is a slight processing overhead in managing the variable-length field. Therefore, there is some advantage to using CHAR when it applies. Many people in the industry use both datatypes; however, this is still a topic of debate.

NOTE

In this scenario, JustLee Books decided to use the AmID code consisting of characters and digits because this is already being used in the payroll system, and the company hopes to integrate all the systems. It is important to consider codes already in use when trying to determine an appropriate primary key value. If JustLee Books did not have any existing values in use to identify account managers, a system-generated value would most likely be used. Typically, these columns would be numeric to enable sequences to be used to populate the column. Sequences are covered in Chapter 6.

The second and third columns of the ACCTMANAGER table will be used to store the first and last name of each account manager. If you store first names and last names in individual columns, you can perform simple searches on each part of a manager's name. Because each account manager's name will consist of characters, these columns will be assigned the datatype of VARCHAR2. Name values vary greatly in length. Therefore, the variable-length datatype is more appropriate than CHAR. Generally, you will define the width so it can hold the largest value that could ever be entered into that column. However, the possibility exists that an account manager will be hired in the future whose first or last name is extremely long. It's relatively easy to increase a column's width at a later time. Therefore, the assumption is that a column width of 12 characters is sufficient to store the names of the current account managers. The columns will be named Amfirst and Amlast.

NOTE

The actual names of the account managers are provided in Chapter 5.

The fourth column of the table will be used to store the employment date of each account manager. Because the datatype will be DATE, you will not have to worry about the length of the column—the length is predetermined by Oracle 10g. All that remains is to choose a name for the column. In this case, Amedate, for "account manager employment date," will be appropriate.

The fifth and final column for the ACCTMANAGER table is the region to which the account manager is assigned. The United States will be divided into six geographical regions, each identified by a two-letter code (such as NE for the Northeast region). The name of the column will be Region and will be defined as a CHAR datatype with a width of two characters. The VARCHAR2 datatype could also be used; however, because the values stored in the Region column will always consist of two characters, the CHAR datatype was selected.

TABLE CREATION

The basic syntax of the SQL command to create a table in Oracle 10g is shown in Figure 3-3.

```
CREATE TABLE [schema] tablename
    (columnname datatype [DEFAULT value],
    [columnname datatype [DEFAULT value], …);
```

FIGURE 3-3   CREATE TABLE syntax

The keywords **CREATE TABLE** instruct Oracle 10g to create a table. The optional **schema** can be included to indicate who will "own" the table. For example, if the person creating the table is also the person who will own the table, the schema can be omitted, and the current user name will be assumed by default. On the other hand, if you were creating the ACCTMANAGER table for someone with the user name of DRAKE, the schema and table name would be entered as DRAKE.ACCTMANAGER to inform Oracle 10g that the table ACCTMANAGER will belong to DRAKE's schema, not yours. The owner of a database object has the right to perform certain operations on that object. In the case of a table, the only way another database user can query or manipulate the data contained in the table is to be given permission from the table's owner or the database administrator. The table name, of course, is the name that will be used to identify the table being created.

> **NOTE**
>
> To create a table for someone else's schema (a table owned by someone else), you must be granted permission, or the privilege, to use the CREATE TABLE command for that user's schema. The different privileges available in Oracle 10g are presented in Chapter 7.

Defining Columns

After the table name has been entered, the columns to be included in the table must be defined. A table can contain a maximum of 1000 columns. *The syntax of the CREATE TABLE command requires the column list to be enclosed within parentheses.* If more than one column will exist in the table, then the name, datatype, and appropriate width are listed for the first column before the next column is defined. Commas separate columns in the list. The CREATE TABLE command also allows a default value to be assigned to a column. The default value is the value that will automatically be stored by Oracle 10g if the user makes no entry into that column. For example, imagine users entering order data. Rather than have the user spend time entering the current date as the order date, have the system do it automatically by setting the DEFAULT value to the current date on the system.

Using the syntax presented in Figure 3-3, the SQL command in Figure 3-4 shows the creation of the ACCTMANAGER table.

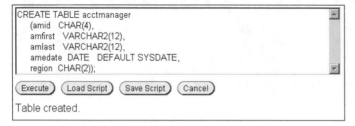

```
CREATE TABLE acctmanager
    (amid   CHAR(4),
     amfirst   VARCHAR2(12),
     amlast   VARCHAR2(12),
     amedate DATE   DEFAULT SYSDATE,
     region  CHAR(2));
```

(Execute) (Load Script) (Save Script) (Cancel)

Table created.

FIGURE 3-4   The creation of the ACCTMANAGER table

In the command given in Figure 3-4, the name of the table is ACCTMANAGER. It is entered here in lower case letters to distinguish it from the CREATE TABLE keywords. Oracle 10*g* SQL commands are not case sensitive; however, for clarity, in this text all keywords are in uppercase and all user-supplied values are in lower case. Because the user who creates the table will also be the owner of the table, the schema has been omitted.

The five columns to be created are listed next, within parentheses. Each column is defined on a separate line simply to improve readability; this is not an Oracle 10*g* requirement. Notice the definition for the Amedate column; it has been assigned a default value of SYSDATE. This instructs Oracle 10*g* to insert the current date (according to the Oracle 10*g* server) if the user enters a new account manager without including the individual's date of employment. Of course, this is only beneficial if the account manager's record is created on the same date the person is hired; otherwise, this makes an incorrect entry if the Amedate column is left blank. In other words, assign defaults with caution!

NOTE

A user cannot have two tables with the same name. If you attempt to create a table with the same name as another table in your schema, Oracle 10*g* displays an ORA-00955 error message. Similarly, if you create a table and then want to create another table using the same table name, you must first delete the existing table using the command DROP TABLE *tablename*;. (The DROP TABLE command is shown later in this chapter.)

Notice that after the command has been executed, Oracle 10*g* returns the message "Table created" to let the user know the table was created successfully. The message does not contain any reference to rows being created. At this point, the table does not contain any data; only the structure of the table has been created (in other words, the table has been defined in terms of a table name and the type of data it will contain). The data, or rows, must be entered into the table as a separate step using the INSERT command. You will enter all the data for the account managers in Chapter 5.

Viewing List of Tables: USER_TABLES

A data dictionary is a typical component of a DBMS that maintains information about database objects. You can query the data dictionary to verify all the tables that exist in your schema. The USER_TABLES data dictionary object maintains information regarding all your tables. Figure 3-5 demonstrates querying the data dictionary to generate a list of all your table names. Imagine how important this can be when you start a new job and need to explore an existing database!

FIGURE 3-5 Listing names of all tables

Viewing Table Structures: DESCRIBE

To determine whether the structure of the table was created correctly, you can use the SQL*Plus command DESCRIBE *tablename* to display the structure of the table, as shown in Figure 3-6. Because the DESCRIBE command is a SQL*Plus command rather than a SQL command, it can be abbreviated and an ending semicolon is not required. The abbreviation for the DESCRIBE command is DESC.

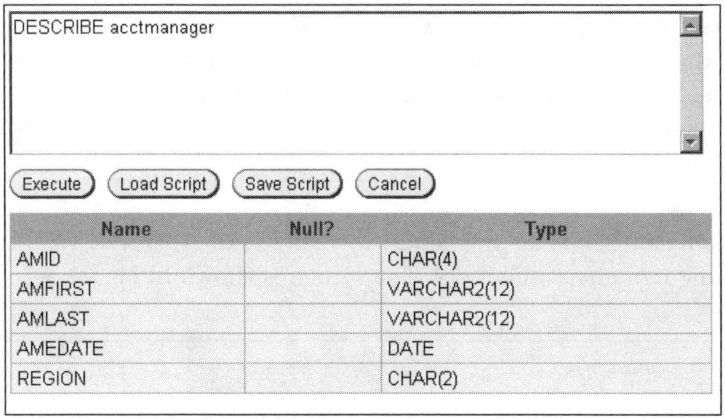

```
DESCRIBE acctmanager
```

(Execute) (Load Script) (Save Script) (Cancel)

Name	Null?	Type
AMID		CHAR(4)
AMFIRST		VARCHAR2(12)
AMLAST		VARCHAR2(12)
AMEDATE		DATE
REGION		CHAR(2)

FIGURE 3-6   The DESCRIBE command

When you issue the **DESCRIBE** command, all columns defined for the ACCTMAN-AGER table are listed. For each column name, you can also check whether the column allows NULL values and the column's datatype. Notice that the results do not display a "NOT NULL" requirement for the AmId column—it is blank. Because this column will be the primary key for the table, it should not be allowed to contain any blank entries. (This problem is corrected in the next chapter.) Thus, if the four columns have the correct name, datatype, and width—*and* your CREATE TABLE command executed—you now have a table ready to accept account managers' data.

TABLE CREATION THROUGH SUBQUERIES

In the previous section, a table was created "from scratch." However, sometimes you might need to create a table based on data contained in existing tables. For example, suppose that management wants each account manager to have a table containing her customers' names. In addition, the table should include the book titles ordered by each customer. A word of caution is in order, however. You generally would not want multiple copies of data within the same database—it could lead to data redundancy. However, in this case, account managers would have the flexibility of adding columns for making comments or changes that should not be reflected in the main database tables. Because the account managers' tables would not be relevant to the transaction-processing activities for JustLee Books (such as order processing), allowing account managers to have tables containing data that already exist in other tables would not be a cause for alarm.

A **subquery** is a SELECT statement that is used within another SQL command. Any type of action that you can perform with a SELECT statement (such as filtering rows, filtering columns, and calculating aggregate amounts) can be performed when creating a table with a subquery. At this point, this text has covered only basic queries, so the example is limited to the SELECT statement features that are already familiar. After you have become familiar with all the features associated with the SELECT statement, you will understand the expanded explanation of this topic in Chapter 12.

The JustLee Books Marketing Department needs to analyze customer data, and management does not want the analysis queries to slow down the production server on which orders are entered. A copy of the customer table is required. The name and street address columns are not needed for the analysis. The Marketing Department requested that the table be named CUST_MKT.

CREATE TABLE...AS

To create a table that will contain data from existing tables, you can use the CREATE TABLE command with an AS clause that contains a subquery. The syntax is shown in Figure 3-7.

```
CREATE TABLE tablename [(columnname, …)]
AS (subquery);
```

FIGURE 3-7   CREATE TABLE...AS command syntax

The command syntax shown in Figure 3-7 uses the CREATE TABLE keywords to instruct Oracle 10g to create a table. The name of the new table is then provided.

If you need to give the columns in the new table names that are different from the column names in the existing table, the new column names must be listed after the table name, inside parentheses. However, if you do not want to change any of the column names, the column list in the CREATE TABLE clause is omitted. If you do provide a column list in the CREATE TABLE clause, it must contain a name for *every* column to be returned by the query—including those names that will remain the same. In other words, if five columns are going to be returned from the subquery, five columns must be listed in the CREATE TABLE clause, or Oracle 10g returns an error message, and the statement fails. In addition, the column list must be in the *same order* as the columns listed in the SELECT clause of the subquery, so Oracle 10g knows which column from the subquery is assigned to which column in the new table.

The AS keyword instructs Oracle 10g that the columns in the new table will be based on the columns returned by the subquery. The AS keyword must precede the subquery. The columns in the new table will be created based on the same datatype and width the columns had in the existing table(s). To distinguish the subquery from the rest of the CREATE TABLE command, *the subquery must be enclosed within parentheses.*

The creation of the CUST_MKT table based on a subquery is shown in Figure 3-8. To verify the results in regard to table structure and contents, execute the DESCRIBE and SELECT commands, as shown in Figures 3-9 and 3-10, respectively.

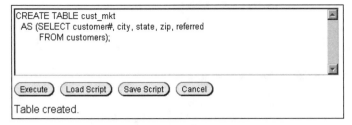

```
CREATE TABLE cust_mkt
  AS (SELECT customer#, city, state, zip, referred
      FROM customers);
```

Execute | Load Script | Save Script | Cancel

Table created.

FIGURE 3-8 Creating a table based on a subquery

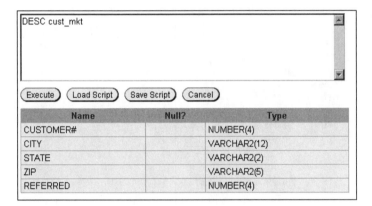

```
DESC cust_mkt
```

Execute | Load Script | Save Script | Cancel

Name	Null?	Type
CUSTOMER#		NUMBER(4)
CITY		VARCHAR2(12)
STATE		VARCHAR2(2)
ZIP		VARCHAR2(5)
REFERRED		NUMBER(4)

FIGURE 3-9 Using DESCRIBE to verify new table structure

```
SELECT *
 FROM cust_mkt;
```
(Execute) (Load Script) (Save Script) (Cancel)

CUSTOMER#	CITY	STATE	ZIP	REFERRED
1001	EASTPOINT	FL	32328	
1002	SANTA MONICA	CA	90404	
1003	TALLAHASSEE	FL	32306	
1004	BOISE	ID	83707	
1005	SEATTLE	WA	98115	
1006	ALBANY	NY	12211	
1007	AUSTIN	TX	78710	1003
1008	CHEYENNE	WY	82003	
1009	BURBANK	CA	91510	1003
1010	ATLANTA	GA	30314	
1011	CHICAGO	IL	60606	
1012	BOSTON	MA	02110	
1013	CLERMONT	FL	34711	1006
1014	CODY	WY	82414	
1015	MIAMI	FL	33111	
1016	BURBANK	CA	91508	1010
1017	KALMAZOO	MI	49006	
1018	MACON	GA	31206	
1019	MORRISTOWN	NJ	07962	1003
1020	TRENTON	NJ	08607	

20 rows selected.

FIGURE 3-10 Using SELECT to verify new table contents

MODIFYING EXISTING TABLES

There might be times when you need to make structural changes to a table. For example, you might need to add a column, delete a column, or simply change the size of a column. Each of these changes is accomplished through the **ALTER TABLE** command. A useful feature of Oracle 10g is that a table can be modified without having to shut down the database. Even if users are accessing a table, it can still be modified without disruption of service. The syntax for the ALTER TABLE command is shown in Figure 3-11.

```
ALTER TABLE tablename
ADD|MODIFY|DROP COLUMN| columnname [definition];
```

FIGURE 3-11 Syntax for the ALTER TABLE command

Whether you should use an ADD, MODIFY, or DROP COLUMN clause depends on the type of change being made. First, let's look at the ADD clause.

ALTER TABLE...ADD Command

Using an **ADD** clause with the ALTER TABLE command allows a user to add a new column to a table. The same rules that apply to creating a column in a new table apply to adding a column to a table that already exists. The new column must be defined by a column name and datatype (and width, if applicable). A default value can also be assigned. The difference is that the new column will be added at the end of the existing table—it will be the last column. The syntax for the ALTER TABLE command with the ADD clause is shown in Figure 3-12.

```
ALTER TABLE tablename
ADD (columnname datatype, [DEFAULT] …);
```

FIGURE 3-12   Syntax for the ALTER TABLE...ADD command

Suppose that after the PUBLISHER table was created, management requests that the table also contain a column for a telephone number extension. The column name should be Ext. The column can consist of a maximum of four numeric digits so the column is defined as a NUMBER datatype with a precision of four. To make this change to the PUBLISHER table, issue the command shown in Figure 3-13.

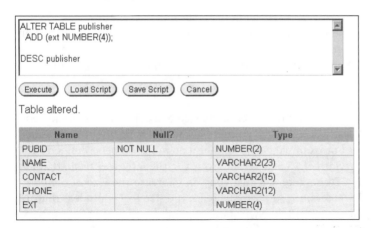

FIGURE 3-13   The ALTER TABLE...ADD command

Oracle 10g returns the message "Table altered" to indicate that the command was completed successfully. To double-check that the column was added with the correct datatype, issue the DESCRIBE command, as shown in Figure 3-13.

NOTE

When you add a column to a table that contains rows of data, the new column is empty for all the existing rows. Perform a SELECT command on the PUBLISHER table to confirm that the new Ext column is empty.

If you need to add more than one column to the ACCTMANAGER table, list the additional column(s) in a column list, and separate each column and its datatype from the other columns with a comma, using the same format as the CREATE TABLE command.

ALTER TABLE...MODIFY Command

A **MODIFY** clause can be used with the ALTER TABLE command to change the definition of an existing column. The changes that can be made to a column include the following:

- Changing the size of a column (increase or decrease)
- Changing the datatype (such as VARCHAR2 to CHAR)
- Changing or adding the default value of a column (such as DEFAULT SYSDATE)

The syntax for the ALTER TABLE...MODIFY command is shown in Figure 3-14.

```
ALTER TABLE tablename
MODIFY (columnname datatype [DEFAULT],…);
```

FIGURE 3-14 Syntax of the ALTER TABLE...MODIFY command

You should be aware of three rules when modifying existing columns.

- A column must be as wide as the data fields it already contains.
- If a NUMBER column already contains data, you can't decrease the precision or scale of the column.
- Changing the default value of a column does not change the values of data already in the table.

Next, let's look at each of these rules.

The first rule applies when you want to decrease the size of a column that already contains data. You can only decrease the size of a column to a size that is no less than the largest width of the existing data. For example, suppose that a column has been defined as a VARCHAR2 datatype with a width of 15 characters. However, the largest entry in that particular column contains only 12 characters. Thus, you can decrease the size of the column only to a width of 12 characters. If you try to decrease the size to a width less than 12, Oracle 10g returns an error message, and the SQL statement fails. As shown in Figure 3-15, when a user attempts to decrease the width of the column to a size that will not accommodate the data it already contains, Oracle 10g returns an ORA-01441 error message, and the table is not altered.

Second, Oracle 10g does not allow you to decrease the precision or scale of a NUMBER column if the column contains any data. Regardless of whether the current values stored in a NUMBER column will be affected, Oracle 10g returns an ORA-01440 error message, and the statement fails unless the column is empty. As shown in Figure 3-16, if you attempt to change the size of the Retail column in the BOOKS table, an error message is displayed, and the table is not altered.

The third rule applies when you modify existing columns and decide to change the default value assigned to a column. When the default value of a column is changed, it changes only the value assigned to *future rows* inserted into the table. The default value

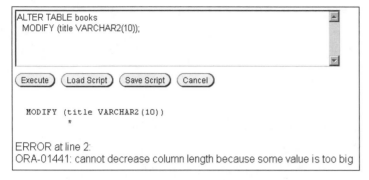

```
ALTER TABLE books
 MODIFY (title VARCHAR2(10));
```

Execute   Load Script   Save Script   Cancel

```
 MODIFY (title VARCHAR2(10))
              *
```

ERROR at line 2:
ORA-01441: cannot decrease column length because some value is too big

FIGURE 3-15 Modify column size error

```
ALTER TABLE books
 MODIFY (retail NUMBER(5,1));
```

Execute   Load Script   Save Script   Cancel

```
 MODIFY (retail NUMBER(5,1))
              *
```

ERROR at line 2:
ORA-01440: column to be modified must be empty to decrease precision or scale

FIGURE 3-16 Modify numeric column error

assigned to rows that already exist in the table remain the same. Thus, if the default value contained in existing rows must be changed, those changes must be performed manually. (Chapter 5 discusses how to change existing values in a row.) Let's assume JustLee Books has decided to assign a service rating code to every publisher that reflects the publisher's service promptness. The code should initially be set to 'N' when a new publisher is added. To demonstrate the effect of adding a DEFAULT option to an existing column, the new column is added first and then it is modified to set the default value. Figure 3-17 displays the ALTER TABLE command to accomplish this column addition task. Notice that the SELECT statement shows that the new Rating column is empty for the existing rows. Second, the DEFAULT option is added on the new column, as shown in Figure 3-18. Notice that the SELECT statement shows that the new Rating column is still empty for the existing rows—the DEFAULT option takes effect only on new rows added.

Now return your attention to the ACCTMANAGER table created earlier in this chapter. A common issue with tables is the need to widen columns to accommodate data values of greater length. After creating the ACCTMANAGER table, suppose that you find out that one of the account managers has a long name that requires more than the 12 spaces you assigned to the Amlast column. To accommodate the name, the Amlast column must be increased to a width of 18 characters. What command should you use to achieve this?

FIGURE 3-17 Adding the Rating column to the PUBLISHER table

NOTE

Column headings may be truncated based on available linespace. Your column headers may vary from those shown in the fgures in this textbook. Formatting column widths in output will be covered later in this book.

FIGURE 3-18 Adding the DEFAULT option to the new column

Figure 3-19 displays the ALTER TABLE command to widen the Amlast column. Notice that the MODIFY clause of the command does not explicitly state that the Amlast column should be increased by six characters. Instead, the datatype and new width are stated with the width increased to the total desired width for the column. After the command is processed, Oracle 10g returns the message "Table altered," as shown in Figure 3-19. To make certain that change is made, you might also use the DESCRIBE command to view the new table structure.

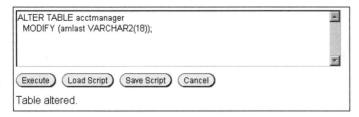

```
ALTER TABLE acctmanager
 MODIFY (amlast VARCHAR2(18));
```

Execute   Load Script   Save Script   Cancel

Table altered.

FIGURE 3-19   The ALTER TABLE...MODIFY command used to increase the width of the Amlast
column

ALTER TABLE...DROP COLUMN Command

The **DROP COLUMN** clause can be used with the ALTER TABLE command to delete an existing column from a table. The clause deletes both the column and its contents, so it should be used with extreme caution. The syntax for the ALTER TABLE...DROP COLUMN command is shown in Figure 3-20.

```
ALTER TABLE tablename
DROP COLUMN columnname;
```

FIGURE 3-20   Syntax for the ALTER TABLE...DROP COLUMN command

You should keep the following in mind when using the DROP COLUMN clause:

- Unlike using an ALTER TABLE command with the ADD or MODIFY clauses, a DROP COLUMN clause can reference only *one* column.
- If you drop a column from a table, the deletion is permanent. You cannot "undo" the damage if you accidentally delete the wrong column from a table. The only option is to add the column back to the table and then manually reenter all the data that it previously contained.
- You cannot delete the last remaining column in a table. If a table contains only one column, and you try to delete the column, the command fails, and Oracle 10g returns an error message.
- A primary key column cannot be dropped from a table.

Previously, you added the Ext column to store the telephone extension of each publisher. However, management has now decided that the extension is not necessary and does not want to waste the storage space that would be taken up by that column. Therefore, the Ext column needs to be deleted from the PUBLISHER table. The command to delete the Ext column is shown in Figure 3-21.

After the command is processed, the Ext column of the PUBLISHER table is removed. To verify that the column no longer exists, use the DESCRIBE command to list the structure of the PUBLISHER table, as shown in Figure 3-22.

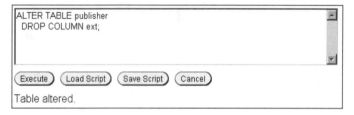

FIGURE 3-21 The ALTER TABLE...DROP COLUMN command

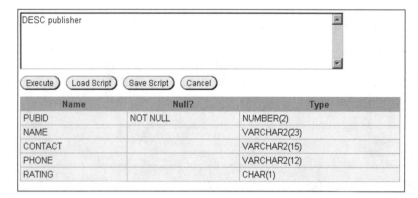

FIGURE 3-22 Verifying the column is dropped

ALTER TABLE...SET UNUSED/DROP UNUSED COLUMNS Command

When the Oracle 10*g* server drops a column from a very large table, this can slow down the processing of queries or other SQL commands from users. To avoid this problem, a **SET UNUSED** clause can be included in the ALTER TABLE command to mark the column for deletion at a later time. If a column is marked for deletion, it is unavailable and will not be displayed in the table structure. Because the column is unavailable, it will not appear in the results of any queries, nor can any other operation be performed on the column except the ALTER TABLE...DROP UNUSED command. In other words, after a column is set as "unused," the column and all its contents are no longer available and cannot be recovered at a future time. It simply postpones the physical erasing of the data from the storage device until a later time—usually when the server is processing fewer queries, such as after business hours. A **DROP UNUSED** clause is used with the ALTER TABLE command to complete the deletion process for any column that has been marked as unused.

The syntax for the ALTER TABLE...SET UNUSED command is shown in Figure 3-23.

```
ALTER TABLE tablename
SET UNUSED (columnname);
     OR
ALTER TABLE tablename
SET UNUSED COLUMN columnname;
```

FIGURE 3-23 Syntax for the ALTER TABLE...SET UNUSED command

As shown in Figure 3-23, the syntax for the ALTER TABLE...SET UNUSED command has two options for the SET UNUSED clause. Regardless of the syntax used, only one column per command can be marked for deletion. The syntax to drop a column previously identified as unused is shown in Figure 3-24.

```
ALTER TABLE tablename
DROP UNUSED COLUMNS;
```

FIGURE 3-24 Syntax for the ALTER TABLE...DROP UNUSED COLUMNS command

When the DROP UNUSED COLUMNS clause is used, any column that has previously been set as "unused" is deleted, and any storage previously occupied by data contained in the column(s) becomes available.

Suppose that the JustLee Books Marketing Department has decided that they do not need to see the customer state data in the CUST_MKT table. To delete that column from the CUST_MKT table, you could use the DROP COLUMN option of the ALTER TABLE command. However, suppose that the table contains thousands of records and deleting a column would slow down operations for other Oracle 10g users. In this case, you can mark the State column of the CUST_MKT table as unused with the command shown in Figure 3-25.

```
ALTER TABLE cust_mkt
  SET UNUSED (state);




( Execute )  ( Load Script )  ( Save Script )  ( Cancel )
Table altered.
```

FIGURE 3-25 The ALTER TABLE...SET UNUSED command

To make certain that the State column was correctly marked for deletion, the DESCRIBE command can be used to check that the column is no longer available, as shown in Figure 3-26.

You can also reference the data dictionary to determine if any columns are marked as unused. The data dictionary object named USER_UNUSED_COL_TABS contains information on which tables have unused columns and how many columns are set to unused. Figure 3-27 displays a query on this object.

After the column has been set as unused, the storage space taken up by data contained within the State column can be reclaimed, using the command shown in Figure 3-28.

Renaming a Table

Oracle 10g allows you to change the name of any table you own, using the **RENAME...TO** command. The syntax for the RENAME...TO command is shown in Figure 3-29.

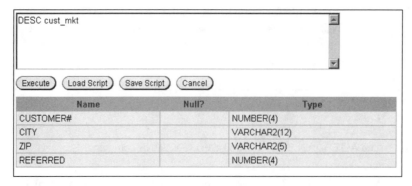

FIGURE 3-26   Verifying the column can no longer be used

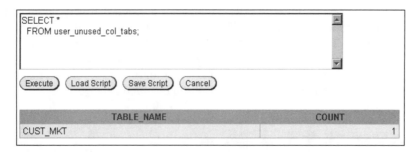

FIGURE 3-27   Listing tables with columns marked as unused

FIGURE 3-28   The ALTER TABLE...DROP UNUSED COLUMNS command

```
RENAME oldtablename TO newtablename;
```

FIGURE 3-29   Syntax for the RENAME...TO command

In a previous section, a table named CUST_MKT was created. However, the Marketing Department wants to maintain a series of snapshots for customer data and, therefore, wants the table name to reflect the month and year the table was created. Assume the current CUST_MKT table was created in September 2005. The command to make the name change is shown in Figure 3-30.

```
RENAME cust_mkt TO cust_mkt_092005;
```

(Execute) (Load Script) (Save Script) (Cancel)

Table renamed.

FIGURE 3-30   The RENAME...TO command

After the RENAME...TO command is executed, any queries directed to the original table named CUST_MKT result in an error message. The table can now only be referenced as the CUST_MKT_092005 table. The SELECT statements shown in Figure 3-31 demonstrate this point.

Truncating a Table

When a table is truncated, all the rows contained in the table are removed, but the table itself remains. In other words, the columns still exist even though no values are stored in them. This is basically the same as deleting all the rows in a table. However, if you simply delete all rows in a table, the storage space occupied by those rows will still be allocated to the table. To delete the rows stored within a table *and* free up the storage space that was occupied by those rows, use the **TRUNCATE TABLE** command. The syntax for the command is shown in Figure 3-32.

Assume that the CUST_MKT_092005 table was originally created a day early and it did not include the customers added on the last day of September. The database administrator is going to repopulate the table; however, all the current rows need to be removed before the repopulation can occur. Keep in mind that the TRUNCATE command will keep the table structure intact, so that it can be reused in the future. TRUNCATE will delete the rows currently contained in the CUST_MKT_092005 table and release the storage space they occupy. The command to perform the truncation is shown in Figure 3-33.

To verify that all rows of the CUST_MKT_092005 table have been removed, perform a query to see all the rows in the table. If the table still exists but contains no rows, Oracle 10g displays the message "no rows selected."

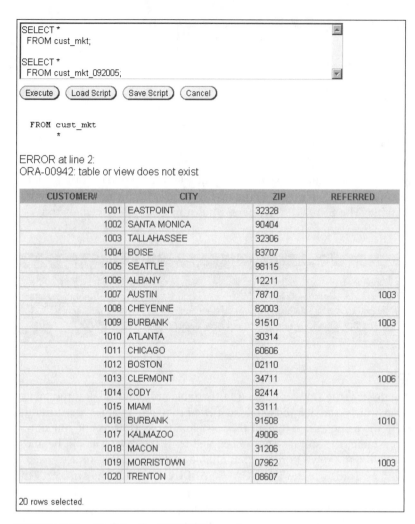

```
SELECT *
 FROM cust_mkt;

SELECT *
 FROM cust_mkt_092005;
```

(Execute) (Load Script) (Save Script) (Cancel)

```
 FROM cust_mkt
      *
```

ERROR at line 2:
ORA-00942: table or view does not exist

CUSTOMER#	CITY	ZIP	REFERRED
1001	EASTPOINT	32328	
1002	SANTA MONICA	90404	
1003	TALLAHASSEE	32306	
1004	BOISE	83707	
1005	SEATTLE	98115	
1006	ALBANY	12211	
1007	AUSTIN	78710	1003
1008	CHEYENNE	82003	
1009	BURBANK	91510	1003
1010	ATLANTA	30314	
1011	CHICAGO	60606	
1012	BOSTON	02110	
1013	CLERMONT	34711	1006
1014	CODY	82414	
1015	MIAMI	33111	
1016	BURBANK	91508	1010
1017	KALMAZOO	49006	
1018	MACON	31206	
1019	MORRISTOWN	07962	1003
1020	TRENTON	08607	

20 rows selected.

FIGURE 3-31   Verifying the RENAME operation

```
TRUNCATE TABLE tablename;
```

FIGURE 3-32   Syntax for the TRUNCATE TABLE command

```
TRUNCATE TABLE cust_mkt_092005;
```

(Execute) (Load Script) (Save Script) (Cancel)

Table truncated.

FIGURE 3-33   The TRUNCATE TABLE command

DELETING A TABLE

A table can be removed from an Oracle 10g database by issuing the **DROP TABLE** command. The syntax for the DROP TABLE command is shown in Figure 3-34.

```
DROP TABLE tablename [PURGE];
```

FIGURE 3-34 Syntax for the DROP TABLE command

Suppose that after truncating the CUST_MKT_092005 table, you realize that you will no longer need the table (or that numerous modifications will have to be made to the table structure so that it is not worth the trouble to make the changes). The CUST_MKT_092005 table can be deleted using the DROP TABLE command, shown in Figure 3-35.

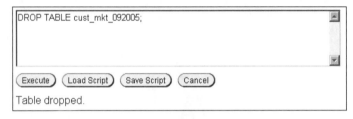

FIGURE 3-35 Using the DROP TABLE command to remove the CUST_MKT_092005 table

After the table has been dropped, the table name will no longer be valid, and the table cannot be accessed by any commands. To verify that the correct table was deleted, you can use the DESCRIBE command to see the structure of the CUST_MKT_092005 table. If the table no longer exists, Oracle 10g returns an error message, as shown in Figure 3-36.

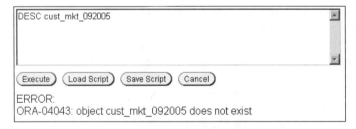

FIGURE 3-36 Using DESCRIBE to verify dropped table

Oracle 10g introduced a new feature related to the DROP TABLE command. In previous Oracle versions, the DROP TABLE command was permanent, and the only method to recover a dropped table was to restore the data from backups. In Oracle 10g, a dropped table is placed into a recycle bin and can be restored—both table structure and data! Now that the CUST_MKT_092005 table has been dropped, you should check the recycle bin with the command shown in Figure 3-37. Notice a new name beginning with BIN has been assigned to the table. This new name is assigned by the system, so results may vary.

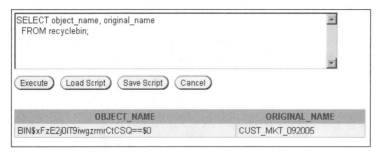

FIGURE 3-37 Checking the recycle bin

Now that you have a table in the recycle bin, how do you recover the table to use it again? You need to use the FLASHBACK command, as shown in Figure 3-38. Keep in mind that the entire table structure and all the data that was in the table at the time of the drop will be restored. After performing the FLASHBACK command, execute a DESCRIBE and SELECT on the table to verify that it is restored.

FIGURE 3-38 Using FLASHBACK to restore a dropped table

Obviously, if the dropped table is actually just moved to a recycle bin, the storage space required for the table is still being used. If you have limited storage space, this might not be desirable. And, in other cases, you know that you want to permanently delete a table. If the table has already been dropped and is now in the recycle bin, you need to remove the table from the recycle bin to permanently delete the table and clear the storage space being used. Drop the CUST_MKT_092005 table again and verify its existence in the recycle bin with the commands shown in Figure 3-39.

Now you can remove the table from the recycle bin using the PURGE TABLE command and referencing the name the table has been assigned in the recycle bin. Figure 3-40 demonstrates removing the table from the recycle bin.

NOTE

Keep in mind that the system assigns the table name in the recycle bin, so you will most likely have a different table name than what is shown in Figure 3-40. Also, if you want to remove all the tables in the recycle bin, you can do so by issuing a PURGE RECYCLE BIN command.

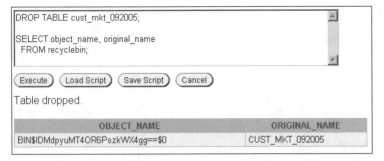

```
DROP TABLE cust_mkt_092005;

SELECT object_name, original_name
  FROM recyclebin;
```

(Execute) (Load Script) (Save Script) (Cancel)

Table dropped.

OBJECT_NAME	ORIGINAL_NAME
BIN$IDMdpyuMT4OR6PszkWX4gg==$0	CUST_MKT_092005

FIGURE 3-39 Dropping a table and check the recycle bin

```
PURGE TABLE "BIN$IDMdpyuMT4OR6PszkWX4gg==$0";
```

(Execute) (Load Script) (Save Script) (Cancel)

Table purged.

FIGURE 3-40 Removing a table from the recycle bin

If you are sure you want to permanently delete a table, you can bypass moving the table to the recycle bin by using the PURGE option in the DROP TABLE statement. For example, you could have initially dropped the CUST_MKT_092005 table using the command shown in Figure 3-41 and then it would not have been moved to the recycle bin. This means the table cannot be recovered.

```
DROP TABLE cust_mkt_092005 PURGE;
```

FIGURE 3-41 Dropping a table with the PURGE option

CAUTION

Always use caution when deleting with the PURGE option. After a table is deleted with PURGE, the table and all the data it contains are gone. In addition, any index that has been created based on that table is also dropped. (Indexes are presented in Chapter 6.)

Table Creation and Management

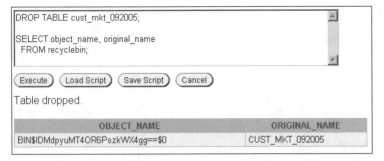

Chapter Summary

- You create a table with the CREATE TABLE command. Each column to be contained in the table must be defined in terms of the column name, datatype, and for certain datatypes, the width.

- A table can contain up to 1000 columns.

- Each column name within a table must be unique.

- Table and column names can contain as many as 30 characters. The names must begin with a letter and cannot contain any blank spaces.

- To create a table based on the structure and data contained in existing tables, use the CREATE TABLE...AS command to use a subquery to extract the necessary data from the existing table.

- You can change the structure of a table with the ALTER TABLE command. Columns can be added, resized, and even deleted with the ALTER TABLE command.

- When using the ALTER TABLE command with the DROP COLUMN clause, only one column can be specified for deletion.

- You can use the SET UNUSED clause to mark a column so its storage space can be freed up at a later time.

- Tables can be renamed with the RENAME...TO command.

- To delete all the rows in a table, use the TRUNCATE TABLE command.

- To remove both the structure of a table and all its contents, use the DROP TABLE command.

- A dropped table is moved to the recycle bin and can be recovered using the FLASH-BACK TABLE command.

- Using the PURGE option in a DROP TABLE command permanently removes the table, meaning you cannot recover it from the recycle bin.

Chapter 3 Syntax Summary

The following table presents a summary of the syntax that you have learned in this chapter. You can use the table as a study guide and reference.

SYNTAX GUIDE		
Commands and Clauses	Description	Example
Creating Tables		

`CREATE TABLE`	Creates a new table in the database—user names columns; defaults and datatypes define/ limit columns. To view the table, use the SQL*PLUS command DESCRIBE.	`CREATE TABLE acctmanager`
		`(amid    VARCHAR2(4),`
		`amname   VARCHAR2(20),`
		`amedate  DATE DEFAULT SYSDATE,`
		`region   CHAR(2));`

`CREATE TABLE ... AS (...)`	Creates a table from existing database tables, using the AS clause and subqueries.	`CREATE TABLE secustomerorder`
		`  AS (SELECT customer#, state,`
		`      ISBN, category, quantity,`
		`      cost, retail`
		`      FROM customers NATURAL JOIN`
		`           orders NATURAL JOIN`
		`           orderitems NATURAL JOIN`
		`           books`
		`      WHERE state IN ('FL', 'GA',`
		`                       'AL'));`

Modifying Tables		
`ALTER TABLE...` `ADD`	Adds a column to a table.	`ALTER TABLE acctmanager` `  ADD (ext NUMBER(4));`
`ALTER TABLE...` `MODIFY`	Changes a column size, datatype, or default value.	`ALTER TABLE acctmanager` `  MODIFY (amname VARCHAR2(25));`
`ALTER TABLE...` `DROP COLUMN`	Deletes one column from a table.	`ALTER TABLE acctmanager` `  DROP COLUMN ext;`

`ALTER TABLE...`
`SET UNUSED`
`or`
`SET UNUSED COLUMN`
	Marks a column for deletion at a later time.	`ALTER TABLE cust_mkt`
		`  SET UNUSED (state);`

`DROP UNUSED` `COLUMNS`	Completes the deletion of a column marked with SET UNUSED.	`ALTER TABLE cust_mkt` `  DROP UNUSED COLUMNS;`
RENAME...TO	Changes a table name.	`RENAME cust_mkt TO cust_mkt_` `092005;`

Table Creation and Management

SYNTAX GUIDE		
Commands and Clauses	Description	Example
Modifying Tables		
TRUNCATE TABLE	Deletes table rows, but table name and column structure remain.	TRUNCATE TABLE cust_mkt_092005;
Deleting Tables		
DROP TABLE . . . PURGE	Removes an entire table permanently from the Oracle 10g database with the PURGE option.	DROP TABLE cust_mkt_092005 PURGE;
PURGE TABLE	Permanently deletes a table in the recyclebin.	PURGE TABLE "BIN$IDMdosJceWxgg041==$0";
Recovering Tables		
FLASHBACK TABLE . . . TO BEFORE DROP	Recovers a dropped table if option not used when table dropped.	PURGE FLASHBACK TABLE cust_mkt_ 092005 TO BEFORE DROP;

Review Questions

To answer the following questions, refer to the tables in Appendix A.

1. Which command is used to create a table based upon data already contained in an existing table?

2. List four datatypes supported by Oracle 10*g*, and provide an example of data that could be stored by each datatype.

3. What are the guidelines you should follow when naming tables and columns in Oracle 10*g*?

4. What is the difference between dropping a column and setting a column as unused?

5. How many columns can be dropped in one ALTER TABLE command?

6. What happens to the existing rows of a table if the DEFAULT value of a column is changed?

7. Explain the difference between truncating a table and deleting a table.

8. If you add a new column to an existing table, where will the column appear relative to the existing columns?

9. What happens if you try to decrease the scale or precision of a NUMBER column to a value less than the data already stored in the field?

10. Is a table and the data contained in the table permanently erased from the system if a DROP TABLE command is issued on the table?

Multiple Choice

To answer the following questions, refer to the tables in Appendix A.

1. Which of the following is a correct statement?
    a. You can restore the data deleted with the DROP COLUMN clause, but not the data deleted with the SET UNUSED clause.
    b. You cannot create empty tables—all tables must contain at least three rows of data.
    c. A table can contain a maximum of 1000 columns.
    d. The maximum length of a table name is 265 characters.

2. Which of the following is a valid SQL statement?
    a. ALTER TABLE secustomersspent ADD DATE lastorder;
    b. ALTER TABLE secustomerorders DROP retail;
    c. CREATE TABLE newtable AS (SELECT * FROM customers);
    d. ALTER TABLE drop column *;

3. Which of the following is not a correct statement?
    a. A table can be modified only if it does not contain any rows of data.
    b. The maximum number of characters in a table name is 30.
    c. You can add more than one column at a time to a table.
    d. You cannot recover data contained in a table that has been truncated.

4. Which of the following is not a valid SQL statement?
    a. CREATE TABLE anothernewtable (newtableid VARCHAR2(2));
    b. CREATE TABLE anothernewtable (date, anotherdate) AS (SELECT orderdate, ship-date FROM orders);
    c. CREATE TABLE anothernewtable (firstdate, seconddate) AS (SELECT orderdate, ship-date FROM orders);
    d. All of the above are valid statements.

5. Which of the following is true?
    a. If you truncate a table, you cannot add new data to the table.
    b. If you change the default value of an existing table, any previously stored default value will be changed to a null value.
    c. If you delete a column from a table, you cannot add a column to the table using the same name as the previously deleted column.
    d. If you add a column to an existing table, it will always be added as the last column of that table.

6. Which of the following will create a new table containing the order number, book title, quantity ordered, and retail price of every book that has been sold?

   a. CREATE TABLE newtable AS(SELECT order#, title, quantity, retail FROM orders);

   b. CREATE TABLE newtable AS(SELECT * FROM orders);

   c. CREATE TABLE newtable AS(SELECT order#, title, quantity, retail FROM orders NATURAL JOIN orderitems);

   d. CREATE TABLE newtable AS(SELECT order#, title, quantity, retail FROM orders NATURAL JOIN orderitems NATURAL JOIN books);

7. Which of the following commands will drop any columns marked as unused from the SECUSTOMERORDERS table?

   a. DROP COLUMN FROM SECUSTOMERORDERS WHERE column_status = UNUSED;

   b. ALTER TABLE SECUSTOMERORDERS DROP UNUSED COLUMNS;

   c. ALTER TABLE SECUSTOMERORDERS DROP (unused);

   d. DROP UNUSED COLUMNS;

8. Which of the following statements is correct?

   a. A table can only contain a maximum of one column that is marked as unused.

   b. You can delete a table by removing all the columns within the table.

   c. Using the SET UNUSED clause allows you to free up all the storage space used by a column.

   d. None of the above statements are correct.

9. Which of the following commands will remove all the data from a table but leave the table's structure intact?

   a. ALTER TABLE secustomerorders DROP UNUSED COLUMNS;

   b. TRUNCATE TABLE secustomerorders;

   c. DELETE TABLE secustomerorders;

   d. DROP TABLE secustomerorders;

10. Which of the following commands would change the name of a table from OLDNAME to NEWNAME?

    a. RENAME oldname TO newname;

    b. RENAME table FROM oldname TO newname;

    c. ALTER TABLE oldname MODIFY TO newname;

    d. CREATE TABLE newname (SELECT * FROM oldname);

11. The default width of a VARCHAR2 field is:

    a. 1

    b. 30

    c. 255

    d. None—there is no default width for a VARCHAR2 field.

12. Which of the following is NOT a valid statement?

    a. You can change the name of a table only if it does not contain any data.

    b. You can change the length of a column that does not contain any data.

    c. You can delete a column that does not contain any data.

    d. You can add a column to a table.

13. Which of the following can be used in a table name?

    a. _

    b. (

    c. %

    d. !

14. Which of the following is true?

    a. When you execute an ALTER TABLE command, other users cannot use the database.

    b. You can add a column to a table as long as the table is not being queried by another user.

    c. All data contained in a table will be lost if the table is dropped.

    d. All of the above statements are true.

15. Which of the following commands is valid?

    a. RENAME customer# TO customernumber FROM customers;

    b. ALTER TABLE customers RENAME customer# TO customernum;

    c. DELETE TABLE customers;

    d. ALTER TABLE customers DROP UNUSED COLUMNS;

16. Which of the following commands will create a new table containing two columns?

    a. CREATE TABLE newname (col1 DATE, col2 VARCHAR2);

    b. CREATE TABLE newname AS(SELECT title, retail, cost FROM books);

    c. CREATE TABLE newname (col1, col2);

    d. CREATE TABLE newname (col1 DATE DEFAULT SYSDATE, col2 VARCHAR2(1));

17. Which of the following is a valid table name?

    a. 9NEWTABLE

    b. DATE9

    c. NEW"TABLE

    d. None of the above are valid table names.

18. Which of the following is a valid datatype?

    a. CHAR3

    b. VARCHAR4(3)

    c. NUMERIC

    d. NUMBER

19. Which of the following will create a table listing books that have a retail price of at least $30.00?

    a. SELECT * FROM books WHERE retail >= 30;

    b. CREATE newtable AS (SELECT * FROM books WHERE retail >= 30);

    c. CREATE newtable AS (SELECT * FROM books WHERE retail > 30);

    d. CREATE newtable AS (SELECT * FROM books) WHERE retail >= 30;

    e. none of the above

20. Which of the following SQL statements will change the size of the Title column in the BOOKS table from the current length of 30 characters to the needed length of 35 characters?

    a. ALTER TABLE books CHANGE title VARCHAR(35);

    b. ALTER TABLE books MODIFY (title VARCHAR2(35));

    c. ALTER TABLE books MODIFY title (VARCHAR2(35));

    d. ALTER TABLE books MODIFY (title VARCHAR2(+5));

Hands-On Assignments

To perform the following activities, refer to the tables in Appendix A.

1. Create a new table that contains the category code and description for the categories of books sold by JustLee Books. The table should be called Category. The columns should be called CatCode and CatDesc. The CatCode column should store a maximum of two characters, and the CatDesc column should store a maximum of 10 characters.

2. Create a new table that contains the following four columns: Emp#, Lastname, Firstname, Job_class. The name of the table should be EMPLOYEES. The Job_class column should be able to store character strings up to a maximum length of four. The column values should not be padded if the value has less than four characters. The Emp# column will contain a numeric ID and should allow a five-digit number.

3. Add two columns to the EMPLOYEES table. One will contain the date of employment for each employee. The default value of the column should be the system date. The new column should be named EmpDate. The second column will be used to contain the date of termination and should be named EndDate.

4. Modify the Job_class column of the EMPLOYEES table so that it will allow a maximum width of two characters to be stored in the column.

5. Delete the EndDate column from the EMPLOYEES table.

6. Rename the EMPLOYEES table to JL_EMPS.

7. Create a new table that contains the following four columns from the existing BOOKS table: ISBN, Cost, Retail, and Category. The name of the ISBN column should be ID, and the other columns should keep their original names. Name the new table BOOK_PRICING.

8. Mark the Category column of the BOOK_PRICING table as unused. Verify that the column is no longer available.

9. Truncate the BOOK_PRICING table and then verify that the table still exists but no longer contains any data.

10. Delete the BOOK_PRICING table permanently so it is not moved to the recycle bin. Delete the JL_EMPS table so it can be restored. Restore the JL_EMPS table and verify that it is now available again.

Advanced Challenge

To perform this activity, refer to the tables in Appendix A.

1. The management of JustLee Books has approved the implementation of the new commission policy for the account managers. The following changes need to be made to the existing database:

   - A new column must be added to the CUSTOMERS table to indicate the region in which each customer lives. The column should be able to store variable-length data to a maximum length of two characters. It should be named Region.

   - A new table, COMMRATE, must be created to store the commission rate schedule. The following columns must exist in the table:
     1. Rate: a numeric field that can store two decimal digits (such as .01 or .03)
     2. Minprice: a numeric field that can store the lowest retail price for a book in that price range of the commission rate
     3. Maxprice: a numeric field that can store the highest retail price for a book in that price range of the commission rate

   Required: Make the necessary changes to the JustLee Books database to support the implementation of the new commission policy.

Case Study: *City Jail*

In the Chapter 1 case study, you designed the new database for City Jail. Now you need to create all the tables needed for the database. First, create all the tables using the information outlined below Section A that follows. Second, make the modifications outlined in Section B. Save all SQL statements used to accomplish these tasks.

Section A

TABLE	COLUMN	DATA DESCRIPTION	LENGTH	SCALE	DEFAULT VALUE
Aliases	Alias_id	Numeric	6		
	Criminal_id	Numeric	6	0	
	Alias	Variable Character	10		
Criminals	Criminal_id	Numeric	6	0	
	Last	Variable Character	15		
	First	Variable Character	10		
	Street	Variable Character	30		

TABLE	COLUMN	DATA DESCRIPTION	LENGTH	SCALE	DEFAULT VALUE
	City	Variable Character	20		
	State	Fixed Character	2		
	Zip	Fixed Character	5	0	
	Phone	Fixed Character	10	0	
	V_status	Fixed Character	1		N (for No)
	P_status	Fixed Character	1		N (for No)
Crimes	Crime_id	Numeric	9	0	
	Criminal_id	Numeric	6	0	
	Classification	Fixed Character	1		
	Data_charged	Date			
	Status	Fixed Character	2		
	Hearing_date	Date			
	Appeal_cut_date	Date			
Sentences	Sentence_id	Numeric	6		
	Criminal_id	Numeric	6		
	Type	Fixed Character	1		
	Prob_id	Numeric	5		
	Start_date	Date			
	End_date	Date			
	Violations	Numeric	3		
Prob_officers	Prob_id	Numeric	5		
	Last	Variable Character	15		
	First	Variable Character	10		
	Street	Variable Character	30		
	City	Variable Character	20		
	State	Fixed Character	2		
	Zip	Numeric	5	0	
	Phone	Numeric	10	0	
	Email	Variable Character	30		
	Status	Fixed Character	1		A (for Active)

TABLE	COLUMN	DATA DESCRIPTION	LENGTH	SCALE	DEFAULT VALUE
Crime_charges	Charge_id	Numeric	10	0	
	Crime_id	Numeric	9	0	
	Crime_code	Numeric	3	0	
	Charge_status	Fixed Character	2		
	Fine_amount	Numeric	7	2	
	Court_fee	Numeric	7	2	
	Amount_paid	Numeric	7	2	
	Pay_due_date	Date			
Crime_officers	Crime_id	Numeric	9	0	
	Officer_id	Numeric	8	0	
Officers	Officer_id	Numeric	8	0	
	Last	Variable Character	15		
	First	Variable Character	10		
	Precinct	Fixed Character	4		
	Badge	Variable Character	14		
	Phone	Numeric	10	0	
	Status	Fixed Character	1		A (for Active)
Appeals	Appeal_id	Numeric	5		
	Crime_id	Numeric	9	0	
	Filing_date	Date			
	Hearing_date	Date			
	Status	Fixed Character	1		P (for Pending)
Crime_codes	Crime_code	Numeric	3	0	
	Code_description	Variable Character	30		

Coding Key for selected columns

TABLE	COLUMN	POSSIBLE VALUES
Criminals	v_status	Y (Yes), N (No)
Criminals	p_status	Y (Yes), N (No)
Crimes	classification	F (Felony), M (Misdemeanor), O (Other), U (Undefined)
Crimes	status	CL (Closed), CA (Can Appeal), IA (In Appeal)
Sentences	Type	J (Jail Period), H (House Arrest), P (Probation)
Prob_officers	status	A (Active), I (Inactive)
Crime_charges	charge_status	PD (Pending), GL (Guilty), NG (Not Guilty)
Officers	status	A (Active), I (Inactive)
Appeals	status	P (Pending), A (Approved), D (Disapproved)

Section B

- Add a default value of 'U' for the Classification column of the CRIMES table.

- Add a column named Date_Recorded to the CRIMES table. This column will need to hold date values and should initially be set to the current date by default.

- Add a column to the PROB_OFFICERS table to contain the pager number for each officer. The column needs to accommodate a phone number including area code. Name the column Pager#.

- Change the Alias column in the ALIASES table to accommodate up to 20 characters.

CHAPTER 4

CONSTRAINTS

LEARNING OBJECTIVES

After completing this chapter, you should be able to do the following:

- Explain the purpose of constraints in a table
- Distinguish among PRIMARY KEY, FOREIGN KEY, UNIQUE, CHECK, and NOT NULL constraints and understand the appropriate use of each constraint
- Understand how constraints can be created when creating a table or when modifying an existing table
- Distinguish between creating constraints at the column level and at the table level
- Create PRIMARY KEY constraints for a single column and a composite primary key
- Create a FOREIGN KEY constraint
- Create a UNIQUE constraint
- Create a CHECK constraint
- Create a NOT NULL constraint using the ALTER TABLE . . . MODIFY command
- Include constraints during table creation
- Use the DISABLE and ENABLE commands
- Use the DROP command

INTRODUCTION

In Chapter 3, you learned how to create tables by using SQL commands. In this chapter, you will learn how

to add constraints to existing tables and how to include constraints during the table creation process.

Constraints are rules used to enforce business rules, practices, and policies—and to ensure the

accuracy and integrity of data. For example, suppose that a customer places an order on April 2, 2005.

However, when the order is shipped, the ship date is entered as March 31, 2005. Shipping an order before the order is placed is impossible, and it indicates a problem with data integrity. If such errors exist in the database, management cannot rely on it for decision making or even to support day-to-day operations. Constraints help solve such problems by not allowing data to be added to tables if the data violates certain rules.

This chapter examines constraints, listed in Figure 4-1, that can prevent errors from being entered into a database.

CONSTRAINT	DESCRIPTION
PRIMARY KEY	Determines which column(s) uniquely identifies each record. The primary key cannot be NULL, and the data value(s) must be unique.
FOREIGN KEY	In a one-to-many or parent-child relationship, the constraint is added to the "many" table. The constraint ensures that if a value is entered into a specified column, it must already exist in the "one" table, or the record is not added.
UNIQUE	Ensures that all data values stored in a specified column are unique. The UNIQUE constraint differs from the PRIMARY KEY constraint in that it allows NULL values.
CHECK	Ensures that a specified condition is true before the data value is added to a table. For example, an order's ship date cannot be earlier than its order date.
NOT NULL	Ensures that a specified column cannot contain a NULL value. The NOT NULL constraint can be created *only* with the column-level approach to table creation.

FIGURE 4-1   List of constraint types

TIP

The examples in this chapter assume you have already created the JustLee Books database as instructed in Chapter 2.

CREATING CONSTRAINTS

You can add constraints during table creation as part of the CREATE TABLE command, or you can do so after the table is created by using the ALTER TABLE command. When creating a constraint, you can choose one of the following options:

1. Name the constraint using the same rules as for tables and columns.
2. Omit the constraint name and allow Oracle 10g to generate the name.

If the Oracle 10g server names the constraint, it follows the format of SYS_Cn, where n is an assigned numeric value to make the name unique. It is a better practice to provide a name for a constraint so you can easily identify it in the future, if necessary. Any constraint violation errors reference the constraint name, so an intuitive name indicating the table, column, and type of constraint is quite helpful.

Industry convention is to use the format *tablename_columnname_constrainttype* for the constraint name. Constraint types are designated by abbreviations, as shown in Figure 4-2.

CONSTRAINT	ABBREVIATION
PRIMARY KEY	_pk
FOREIGN KEY	_fk
UNIQUE	_uk
CHECK	_ck
NOT NULL	_nn

FIGURE 4-2 Constraint abbreviations

> **NOTE**
>
> Most development groups have a set of coding conventions, which include guidelines for naming database objects, including constraints. Upon joining a new development group or company, you should request and review their coding convention guide.

You can create a constraint in two ways: at the column level or at the table level. Creating a constraint at the column level means the definition of the constraint is included as part of the column definition, similar to assigning a default value to a column. Creating a constraint at the table level means the definition of the constraint is separate from the column definition.

Creating Constraints at the Column Level

When you create constraints at the column level, the constraint applies to the column specified. The optional **CONSTRAINT** keyword is used *if* you want to give the constraint a specific name. The constraint type uses the following keywords to identify the type of constraint you are creating:

- PRIMARY KEY
- FOREIGN KEY
- UNIQUE
- CHECK
- NOT NULL

> **NOTE**
>
> A **NULL** value means that no entry is made.

You can create any type of constraint at the column level—unless the constraint is being defined for more than one column (for example, a composite primary key). *If the constraint applies to more than one column, you must create the constraint at the table level.*

The general syntax for creating a constraint at the column level is shown in Figure 4-3.

```
columnname [CONSTRAINT constraintname] constrainttype,
```

FIGURE 4-3   Syntax for creating a column-level constraint

As you will see later in this chapter, a NOT NULL constraint can be created only at the column level.

Creating Constraints at the Table Level

The syntax for creating a constraint at the table level is shown in Figure 4-4.

```
[CONSTRAINT constraintname] constrainttype
(columnname, ...),
```

FIGURE 4-4   Syntax for creating a table-level constraint

When you create a constraint at the table level, the constraint definition is separate from any column definitions. If you create the constraint at the same time that you are creating a table, you list the constraint *after* all the columns are defined. In fact, *the main difference in the syntax of a column-level constraint and a table-level constraint is that you provide the column name(s) for the table-level constraint at the end of the constraint definition in a set of parentheses, rather than at the beginning of the constraint definition.* You can use the table-level approach to create any type of constraint except a NOT NULL constraint. As stated previously, you can create a NOT NULL constraint only using the column-level approach.

NOTE

Although a constraint can be created at the column level or at the table level, the constraint is *always* enforced at the table level, which means the entire row cannot be added, modified, or deleted if it violates a constraint.

To simplify the examples for the different types of constraints, the following sections demonstrate how to add constraints to an existing table. After you have learned how to add constraints by using the ALTER TABLE command, you will learn how to include constraints at both the column level and the table level during table creation.

USING THE PRIMARY KEY CONSTRAINT

A **PRIMARY KEY constraint** is used to enforce the primary key requirements for a table. Although you can create a table in Oracle 10*g* without specifying a primary key, the constraint makes certain that the column(s) identified as the table's primary key is unique and *does not contain NULL values*. As stated previously, a **NULL** value means that no entry is made. It is not equivalent to entering a zero or a blank. The syntax of the ALTER TABLE command to add a PRIMARY KEY constraint to an existing table is shown in Figure 4-5.

```
ALTER TABLE tablename
ADD [CONSTRAINT constraintname] PRIMARY KEY (columnname);
```

FIGURE 4-5 Syntax of the ALTER TABLE command to add a PRIMARY KEY constraint

NOTE

You used the bookscript.sql file in Chapter 2 to create the JustLee database. Review the CREATE TABLE statements in this file. The only constraint included on some tables is a PRIMARY KEY. You will be adding constraints to these tables in the chapter examples.

Let's look at an example. The PROMOTION table stores data regarding the gifts customers will receive during JustLee Books' annual marketing promotion. Upon order completion, each book price will be compared to the Minretail and Maxretail columns to determine which gift the customer will receive for that purchase. Notice that the PROMOTION table does not have a PRIMARY KEY constraint. By not having a primary key designated for the PROMOTION table, a user could mistakenly enter a bookmark as a gift for multiple retail value ranges. Therefore, the purchase of a $10 book and a $50 book could both accidentally be assigned a gift of a bookmark rather than a bookmark and book cover, respectively. This would defeat the purpose of the promotion, which offers more valuable gifts for more expensive purchases. Adding a primary key constraint to the Gift column ensures that a particular gift will be entered only for one value range. To designate the Gift column as the primary key for the PROMOTION table, issue the ALTER TABLE command shown in Figure 4-6.

```
ALTER TABLE promotion
ADD CONSTRAINT promotion_gift_pk PRIMARY KEY (gift);
```

FIGURE 4-6   Adding a PRIMARY KEY constraint

Note the following elements in Figure 4-6:

- The ADD clause instructs Oracle 10g that a constraint will be added to the PROMOTION table (listed after the ALTER TABLE keywords).
- The user has chosen the constraint name, rather than having it assigned by Oracle 10g, and the constraint name is added as **promotion_gift_pk**.
- Because PRIMARY KEY is provided as the constraint type, Oracle 10g makes the Gift column the primary key for the PROMOTION table.

Upon execution of the command, you will receive the message "Table altered," as shown in Figure 4-7. After this command has been executed, all rows must have an entry in the Gift column, and each entry must be unique.

FIGURE 4-7   Results of adding a PRIMARY KEY constraint

You can create only *one* PRIMARY KEY constraint for each table. If the primary key consists of more than one column (a composite primary key), you must create it at the table level. For example, the ORDERITEMS table uses two columns to uniquely identify each item on an order: Order# and Item#. To indicate that the primary key for a table consists of more than one column, simply list the column names within parentheses after the constraint type. You must use commas to separate the list of column names. This is shown in Figure 4-8.

FIGURE 4-8 Adding a composite PRIMARY KEY constraint

After the constraint shown in Figure 4-8 is added to the ORDERITEMS table, all entries in the Order# and Item# columns must create a unique combination in the table, and neither column can contain a NULL value.

> **N O T E**
>
> Because a table can have only one PRIMARY KEY constraint, the name assigned to the constraint (**orderitems_pk**) in Figure 4-8 does not include the names of the columns used to create the composite primary key. However, if the user wanted to include the name of the columns, the constraint name could be assigned as **orderitems_order#item#_pk.**

Although Oracle 10g has already returned the message "Table altered," you can make certain that the two constraints just created actually exist by selecting the names of all constraints that you have created or own from the **USER_CONSTRAINTS** view. The SQL statement to view the names of existing constraints is shown in Figure 4-9.

```
SELECT constraint_name
FROM user_constraints;
```

FIGURE 4-9 SQL command to list existing constraints

After the SQL command is executed, a list of constraint names appears. In Figure 4-10, the two constraints that were previously created, **promotion_gift_pk** and **orderitems_pk** are displayed in the middle of the list. Notice, however, that there are also other constraints in the results. The output will vary, depending on the tables contained within your schema. In this case, thirteen constraints are displayed in the output. Because the two constraints you have created are listed in the output, you can be assured that they were successfully created. You'll learn more about this view and its contents in a later section.

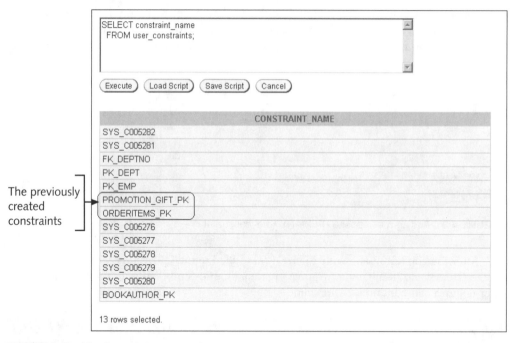

The previously created constraints

FIGURE 4-10   Viewing existing constraint names

TIP

Notice the constraints that begin with **SYS_C**. Oracle 10*g* named these constraints, not the user. As previously mentioned, the constraint names assigned by the software are not very informative, and it is difficult to determine their purpose, or even which table(s) is affected by the constraints. Also notice that there are constraints which have a name that first identifies the type of constraint (**pk** or **fk**). However, it is unclear whether the remainder of the constraint names identify a column or a table—unless you are familiar with the contents of the database. Most organizations have some type of standard naming convention that they use for constraints. Ideally, that naming convention would consist of the table name, column name, and constraint type. By using a standardized naming convention, it is much easier to identify a specific constraint for future reference.

USING THE FOREIGN KEY CONSTRAINT

Suppose that an order is placed with JustLee Books by a new customer who does not exist in the CUSTOMERS table. If the customer information was not collected, the customer's name and billing address would not be contained in the database. This would make it difficult to bill the customer for the order—not exactly what one might consider "good business practice." Or, perhaps there is a book in the BOOKS table that is published with the publisher ID of 9—a publisher ID that does not exist in the PUBLISHER table. (The publisher ID could simply be a typo, or perhaps someone neglected to add the publisher to the PUBLISHER table.)

You can prevent these problems by using a FOREIGN KEY constraint. To prevent someone from entering an order from a customer who does not have a record in the CUSTOMERS table, you can create a constraint that compares any entry made into the Customer# column of the ORDERS table with all customer numbers existing in the CUSTOMERS table. Thus, if a customer service representative enters a customer number not found in the CUSTOMERS table, the corresponding entry in the ORDERS table is rejected. This requires the customer service representative first to collect and enter the customer's information into the CUSTOMERS table, and then enter the order into the ORDERS table.

The syntax to add a FOREIGN KEY constraint to a table is shown in Figure 4-11.

```
ALTER TABLE tablename
ADD [CONSTRAINT constraintname] FOREIGN KEY (columnname)
REFERENCES referencedtablename (referencedcolumnname);
```

FIGURE 4-11 Syntax of the ALTER TABLE command to add a FOREIGN KEY constraint

The keywords **FOREIGN KEY** are used to identify a column that, if it contains a value, must match data contained in another table. The name of the column identified as the foreign key is contained within a set of parentheses after the FOREIGN KEY keywords. The keyword **REFERENCES** refers to **referential integrity**, which means that the user is referring to something that exists in another table. For example, the value entered into the Customer# column of the ORDERS table references a value in the Customer# column of the CUSTOMERS table. The REFERENCES keyword is used to identify the table and column that must already contain the data being entered. The column referenced must be a primary key column. In this case, the Customer# column of the CUSTOMERS table must be defined as the primary key column of that table.

To create a FOREIGN KEY constraint on the Customer# column of the ORDERS table that ensures that any customer number entered also exists in the CUSTOMERS table before the order is accepted, use the command shown in Figure 4-12.

```
ALTER TABLE orders
ADD CONSTRAINT orders_customer#_fk FOREIGN KEY (customer#)
REFERENCES customers (customer#);
```

FIGURE 4-12 Command to add a FOREIGN KEY constraint to the ORDERS table

This command instructs Oracle 10g to add a FOREIGN KEY constraint on the Customer# column of the ORDERS table. The name chosen for the constraint is **orders_customer#_fk**. This constraint makes sure that an entry for the Customer# column of the ORDERS table matches a value that is stored in the Customer# column of the CUSTOMERS table. Upon execution of the command, the message "Table altered" is returned, as shown in Figure 4-13.

```
ALTER TABLE orders
  ADD CONSTRAINT orders_customer#_fk FOREIGN KEY (customer#)
  REFERENCES customers (customer#);



  Execute    Load Script    Save Script    Cancel
Table altered.
```

FIGURE 4-13   Adding a FOREIGN KEY constraint

The syntax for the FOREIGN KEY constraint is more complex than for the PRIMARY KEY constraint because two tables are involved in the constraint. The CUSTOMERS table is the referenced table (it is the "one" side of the one-to-many relationship between the CUSTOMERS and ORDERS table—each order can be placed by only one customer, but one customer can place many orders). Thus, the CUSTOMERS table is considered the parent table for the constraint; the ORDERS table is considered the child table.

When a FOREIGN KEY constraint exists between two tables, by default, a record cannot be deleted from the parent table if matching entries exist in the child table. This means that in the JustLee Books example, *you cannot delete a customer from the CUSTOMERS table if there are orders in the ORDERS table for that customer*.

However, suppose that you do need to delete a customer from the CUSTOMERS table. Perhaps the customer has not paid for previous orders, or perhaps the customer has passed away. Your goal is to remove the customer from the database to make certain no one places an order using that customer's information. The FOREIGN KEY constraint requires that you first delete all of that customer's orders from the ORDERS table (the child table) and then delete the customer from the CUSTOMERS table (the parent table). If the customer has placed many orders, this process could take some time.

There is a much simpler solution, however: You can add the keywords ON DELETE CASCADE to the end of the command issued in Figure 4-12. If the **ON DELETE CASCADE** keywords are included in the constraint definition and a record is deleted from the parent table, any corresponding records in the child table are also automatically deleted. Figure 4-14 shows a FOREIGN KEY constraint with the ON DELETE CASCADE option.

```
ALTER TABLE orders
ADD CONSTRAINT orders_customer#_fk FOREIGN KEY
             (customer#)
REFERENCES customers (customer#) ON DELETE CASCADE;
```

FIGURE 4-14   FOREIGN KEY constraint with the ON DELETE CASCADE option

N O T E

If you attempt the command shown in Figure 4-14 and receive an error message, this might be because a FOREIGN KEY constraint already exists with the same name. Enter the following command:
ALTER TABLE orders
DROP CONSTRAINT orders_customer#_fk;
After you've removed the previous constraint from the database, you can then reenter the command from Figure 4-14 without receiving an error message.

N O T E

Thus, using the example in Figure 4-14, if a customer who has placed 20 orders is deleted from the CUSTOMERS table, all orders placed by that customer are deleted from the ORDERS table. Clearly, you must be very cautious with the ON DELETE CASCADE option. It could create a problem for unsuspecting users who unintentionally delete outstanding orders. Make absolutely certain that any records that might get deleted from the child table through this option will not be needed in the future. If that possibility exists—even remotely—do not include the ON DELETE CASCADE keywords, and force the user to explicitly delete the entries in the child table before removing the parent record.

If a record in a child table has a NULL value for a column that has a FOREIGN KEY constraint, the record is accepted. For example, all the constraint ensures is that the customer number is a valid number, *not* that a customer number has been entered for an order. Basically, this means that an order could be entered into the ORDERS table without an entry in the Customer# column, and the order would still be accepted. To force the user to enter a customer number for an order, you should also add a NOT NULL constraint for the Customer# column in the ORDERS table. (You will add such a constraint in a later section.)

N O T E

A FOREIGN KEY constraint cannot reference a column in a table that has not already been designated as the primary key for the referenced table.

A FOREIGN KEY constraint is unlike the other constraints in that it involves more than one table. Because a FOREIGN KEY affects more than one table, it can impact a DROP TABLE command on the parent table of the relationship established by the constraint. For example, we just created a FOREIGN KEY constraint relating the CUSTOMERS and ORDERS tables. If you attempt to delete the parent table (CUSTOMERS), an error referencing the FOREIGN KEY is raised, as shown in Figure 4-15.

Two options exist to allow the deletion of the parent table:

- DROP the child table and then DROP the parent table.
- DROP the parent table with the CASCADE CONSTRAINTS option, as shown in Figure 4-16. This option deletes the FOREIGN KEY constraint in the child table and then deletes the parent table.

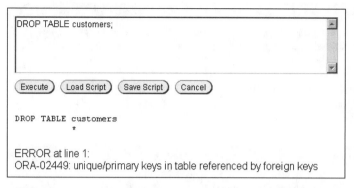

FIGURE 4-15 DROP TABLE error due to FOREIGN KEY

```
DROP TABLE customers
  CASCADE CONSTRAINTS;
```

FIGURE 4-16 Using the CASCADE CONSTRAINTS option

USING THE UNIQUE CONSTRAINT

The purpose of a **UNIQUE** constraint is to ensure that two records cannot have the same value stored in the same column. Although this sounds like a PRIMARY KEY constraint, there is one major difference. *A UNIQUE constraint allows NULL values*, which are not permitted with a PRIMARY KEY constraint. Therefore, the UNIQUE constraint performs a check on the data only if a value is entered for the column. The syntax to add a UNIQUE constraint to an existing table is shown in Figure 4-17.

```
ALTER TABLE tablename
ADD [CONSTRAINT constraintname] UNIQUE (columnname);
```

FIGURE 4-17 Syntax for adding a UNIQUE constraint to a table

As shown in Figure 4-17, the syntax to add a UNIQUE constraint is the same as the syntax for adding a PRIMARY KEY constraint, except the UNIQUE keyword is used to indicate the type of constraint being created.

For example, suppose that JustLee Books wants to make certain that each book in inventory has a different title entry to assist customers and employees to differentiate books with titles that are, in fact, the same. The company may add subject or author information at the end of the title so customers will not accidentally select the wrong book to purchase. To create a UNIQUE constraint on the Title column of the BOOKS table, issue the command shown in Figure 4-18.

After the command is successfully issued, Oracle 10g does not allow any entry into the Title column of the BOOKS table that will duplicate an existing entry. If multiple books have the same title, some modification is now required to differentiate the titles of the two

```
ALTER TABLE books
  ADD CONSTRAINT books_title_uk UNIQUE (title);
```

(Execute) (Load Script) (Save Script) (Cancel)

Table altered.

FIGURE 4-18   Creating a UNIQUE constraint

books. For example, if a second edition of a book is published with the same title as the first edition, the user must include the edition number in the title to make it different from the record for the previous edition.

USING THE CHECK CONSTRAINT

In this chapter's introduction, we presented a scenario in which an order's ship date was earlier than its order date. You can prevent data entry errors of this type through the use of a CHECK constraint. A **CHECK** constraint requires that a specific condition be met before a record is added to a table. With a CHECK constraint, you can, for example, make certain that a book's cost is greater than zero, or that its retail price is less than $200.00, or that a seller's commission rate is less than 50%. The condition included in the constraint cannot reference certain built-in functions, such as SYSDATE, or refer to values stored in other rows (though it can be compared to values within the *same* row). For instance, you could use the condition that the order date must be earlier than or equal to the ship date. However, you could not add a CHECK constraint that requires the ship date for an order to be the same as the current system date, because you would have to reference the SYSDATE function. The syntax for adding a CHECK constraint to an existing table is shown in Figure 4-19.

```
ALTER TABLE tablename
ADD [CONSTRAINT constraintname] CHECK (condition);
```

FIGURE 4-19   Syntax for adding a CHECK constraint to an existing table

NOTE

The SYSDATE function was introduced earlier in this text. Other functions are covered in Chapter 11.

The syntax to add a CHECK constraint follows the same format as the syntax to add a PRIMARY KEY or UNIQUE constraint. However, rather than list the column name(s) for the constraint, you list the condition that must be satisfied.

To solve the problem of an incorrect ship date being entered into the table, the condition can be stated as **(orderdate<=shipdate)**, so the ship date entered in a record cannot be earlier than the order date. The command to add the CHECK constraint to the ORDERS table is shown in Figure 4-20.

```
ALTER TABLE orders
ADD CONSTRAINT orders_shipdate_ck
CHECK (orderdate<=shipdate);
```

FIGURE 4-20   Adding a CHECK constraint to the ORDERS table

If any records that are already stored in the ORDERS table violate the **orderdate<=shipdate** condition, Oracle 10g returns an error message stating that the constraint has been violated, and the ALTER TABLE command fails. This is true for all the constraint types. If you attempt to add a CHECK constraint and you receive an error message indicating that such a violation exists, issue a SELECT statement and review the data in the table to identify any records preventing the constraint from being added to the table. After those records are identified and corrected, the ALTER TABLE command can be reissued and should be successful. After the CHECK constraint is successfully executed, the message "Table altered" is returned, as shown in Figure 4-21.

```
ALTER TABLE orders
ADD CONSTRAINT orders_shipdate_ck CHECK (orderdate<=shipdate);


  ( Execute )  ( Load Script )  ( Save Script )  ( Cancel )
Table altered.
```

FIGURE 4-21   Execution of statement to add a CHECK constraint

You can use a variety of operators to create the condition needed for your data. Examples are listed as follows:

- Less Than (<) : retail < 200
- Greater Than (>) : cost > 0
- Range (BETWEEN) : retail BETWEEN 0 AND 200
- List of Values (IN) : region IN ('NE','SE','NW','SW')

NOTE

The BETWEEN operator is inclusive. The preceding example is interpreted as the retail column value that could be 0 or 200 or any number between those two values.

USING THE NOT NULL CONSTRAINT

The **NOT NULL** constraint is a special CHECK constraint with the condition of IS NOT NULL. As such, it prevents users from adding a row that contains a NULL value in the specified column. However, a NOT NULL constraint is not added to a table in the same manner as constraints previously presented in this chapter. *A NOT NULL constraint can be added only to an existing column by using the ALTER TABLE...MODIFY command.*

The syntax for adding a NOT NULL constraint is shown in Figure 4-22.

```
ALTER TABLE tablename
MODIFY (columnname [CONSTRAINT constraintname]
NOT NULL);
```

FIGURE 4-22   Syntax for adding a NOT NULL constraint to an existing table

The ALTER TABLE...MODIFY command is the same command used in Chapter 3 to redefine a column. You need to list just the column's name and the keywords NOT NULL. You do not have to list the column's datatype and width or any default value, if one exists. For example, suppose that you want to add a NOT NULL constraint to the PubID column of the BOOKS table. The command and its successful execution, "Table altered," are shown in Figure 4-23.

```
ALTER TABLE books
  MODIFY (pubid CONSTRAINT books_pubid_nn NOT NULL);
```

(Execute) (Load Script) (Save Script) (Cancel)

Table altered.

FIGURE 4-23   Adding a NOT NULL constraint

Because you would expect only one NOT NULL constraint for a particular column, the industry rarely gives constraint names to this type of constraint (although it is still advisable if you ever need to delete the constraint in the future). If you do not want to assign a name to a NOT NULL constraint, simply omit the CONSTRAINT keyword and list the constraint type directly after the column name, as shown in Figure 4-24.

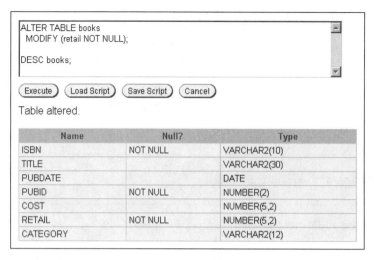

FIGURE 4-24 Adding a NOT NULL constraint without a name

As shown in Figure 4-24, you can use the DESCRIBE command (DESC) to determine whether a column can contain a NULL value. The Null? column in the results of the DESCRIBE command contains the value NOT NULL if the column is not allowed to contain NULL values. The only problem with using this approach is that you cannot tell whether a NOT NULL constraint is being used to prevent NULL values, or if it is due to a PRIMARY KEY constraint. Methods for identifying specific information regarding the constraints for a table are presented later in this chapter using the USER_CONSTRAINTS view.

INCLUDING CONSTRAINTS DURING TABLE CREATION

Now that you've examined adding constraints to existing tables, let's look at adding constraints to tables during table creation. If constraints are included during a table's creation, data that violates those constraints cannot ever be added to the table—unless the user disables the constraints. When the design process for a database is thorough, you identify all needed constraints before creating a table. In this case, the constraints can be included in the CREATE TABLE command, so they will not need to be added at a later time as a separate step.

As previously mentioned, you can use two approaches to define constraints: at the column level or at the table level. If a constraint is created at the column level as part of the CREATE TABLE command, the constraint type is simply listed after the datatype for the column. In Chapter 3, you created the ACCTMANAGER table, using the command shown in Figure 4-25.

```
CREATE TABLE acctmanager
    (amid    CHAR(4),
     amfirst VARCHAR2(12),
     amlast  VARCHAR2(12),
     amedate DATE  DEFAULT SYSDATE,
     region  CHAR(2));
```

FIGURE 4-25   Original command to create the ACCTMANAGER table

However, this command did not include any references to a PRIMARY KEY constraint or any other type of constraint. Because the AmID column was designed to be the primary key for the ACCTMANAGER table, a PRIMARY KEY constraint should be created to ensure that the column will never have any NULL values and that the same ID will not be assigned to more than one account manager. In addition, an account manager's name and the region to which he is assigned should never be left blank. With some minor modifications to the CREATE TABLE command shown in Figure 4-25, these constraints can be included during the creation of the ACCTMANAGER table, as shown in Figure 4-26. Drop the existing ACCTMANAGER table before running the CREATE TABLE statement shown in Figure 4-26.

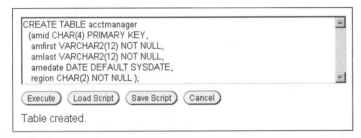

```
CREATE TABLE acctmanager
 (amid CHAR(4) PRIMARY KEY,
  amfirst VARCHAR2(12) NOT NULL,
  amlast VARCHAR2(12) NOT NULL,
  amedate DATE DEFAULT SYSDATE,
  region CHAR(2) NOT NULL );
```

(Execute) (Load Script) (Save Script) (Cancel)

Table created.

FIGURE 4-26   Creating a table with constraints defined at the column level

The modified command in Figure 4-26 adds a PRIMARY KEY constraint to the AmID column and the NOT NULL constraint to the Amname and Region columns. A NOT NULL constraint is not added to the Amedate column because a default value would automatically be assigned if the user does not enter a value for the column. In this example, the constraints are not given a name, so the Oracle server assigns a unique name for each of the three constraints. However, you should remember that this could cause a headache in the future if you ever need to delete the constraint. A name can be assigned to a constraint, using the same format as the ALTER TABLE command, by including the keyword CONSTRAINT followed by the name of the constraint.

NOTE

If you enter the example shown in Figure 4-26 and receive an error message indicating that a table with the same name already exists, you need to issue the **DROP TABLE acctmanager**; command so you can re-create the table.

Constraints

With the exception of the NOT NULL constraint, constraints can also be included in the CREATE TABLE command using the table-level approach. In this case, you list the constraints at the end of the command after all columns are defined, as shown in Figure 4-27.

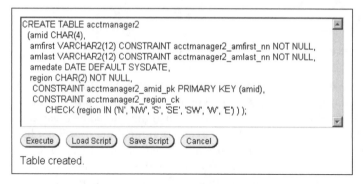

```
CREATE TABLE acctmanager2
  (amid CHAR(4),
   amfirst VARCHAR2(12) CONSTRAINT acctmanager2_amfirst_nn NOT NULL,
   amlast VARCHAR2(12) CONSTRAINT acctmanager2_amlast_nn NOT NULL,
   amedate DATE DEFAULT SYSDATE,
   region CHAR(2) NOT NULL,
   CONSTRAINT acctmanager2_amid_pk PRIMARY KEY (amid),
   CONSTRAINT acctmanager2_region_ck
      CHECK (region IN ('N', 'NW', 'S', 'SE', 'SW', 'W', 'E') ) );
```

Execute   Load Script   Save Script   Cancel

Table created.

FIGURE 4-27   CREATE TABLE command including constraints created with the table-level approach

In the example shown in Figure 4-27, a table named ACCTMANAGER2 has been created with a structure similar to the ACCTMANAGER table. Note the following:

- The NOT NULL constraints for the name columns have been changed to include constraint names.
- The PRIMARY KEY constraint is now being created using the table-level approach.
- A CHECK constraint has been added to the table to make certain that an account manager is assigned to a region using a predetermined code of N for north, NW for Northwest, and so on.
- Notice the three parentheses at the end of the CREATE TABLE command. One parenthesis closes the code list used for the IN logical operator, the second parenthesis closes the condition for the CHECK constraint, and the last parenthesis closes the column list for the CREATE TABLE command.

As shown in Figure 4-27, both the column-level and table-level approaches can be used in the same command, should the need arise. However, the general practice in the industry is to create constraints using the table-level approach. This is not a requirement; it is simply good practice because a column list can become cluttered if a constraint name is provided in the middle of a list that defines all the columns. Therefore, most users define all the columns first, and then include the constraints at the end of the CREATE TABLE command to separate the column definitions from the constraints. It makes it much easier to go back and identify a problem if you receive an error message.

MULTIPLE CONSTRAINTS ON A SINGLE COLUMN

You can assign a column as many constraints as necessary to satisfy all the business rules for that column. For example, we created a UNIQUE constraint on the title column of the BOOKS table earlier, but this might not be enough. To enter a new book, a book title would be required in most cases and, therefore, the column would also need a NOT NULL constraint. In addition, earlier we added a composite PRIMARY KEY to the ORDERITEMS table that included the order# and item#. The order# must also be assigned a FOREIGN KEY constraint that references the ORDERS table to ensure a valid order# is entered. The CREATE TABLE statement for the ORDERITEMS table including both the PRIMARY KEY and FOREIGN KEY is shown in Figure 4-28. Notice that both of the constraints affect the order# column.

```
CREATE TABLE orderitems
(ORDER# NUMBER(4) NOT NULL,
 ITEM# NUMBER(2) NOT NULL,
 ISBN VARCHAR2(10),
 QUANTITY NUMBER(3),
 CONSTRAINT orderitems_pk PRIMARY KEY (order#, item#),
 CONSTRAINT orderitems_order#_fk FOREIGN KEY (order#)
   REFERENCES orders (order#));
```

FIGURE 4-28 Assigning multiple constraints to a column

NOTE

A common error is assigning a NOT NULL constraint to a PRIMARY KEY column. This does not generate an error but does duplicate processing because a PRIMARY KEY constraint does not allow NULL values. Recall that a PRIMARY KEY checks for both uniqueness and no NULL values.

VIEWING CONSTRAINTS

So far in this chapter, you have learned various ways to create different types of constraints. You have also used the USER_CONSTRAINTS view to verify the creation of a constraint. But how, at a later time, do you determine which columns have constraints, and how can you determine the purpose of a constraint? The Oracle 10g server stores information about constraints in its data dictionary. Let's display the constraints that have just been created for the ACCTMANAGER2 table. To view the contents of the portion of the data dictionary that references constraints, use the SELECT statement shown in Figure 4-29. This statement displays information regarding all constraints on the ACCTMANAGER2 table.

```
SELECT constraint_name, constraint_type,
       search_condition
FROM user_constraints
WHERE table_name = 'ACCTMANAGER2';
```

FIGURE 4-29 SELECT statement to view data about existing constraints

In the SELECT clause, note the columns listed:

1. The first column referenced, **constraint_name**, lists the name of any constraint that exists in the ACCTMANAGER2 table.
2. The second column, **constraint_type**, lists the following letters:
   - P if the constraint is a PRIMARY KEY constraint
   - C if the constraint is a CHECK or NOT NULL constraint
   - U if the constraint is a UNIQUE constraint
   - R if the constraint a FOREIGN KEY constraint

   The R might seem a little strange for a FOREIGN KEY constraint; however, the purpose of the constraint is to ensure referential integrity—that you are referencing something that actually exists. Therefore, the assigned code is the letter R.
3. The third column listed in the SELECT clause, **search_condition**, is used to display the condition that is being used by a CHECK constraint. This column is blank for any constraint that is not a CHECK constraint.

Figure 4-30 shows partial output of a SELECT statement listing all table constraints.

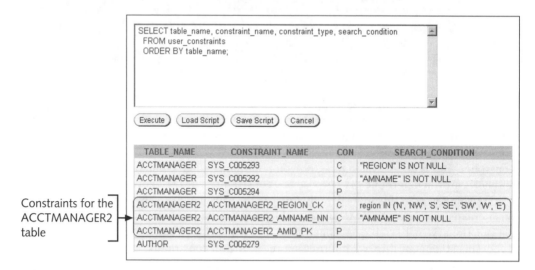

Constraints for the ACCTMANAGER2 table

FIGURE 4-30 Query of constraint information

In Figure 4-30, the results of the SELECT statement list three constraints for the ACCT-MANAGER2 table. The first two constraints are CHECK constraints, as indicated by the **C** in the second column of the output. However, if you look at the search conditions displayed in the third column of the results, the second constraint, **acctmanager2_amname_nn**, is a NOT NULL constraint because the condition specifies a NOT NULL requirement for the column. As indicated by the **P** for the constraint type of the third constraint, this is the PRIMARY KEY constraint.

N O T E

If the results displayed in Figure 4-30 appear to wrap across several lines, you can change the number of characters that appear on one line by issuing the **SET LINESIZE 150** command.

DISABLING AND DROPPING CONSTRAINTS

Sometimes, you will want to temporarily disable or drop a constraint. In this section, you'll examine these options.

Using DISABLE/ENABLE

When a constraint exists for a column(s), each entry made to that column is evaluated to determine whether the value is allowed in that column (that is, it doesn't violate the constraint). If you are adding a large block of data to a table, this validation process can severely slow down the Oracle server's processing speed. If you are certain that the data you're adding adheres to the constraints, you can disable the constraints while adding that particular block of data to the table.

To **DISABLE** a constraint, you issue an ALTER TABLE command and change the status of the constraint to DISABLE. At a later time, you can reissue the ALTER TABLE command and change the status of the constraint back to **ENABLE**. The syntax for using the ALTER TABLE command to change the status of a constraint is shown in Figure 4-31.

```
ALTER TABLE tablename
DISABLE CONSTRAINT constraintname;

ALTER TABLE tablename
ENABLE CONSTRAINT constraintname;
```

FIGURE 4-31   Syntax to disable or enable an existing constraint

For example, suppose that you must set up records for account managers in the ACCT-MANAGER2 table to test some queries or other SQL operations. You know the number range that will be assigned for the IDs of managers and regions. However, you do not know the managers' names, and at this point, you do not want to make up names just for a few test runs. However, you want the NOT NULL constraint for the Amname column to remain for future use. The simplest solution is to disable, or turn off, the constraint temporarily and then enable it when you are finished, as shown in Figure 4-32.

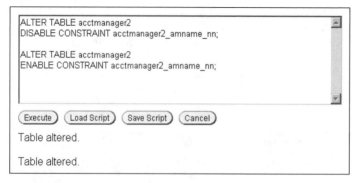

FIGURE 4-32   Disabling and enabling constraints

In the examples provided in Figure 4-32, the first ALTER TABLE command is used to temporarily turn off the **acctmanager2_amname_nn constraint**—note the keyword DISABLE. Upon successful execution of the statement, Oracle 10g returns the message "Table altered." To instruct Oracle 10g to begin enforcement of the constraint again, the same statement is reissued with the keyword ENABLE rather than DISABLE. After the constraint is enabled, data added to the table is again not allowed to contain NULL values in the Amname column.

NOTE

Keep in mind that when a constraint is enabled, all the data relevant to that constraint is checked for validity. Therefore, if data was added that violates a constraint when the constraint is disabled, an error occurs when the constraint is enabled.

DROPPING CONSTRAINTS

If you create a constraint and then decide that it is no longer needed (or if you find an error in the constraint), you can delete the constraint from the table with the DROP (constraintname) command. In addition, if you need to change or modify a constraint, your only option is to delete the constraint and then create a new one. You use the ALTER TABLE command to drop an existing constraint from a table using the syntax shown in Figure 4-33.

```
ALTER TABLE tablename
DROP PRIMARY KEY | UNIQUE (columnname)|
CONSTRAINT constraintname;
```

FIGURE 4-33   Syntax for the ALTER TABLE command to delete a constraint

Note the following guidelines for the syntax shown in Figure 4-33:

- The DROP clause varies depending on the type of constraint being deleted. If the DROP clause references the PRIMARY KEY constraint for the table, using the keywords PRIMARY KEY is sufficient because only one such clause is allowed for each table in the database.
- To delete a UNIQUE constraint, only the column name affected by the constraint is required because a column is referenced by only one UNIQUE constraint.
- Any other type of constraint must be referenced by the constraint's actual name—regardless of whether the constraint name is assigned by a user or the Oracle server.

Figure 4-34 shows the user dropping a NOT NULL constraint.

```
ALTER TABLE acctmanager2
DROP CONSTRAINT acctmanager2_amname_nn;
```

(Execute) (Load Script) (Save Script) (Cancel)

Table altered.

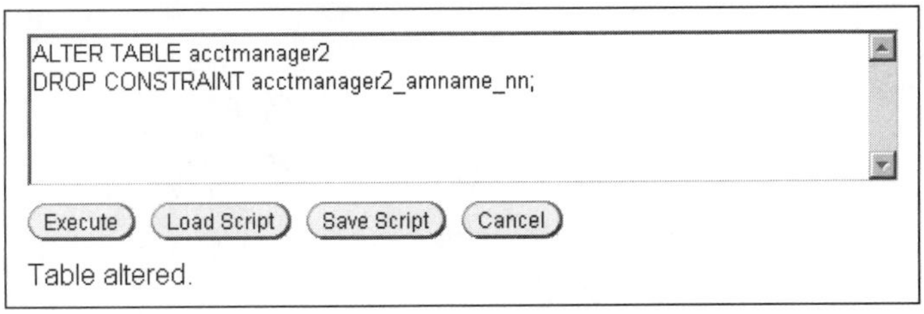

FIGURE 4-34   Dropping a NOT NULL constraint by name

In Figure 4-34, the ALTER TABLE command is issued to delete the NOT NULL constraint for the Amname column of the ACCTMANAGER2 table. If the command is executed successfully, the constraint no longer exists, and the column is allowed to contain NULL values.

The FOREIGN KEY constraint raises a special concern when attempting to drop constraints because it involves a relationship between two tables. Recall that a FOREIGN KEY column references a PRIMARY KEY column of another table. If we attempt to drop the PRIMARY KEY, an error is raised indicating that a FOREIGN KEY reference exists, as shown in Figure 4-35.

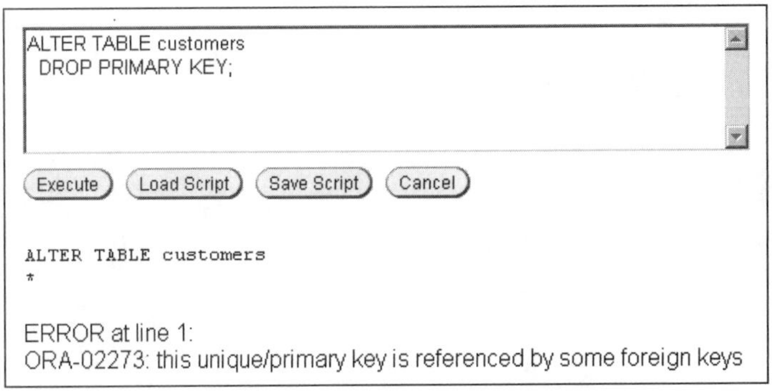

```
ALTER TABLE customers
 DROP PRIMARY KEY;
```

(Execute) (Load Script) (Save Script) (Cancel)

```
ALTER TABLE customers
*
ERROR at line 1:
ORA-02273: this unique/primary key is referenced by some foreign keys
```

FIGURE 4-35   Error dropping a PRIMARY KEY referenced by a FOREIGN KEY

If desired, the associated FOREIGN KEY can be deleted along with the PRIMARY KEY deletion by using a CASCADE option, as shown in Figure 4-36.

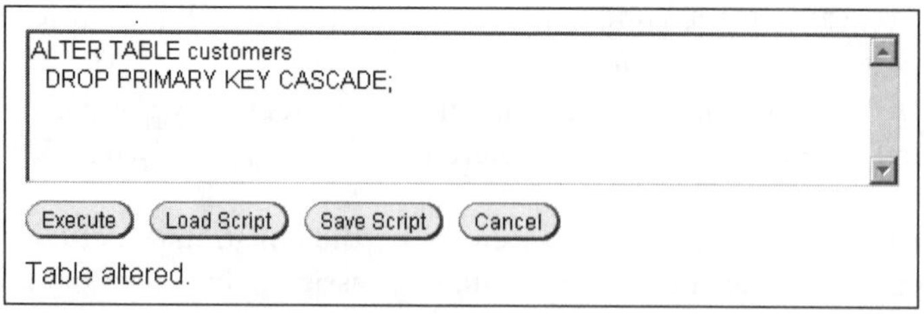

```
ALTER TABLE customers
  DROP PRIMARY KEY CASCADE;
```

(Execute) (Load Script) (Save Script) (Cancel)

Table altered.

FIGURE 4-36   Dropping a PRIMARY KEY referenced by a FOREIGN KEY using a CASCADE option

Chapter Summary

- A constraint is a rule that is applied to data being added to a table. The constraint represents business rules, policies, and/or procedures. Data violating the constraint is not added to the table.

- A constraint can be included during table creation as part of the CREATE TABLE command or added to an existing table using the ALTER TABLE command.

- A constraint that is based on composite columns (more than one column) must be created using the table-level approach.

- A NOT NULL constraint can be created using only the column-level approach.

- A PRIMARY KEY constraint does not allow duplicate or NULL values in the designated column.

- Only one PRIMARY KEY constraint is allowed in a table.

- A FOREIGN KEY constraint requires that the column entry match a referenced column entry in the referenced table or be NULL.

- A UNIQUE constraint is similar to a PRIMARY KEY constraint except it allows NULL values to be stored in the specified column.

- A CHECK constraint ensures that the data meets a given condition before it is added to the table. The condition cannot reference the SYSDATE function or values stored in other rows.

- A NOT NULL constraint is a special type of CHECK constraint. It can be added only to a column using the CREATE TABLE command or the MODIFY clause of the ALTER TABLE command.

- A constraint can be disabled or enabled using the ALTER TABLE command and the DISABLE and ENABLE keywords.

- A constraint cannot be modified. To change a constraint, the constraint must first be dropped with the DROP command and then re-created.

Chapter 4 Syntax Summary

The following table presents a summary of the syntax that you have learned in this chapter. You can use the table as a study guide and reference.

SYNTAX GUIDE		
Constraint	Description	Example

PRIMARY KEY	Determines which column(s) uniquely identifies each record. The primary key cannot be NULL, and the data value(s) must be unique.	*Constraint created during table creation:*
		`CREATE TABLE newtable`
		`(firstcol NUMBER PRIMARY KEY,`
		`secondcol VARCHAR2(20));`
		or
		`CREATE TABLE newtable (firstcol`
		`NUMBER,`
		`secondcol VARCHAR2(20),`
		`CONSTRAINT constraint_name_pk`
		`PRIMARY KEY (firstcol));`
		Constraint created after table creation:
		`ALTER TABLE newtable`
		`ADD CONSTRAINT constraint_name_pk`
		`PRIMARY KEY (firstcol);`

FOREIGN KEY	In a one-to-many relationship, the constraint is added to the "many" table. The constraint ensures that if a value is entered to the specified column, it exists in the table being referred to, or the row is not added.	*Constraint created during table creation:*
		`CREATE TABLE newtable`
		`(firstcol NUMBER,`
		`secondcol VARCHAR2(20) REFERENCES`
		`another table(col1));`
		or
		`CREATE TABLE newtable`
		`(firstcol NUMBER,`
		`secondcol VARCHAR2(20),CONSTRAINT`
		`constraint_name_fk FOREIGN KEY`
		`(secondcol) REFERENCES`
		`anothertable(col1);`
		Constraint created after table creation:
		`ALTER TABLE newtable`
		`ADD CONSTRAINT constraint_name_fk`
		`FOREIGN KEY (secondcol)`

Constraint	Description	Example

UNIQUE	Ensures that all data values stored in the specified column are unique. The UNIQUE constraint differs from the PRIMARY KEY constraint in that it allows NULL values.	*Constraint created during table creation:*
		`CREATE TABLE newtable`
		`(firstcol NUMBER, secondcol`
		`VARCHAR2(20) UNIQUE);`
		or
		`CREATE TABLE newtable`
		`(firstcol NUMBER, secondcol`
		`VARCHAR2(20), CONSTRAINT`
		`constraint_name_uk UNIQUE`
		`(secondcol));`
		Constraint created after table creation:
		`ALTER TABLE newtable`
		`ADD CONSTRAINT constraint_name_uk`
		`UNIQUE (secondcol);`

CHECK	Ensures that a specified condition is TRUE before the data value is added to the table. For example, an order's ship date cannot be "less than" its order date.	*Constraint created during table creation:*
		`CREATE TABLE newtable`
		`(firstcol NUMBER,`
		`secondcol VARCHAR2(20),`
		`thirdcol NUMBER CHECK (BETWEEN 20`
		`AND 30));`
		or
		`CREATE TABLE newtable`
		`(firstcol NUMBER,`
		`secondcol VARCHAR2(20),`
		`thirdcol NUMBER,`
		`CONSTRAINT constraint_name_ck`
		`CHECK (thirdcol BETWEEN 20`
		`AND 80));`
		Constraint created after table creation:
		`ALTER TABLE newtable`
		`ADD CONSTRAINT constraint_name_ck`
		`CHECK (thirdcol BETWEEN 20`
		`AND 80);`

NOT NULL	Requires that the specified column cannot contain a NULL value. It can be created *only* with the column-level approach to table creation.	*Constraint created during table creation:*
		`CREATE TABLE newtable`
		`(firstcol NUMBER,secondcol`
		`VARCHAR2(20),`
		`thirdcol NUMBER NOT NULL);`
		Constraint created after table creation:
		`ALTER TABLE newtable`
		`MODIFY (thirdcol NOT NULL);`

117

Constraints

Review Questions

To answer these questions, refer to the tables in Appendix A.

1. What is the difference between a PRIMARY KEY constraint and a UNIQUE constraint?
2. All constraints are enforced at what level?
3. A table can have a maximum of how many PRIMARY KEY constraints?
4. Which type of constraint can be used to make certain the category for a book is included when a new book is added to inventory?
5. Which type of constraint would be required to ensure that every book has a profit margin between 15% and 25%?
6. How is adding a NOT NULL constraint to an existing table different from adding other types of constraints?
7. When must you define constraints at the table level rather than at the column level?
8. To which table do you add a FOREIGN KEY constraint if you want to make certain every book ordered is contained in the BOOKS table?
9. What is the difference between disabling a constraint and dropping a constraint?
10. What is the simplest way to determine whether a particular column can contain NULL values?

Multiple Choice

To answer the following questions, refer to the tables in Appendix A.

1. Which of the following statements is correct?
   a. A PRIMARY KEY constraint allows NULL values in the primary key column(s).
   b. You can enable a dropped constraint if you need it in the future.
   c. Every table must have at least one PRIMARY KEY constraint, or Oracle 10*g* does not allow the table to be created.
   d. None of the above statements is correct.
2. Which of the following is not a valid constraint type?
   a. PRIMARY KEYS
   b. UNIQUE
   c. CHECK
   d. FOREIGN KEY
3. Which of the following SQL statements is invalid and returns an error message?
   a. ALTER TABLE books
      ADD CONSTRAINT books_pubid_uk UNIQUE (pubid);
   b. ALTER TABLE books
      ADD CONSTRAINT books_pubid_pk PRIMARY KEY (pubid);
   c. ALTER TABLE books
      ADD CONSTRAINT books_pubid_nn NOT NULL (pubid);

    d.   ALTER TABLE books
        ADD CONSTRAINT books_pubid_fk FOREIGN KEY (pubid)
        REFERENCES publisher (pubid);

    e.   All of the above statements are invalid.

4.   What is the maximum number of PRIMARY KEY constraints allowed for a table?

    a.   1

    b.   2

    c.   30

    d.   255

5.   Which of the following is a valid SQL command?

    a.   ALTER TABLE books
        ADD CONSTRAINT UNIQUE (pubid);

    b.   ALTER TABLE books
        ADD CONSTRAINT PRIMARY KEY (pubid);

    c.   ALTER TABLE books
        MODIFY (pubid CONSTRAINT NOT NULL);

    d.   ALTER TABLE books
        ADD FOREIGN KEY CONSTRAINT (pubid)
        REFERENCES publisher (pubid);

    e.   None of the above commands is valid.

6.   How many NOT NULL constraints can be created at the table level using the CREATE TABLE command?

    a.   0

    b.   1

    c.   12

    d.   30

    e.   255

7.   The FOREIGN KEY constraint should be added to which table?

    a.   the table representing the "one" side of a one-to-many relationship

    b.   the parent table in a parent-child relationship

    c.   the child table in a parent-child relationship

    d.   the table that does not have a primary key

8.   What is the maximum number of columns that you can define as a primary key using the column-level approach when creating a table?

    a.   0

    b.   1

    c.   30

    d.   255

9. Which of the following commands can you use to rename a constraint?

   a. RENAME

   b. ALTER CONSTRAINT

   c. MOVE

   d. NEW NAME

   e. None of the above commands can be used.

10. Which of the following is a valid SQL statement?

    a. CREATE TABLE table1
    (col1   NUMBER PRIMARY KEY
    col2   VARCHAR2(20) PRIMARY KEY,
    col3   DATE DEFAULT SYSDATE,
    col4   VARCHAR2(2));

    b. CREATE TABLE table1
    (col1   NUMBER PRIMARY KEY
    col2   VARCHAR2(20),
    col3   DATE,
    col4   VARCHAR2(2) NOT NULL,
    CONSTRAINT table1_col3_ck CHECK (col3 = SYSDATE));

    c. CREATE TABLE table1
    (col1   NUMBER,
    col2   VARCHAR2(20),
    col3   DATE,
    col4   VARCHAR2(2),
    PRIMARY KEY (col1));

    d. CREATE TABLE table1
    (col1   NUMBER,
    col2   VARCHAR2(20),
    col3   DATE DEFAULT SYSDATE,
    col4   VARCHAR2(2);

11. In the initial creation of a table, if a UNIQUE constraint is being included for a composite column that requires that the combination of entries in the specified columns be unique, which of the following statements is correct?

    a. The constraint can be created using only the ALTER TABLE command.

    b. The constraint can be created using only the table-level approach.

    c. The constraint can be created using only the column-level approach.

    d. The constraint can be created using only the ALTER TABLE...MODIFY command.

12. Which type of constraint would you use on a column to allow values only above 100 to be entered?

    a. PRIMARY KEY

    b. UNIQUE

    c. CHECK

    d. NOT NULL

13. Which of the following commands can be used to enable a disabled constraint?

    a.   ALTER TABLE...MODIFY

    b.   ALTER TABLE...ADD

    c.   ALTER TABLE...DISABLE

    d.   ALTER TABLE...ENABLE

14. Which of the following keywords will allow the user to delete a record from a table even if rows in another table reference the record through a FOREIGN KEY constraint?

    a.   CASCADE

    b.   CASCADE ON DELETE

    c.   DELETE ON CASCADE

    d.   DROP

    e.   ON DELETE CASCADE

15. Which of the following data dictionary objects should be used to view information about the constraints in a database?

    a.   USER_TABLES

    b.   USER_RULES

    c.   USER_COLUMNS

    d.   USER_CONSTRAINTS

    e.   None of the above objects should be used.

16. Which of the following types of constraints cannot be created at the table level?

    a.   NOT NULL

    b.   PRIMARY KEY

    c.   CHECK

    d.   FOREIGN KEY

    e.   None of the above constraints can be created at the table level.

17. Suppose that you created a PRIMARY KEY constraint at the same time you created a table and later decide to name the constraint. Which of the following commands can you use to change the name of the constraint?

    a.   ALTER TABLE...MODIFY

    b.   ALTER TABLE...ADD

    c.   ALTER TABLE...DISABLE

    d.   None of the above commands can be used.

18. You are creating a new table consisting of three columns: col1, col2, and col3. The column called col1 should be the primary key and cannot have any NULL values, and each entry should be unique. The column labeled col3 must not contain any NULL values either. How many total constraints will you be required to create?

    a. 1
    b. 2
    c. 3
    d. 4

19. Which of the following types of restrictions can be viewed using the DESCRIBE command?

    a. NOT NULL
    b. FOREIGN KEY
    c. UNIQUE
    d. CHECK

20. Which of the following is the valid syntax for adding a PRIMARY KEY constraint to an existing table?

    a. ALTER TABLE *tablename*
       ADD CONSTRAINT PRIMARY KEY (*columnname*);

    b. ALTER TABLE tablename
       ADD CONSTRAINT (*columnname*)
       PRIMARY KEY *constraintname*;

    c. ALTER TABLE *tablename*
       ADD [CONSTRAINT *constraintname*]
       PRIMARY KEY;

    d. None of the above is valid syntax.

Hands-On Assignments

JustLee Books has become the exclusive distributor for a number of books. The company now needs to assign sales representatives to retail bookstores to handle the new distribution duties. You will create new tables in these assignments to support this functionality.

1. Modify the following SQL command so that the Rep_id column is the PRIMARY KEY for the table and the default value of 'Y' is assigned to the Comm column.

   ```
   CREATE TABLE Store_reps
   (rep_id   NUMBER(5),
   last   VARCHAR2(15),
   first   VARCHAR2(10),
   comm   CHAR(1) );
   ```

2. Change the STORE_REPS table so that NULL values cannot be entered in the name columns.

3. Change the STORE_REPS table so that only a 'Y' or 'N' can be entered in the Comm column.

4. Add a column named Base_salary with a datatype of NUMBER(7,2) to the STORE_REPS table. Ensure that the amount entered is above zero.

5. Create a table named BOOK_STORES to include the columns listed in the table that follows:

COLUMN NAME	DATATYPE	CONSTRAINT COMMENTS
Store_id	NUMBER(8)	Primary Key column
Name	VARCHAR2(30)	Should be unique and not NULL
Contact	VARCHAR2(30)	
Rep_id	VARCHAR2(5)	

6. Add a constraint to ensure the Rep_id value entered into the BOOK_STORES table is a valid value contained in the STORE_REPS table. Consider that the Rep_id columns of both tables were initially created using different datatypes. Does this cause an error when adding the constraint to accomplish this task? Make table modifications as needed to be able to add the required constraint.

7. Change the constraint created in assignment #6 so that associated rows of the BOOK_ STORES table will be deleted automatically if a row in the STORE_REPS table is deleted.

8. Create a table named REP_CONTRACTS that contains the columns listed in the table that follows. A composite PRIMARY KEY including the Rep_id, Store_id, and quarter columns should be assigned. In addition, FOREIGN KEYS should be assigned to both the Rep_id and Store_id columns.

COLUMN NAME	DATATYPE
Store_id	NUMBER(8)
Name	NUMBER(5)
Quarter	CHAR(3)
Rep_id	VARCHAR2(50)

9. Produce a list of information regarding all the constraints existing in your database. Sort the list by table name.

10. Issue the commands to disable and then enable the CHECK constraint on the Base_ salary column.

Advanced Challenge

To perform this activity, refer to the tables in Appendix A and to the ERD for the JustLee Books database in Chapter 1.

Now that you have received training in constraints, your supervisor asks you to examine the CUSTOMERS, ORDERS, ORDERITEMS, BOOKS, PUBLISHER, and BOOKAUTHOR tables to determine whether additional constraints to enforce data relationships should be created for any of the tables. Currently, most of the tables have only primary key constraints.

Create and execute the SQL statements needed to enforce the data relationships among these tables.

Case Study: *City Jail*

In previous chapters, you have designed and created tables for the City Jail database. These tables do not include any constraints. Review the information provided in Chapters 1 and 3 case studies to determine what constraints you might need for the City Jail database.

First, using the format in the table that follows, create a list of constraints needed. Second, create and execute all the SQL statements needed to add these constraints.

TABLE NAME	COLUMN(S)	CONSTRAINT TYPE	CONDITION

CHAPTER 5

DATA MANIPULATION AND TRANSACTION CONTROL

LEARNING OBJECTIVES

After completing this chapter, you should be able to do the following:

- Use the INSERT command to add a record to an existing table
- Understand constraint violations during data manipulation
- Use a subquery to copy records from an existing table
- Use the UPDATE command to modify the existing rows of a table
- Use substitution variables with an UPDATE command
- Delete records
- Manage transactions with transaction control statements COMMIT, ROLLBACK, and SAVEPOINT
- Differentiate between a shared lock and an exclusive lock
- Use the SELECT...FOR UPDATE command to create a shared lock

INTRODUCTION

In Chapters 3 and 4, you issued data definition language (DDL) commands to create, alter, and drop database objects, such as the ACCTMANAGER2 table. This chapter addresses methods for adding new rows and modifying or deleting existing rows. In this chapter, however, all the operations performed differ from DDL statements in that they affect the data contained *within* tables, not to the actual structure of tables. Commands used to modify data are called **data manipulation language (DML)** commands. Figure 5-1 lists the commands you'll use in this chapter.

COMMAND	DESCRIPTION
INSERT	Adds new row(s) to a table; the user can include a subquery to copy row(s) from an existing table
UPDATE	Adds data to, or modifies data within, existing row(s)
COMMIT	Permanently saves changed data in a table
ROLLBACK	Allows the user to "undo" uncommitted changes to data
SAVEPOINT	Enables setting markers within a transaction
DELETE	Removes row(s) from a table
LOCK TABLE	Prevents other users from making changes to a table
SELECT...FOR UPDATE	Creates a shared lock on a table to prevent another user from making changes to data in specified columns

FIGURE 5-1   DML and transaction control commands

NOTE

Go to the Chapter5 folder in your Data Files. Before working through the examples in this chapter, run the script prech5.sql to ensure that all necessary tables and constraints are available. Reference the steps used in the beginning of Chapter 2 for loading and executing a script. Ignore any errors in the DROP TABLE statements at the beginning of the script. An "object does not exist" error indicates merely that the table was not previously created in the schema. If the script is not available, ask your instructor to provide you with a copy.

INSERTING NEW ROWS

As discussed in previous chapters, the management of JustLee Books is implementing a new commission policy for regional account managers. The ACCTMANAGER2 table was created to store data about account managers. Now that the table has been created and all necessary constraints for the table are in place, it is time to add data to the table. The data shown in Figure 5-2 is for those account managers you need to add to the ACCT-MANAGER2 table. (Blank spaces indicate that data has not yet been provided to the data entry clerk.)

ID	NAME	EMPLOYMENT DATE	REGION
T500	Nick Taylor	September 5, 2005	NE
L500	Mandy Lopez	October 1, 2005	
J500	Sammie Jones	Today	NW

FIGURE 5-2   Data for account managers

NOTE

Reference the prech5.sql script used in the beginning of the chapter to review the CREATE TABLE statement for the ACCTMANAGER2 table. This table contains a PRIMARY KEY, CHECK, and NOT NULL constraints. Also, the Amedate column has a default option setting.

INSERT Command

You can add rows to a new or existing table with the **INSERT** command. The syntax for the INSERT command is shown in Figure 5-3.

```
INSERT INTO tablename [(columnname, …)]
VALUES (datavalue, …);
```

FIGURE 5-3 Syntax of the INSERT command

Note the following syntax elements in Figure 5-3.

- The keywords **INSERT INTO** are followed by the name of the table into which the rows will be entered. The table name is followed by the names of the columns that will contain the data.
- The VALUES clause identifies the data values that will be inserted into the table. You list the data in parentheses after the VALUES keyword.
- If the data entered in the VALUES clause is in the same order as the columns in the table, column names can be omitted in the INSERT INTO clause. However, if you enter data for only some of the columns, or if columns are listed in a different order than they are listed in the table, the names of the columns **must** be provided in the INSERT INTO clause in the same order as they will be given in the VALUES clause. You must list the column names within a set of parentheses after the table name in the INSERT INTO clause.
- If more than one column is listed, column names must be separated by commas.
- If more than one data value is entered, they must be separated by commas.
- You must use single quotation marks to enclose data inserted into a column defined for nonnumeric data (the column's datatype is not NUMBER).

NOTE

You can review the order of the columns in a table by using the DESCRIBE *tablename* command.

To insert the first account manager's data (Figure 5-2) into the ACCTMANAGER2 table, use the command shown in Figure 5-4.

```
INSERT INTO acctmanager
VALUES ('T500', 'NICK TAYLOR', '05-SEP-05', 'NE');
```

FIGURE 5-4   The INSERT command

The INSERT INTO clause shown in Figure 5-4 does not contain a list of the column names because the VALUES clause contains a valid entry for every column in the ACCTMANAGER2 table, and the data is given in the same order as the columns are listed in the table.

The character data in the VALUES clause is entered in all uppercase characters because that is used for all the tables in the JustLee Books database. When you enter character data in a table, the data retains the case you use in the INSERT INTO command. For example, if the name of an account manager had been entered in mixed case (upper- and lower-case letters), the table would have stored the name in mixed case. This can make future record searches difficult if the data does not follow a consistent format or case because character data matching is case sensitive. Figure 5-5 shows that the first account manager, Nick Taylor, has been added to the ACCTMANAGER2 table.

FIGURE 5-5   Command to insert a row into the ACCTMANAGER2 table

After you execute the INSERT command, the message "1 row created" appears, indicating that the data has been inserted into the table. To verify that the row was added, a SELECT statement can be used to view the table's contents.

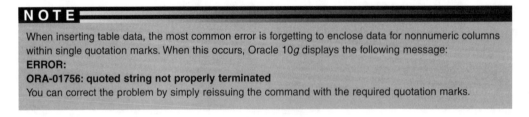

NOTE

When inserting table data, the most common error is forgetting to enclose data for nonnumeric columns within single quotation marks. When this occurs, Oracle 10*g* displays the following message:
ERROR:
ORA-01756: quoted string not properly terminated
You can correct the problem by simply reissuing the command with the required quotation marks.

The next record to enter into the ACCTMANAGER2 table contains data about Mandy Lopez. However, as shown in Figure 5-2, she has not yet been assigned a marketing region, so that column will be left blank. You can take one of the following approaches to indicate that the Region column will contain a NULL value:

1. List all the columns **except** the Region column in the INSERT INTO clause, and provide the data for the listed columns in the VALUES clause.
2. In the VALUES clause, substitute **two single quotation marks** in the position that should contain the account manager's assigned region. Oracle 10g interprets the quotation marks to mean that a NULL value should be stored in that column. Be sure you do not add a blank space between the two single quotation marks—this adds a blank space value rather than a NULL value.
3. In the VALUES clause, include the keyword NULL in the position where the region should be listed. As long as the keyword NULL is not enclosed in single quotation marks, Oracle 10g leaves the column blank. However, if the keyword is mistakenly entered as 'NULL,' the software tries to store the word "NULL" in the column.

The INSERT INTO commands are shown in Figure 5-6, which demonstrates each of the previously listed methods.

```
INSERT INTO acctmanager2 (amid, amfirst, amlast, amedate)
  VALUES ('L500', 'MANDY','LOPEZ','01-OCT-05');

INSERT INTO acctmanager2
  VALUES ('L500', 'MANDY','LOPEZ','01-OCT-05','");

INSERT INTO acctmanager2
  VALUES ('L500', 'MANDY','LOPEZ','01-OCT-05', NULL);
```

FIGURE 5-6 Methods to enter NULL values

Execute the third approach, and Oracle 10g adds Mandy Lopez to the ACCTMANAGER2 table, as shown in Figure 5-7.

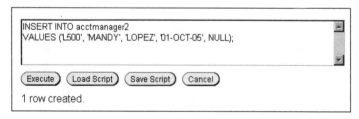

FIGURE 5-7 Insert for Mandy Lopez

Next, you need to enter the record for Sammie Jones into the ACCTMANAGER2 table. Note that Sammie's employment date should be set to the current date. A value of SYS-DATE could be included in the VALUES clause of the INSERT statement to cause Oracle 10g to insert the current date into the Amedate column. Or, in this case, the Amedate column has a default option set to SYSDATE. To instruct the software to use the default option, a column list must be provided in the INSERT INTO clause that omits the Amedate column. Figure 5-8 shows the commands for both of these approaches.

```
INSERT INTO acctmanager2 (amid, amfirst, amlast, region)
   VALUES ('J500', 'Sammie' , 'Jones', SYSDATE, 'NW');

INSERT INTO acctmanager2 (amid, amfirst, amlast, region)
   VALUES ('J500', 'Sammie' , 'Jones', 'NW');
```

FIGURE 5-8 Commands to add Sammie Jones' record

Use Figure 5-9 as a guide to execute the second method that was shown in Figure 5-8 for adding the record. The column list provided in the INSERT INTO clause lists all the columns of the ACCTMANAGER2 table except the Amedate column. Although the columns are listed in the same sequence as they appear in the actual table, this is not a requirement. What is required is that the data listed in the VALUES clause matches the **exact order** of the columns listed in the INSERT INTO clause. The order of the list is how Oracle 10g is able to determine which data value belongs to which column.

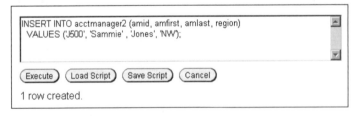

FIGURE 5-9 Use an INSERT statement that will apply a default column option

As shown in Figure 5-10, a display of the current contents of the ACCTMANAGER2 table confirms the addition of the three records. Because the current date was used for Sammie Jones' employment date, yours will vary from what is displayed in Figure 5-10. Note that the name values for Sammie Jones are displayed as mixed case, which illustrates that character values are stored in the case that is used when inserted.

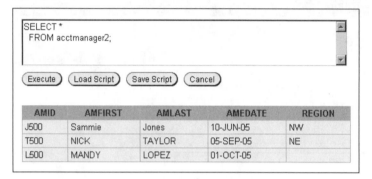

FIGURE 5-10   Contents of the ACCTMANAGER2 table

Constraint Violations

In the last chapter, you added constraints to tables to enforce data integrity rules. When you add or modify table data, the data is checked for compliance with any applicable constraints. You will use the ORDERITEMS table to test the constraints with INSERT statements. The CREATE TABLE statement for the ORDERITEMS table is listed in Figure 5-11. Review the constraints defined.

The rows listed in Figure 5-12 need to be added to the ORDERITEMS table.

The first INSERT is shown in Figure 5-13. When Oracle 10g attempts to execute the command, the values to be entered are compared against all existing constraints. A constraint error is shown upon execution and the INSERT is not successful.

The error indicates that the unique constraint named SCOTT.ORDERITEMS_PK is violated. Recall that the PRIMARY KEY enforces two rules: \ uniqueness and NOT NULL. The constraint name includes the schema name prefix, which, in this case is SCOTT. Notice how helpful a good constraint name can be! In this case, the name indicates the table name

```
CREATE TABLE ORDERITEMS
    ( ORDER# NUMBER(4),
      ITEM# NUMBER(2),
      ISBN VARCHAR2(10),
      QUANTITY NUMBER(3) NOT NULL,
      CONSTRAINT orderitems_pk PRIMARY KEY (order#, item#),
      CONSTRAINT orderitems_order#_fk FOREIGN KEY (order#)
        REFERENCES orders (order#) ,
      CONSTRAINT orderitems_quantity_ck CHECK (quantity > 0) );
```

FIGURE 5-11   ORDERITEMS table constraints

ORDER	ITEM#	ISBN	QUANTITY
1020	1	3437212490	1
1021	1	3437212490	1
1020	2	0401140733	0

FIGURE 5-12   Data for ORDERITEMS

FIGURE 5-13   PRIMARY KEY violation

and the type of constraint violated. If you review the data currently in the ORDERITEMS table, you will discover that a record with the same primary key values already exists in the table.

The second INSERT is shown in Figure 5-14. An integrity constraint violation is indicated on this statement.

This error indicates that the constraint name of ORDERITEMS_ORDER#_FK has been violated and the parent key cannot be found. The FOREIGN KEY constraint on the ORDERITEMS table references the ORDERS table for a valid order number. In this case, the

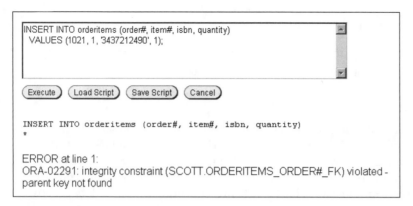

```
INSERT INTO orderitems (order#, item#, isbn, quantity)
 VALUES (1021, 1, 3437212490', 1);
```

Execute   Load Script   Save Script   Cancel

```
INSERT INTO orderitems (order#, item#, isbn, quantity)
*
ERROR at line 1:
ORA-02291: integrity constraint (SCOTT.ORDERITEMS_ORDER#_FK) violated -
parent key not found
```

FIGURE 5-14   FOREIGN KEY violation

order number of 1021 does not exist in the ORDERS table. The ORDERS table (the referenced table) is considered the parent table. The ORDERITEMS table is considered the child table in this relationship.

The last INSERT is shown in Figure 5-15. In this case, the CHECK constraint on the quantity column is violated because the constraint allows only quantity values above zero.

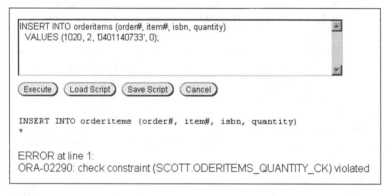

```
INSERT INTO orderitems (order#, item#, isbn, quantity)
 VALUES (1020, 2, 0401140733', 0);
```

Execute   Load Script   Save Script   Cancel

```
INSERT INTO orderitems (order#, item#, isbn, quantity)
*
ERROR at line 1:
ORA-02290: check constraint (SCOTT.ODERITEMS_QUANTITY_CK) violated
```

FIGURE 5-15   CHECK constraint violation

Inserting Data from an Existing Table

In Chapter 3, you learned how to use the CREATE TABLE command with a subquery to create and populate a new table based upon the structure and content of an existing table. However, what if a table already exists, and you need to add to it copies of existing records that are contained in another table? In that case, you would not be able to use the CREATE TABLE command. Because the table already exists, you would need to use the INSERT INTO command with a subquery. The syntax for combining an INSERT INTO command with a subquery is shown in Figure 5-16.

```
INSERT INTO tablename [(columnname, …)]
subquery;
```

FIGURE 5-16   Syntax for the INSERT INTO command with a subquery

Note the following elements in Figure 5-16:

- The main difference between using the INSERT INTO command with data values and with a subquery is that **the VALUES clause is not included when the command is used with a subquery**. The keyword VALUES indicates that the clause contains data values that must be inserted into the indicated table. However, there are no data values being entered by the user—the data is derived from the results of the subquery.
- Also, note that unlike the CREATE TABLE command, the INSERT INTO command does not require the subquery to be enclosed within a set of parentheses, although including parentheses does not generate an error message.

Part of JustLee Books' new commission policy is that an account manager will begin receiving a sales commission after three months of employment. You need to identify the account managers who are eligible to receive a commission for orders placed after December 5, 2005. The eligible account managers need to be added to the ACCTMANAGER table. In this scenario, a condition must be used to copy only those account managers who have an employment date of on or before September 5, 2005. You can use a WHERE clause to provide the condition in the subquery.

The ACCTMANAGER2 table currently contains all account manager data. The account manager commission data must be copied into the ACCTMANAGER table. By copying the relevant data into the second table, SQL statements from the Payroll Department's program that calculates commissions can be tested. This allows the Information Technology Department to have the flexibility of making changes to the original ACCTMANAGER2 table without interfering with Payroll's program testing. The command to copy the relevant rows from the ACCTMANAGER2 table into the ACCTMANAGER table is shown in Figure 5-17.

Note the following elements in Figure 5-17:

- The SELECT clause of the subquery lists the columns to be copied from the ACCTMANAGER2 table, which is identified in the FROM clause. In this example, the INSERT INTO clause does not contain a column list because the columns returned by the subquery are in the same order as the columns in the ACCTMANAGER table.
- The WHERE clause provides the condition that an account manager must have been employed by JustLee Books on or before September 5, 2005 to receive a commission for sales that began on December 5, 2005.

After the command has been executed, the ACCTMANAGER table can be queried to determine which account managers are eligible to receive a commission for orders placed in December. As shown in the results in Figure 5-17, there are currently two eligible account managers, Nick Taylor and Sammie Jones. Results might vary, depending on your computer's system date because SYSDATE was used for Sammie Jones' employment date.

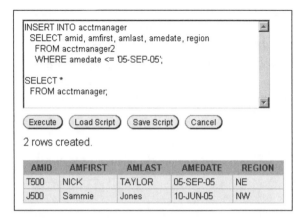

FIGURE 5-17   INSERT INTO command with a subquery

> **NOTE**
>
> Oracle 10g contains several utilities for importing and exporting data to and from external files such as SQL*Loader and Data Pump. This topic is beyond the scope of the text, so it is not addressed in this chapter. Appendix G offers an introduction to SQL*Loader to provide exposure to a bulk import operation.

MODIFYING EXISTING ROWS

There will be many times when you need to change column data values. For example, when customers move, their mailing addresses need to be updated; when the wholesale cost of books changes, retail prices need to be changed. Because the INSERT INTO command can be used only to add new rows to a table, it cannot modify existing data. To alter existing table data, you use the UPDATE command. In this section, you will learn how to perform updates and create interactive update scripts using substitution variables.

UPDATE command

You change the contents of existing rows using the **UPDATE** command. The syntax for the UPDATE command is shown in Figure 5-18.

```
UPDATE tablename
SET columnname = new_datavalue, …
[WHERE condition];
```

FIGURE 5-18   Syntax for the UPDATE command

Note the following elements in Figure 5-18:

- The UPDATE clause identifies the table containing the record(s) to be changed.
- The **SET** clause is used to identify the column(s) to be changed and the new value to be assigned to the column.

- The optional WHERE clause identifies the exact row or set of rows to be changed by the UPDATE command. If the WHERE clause is omitted, then the column specified in the SET clause is updated for **all records** contained in the table.

Let's address several changes that have been requested for data in the ACCTMAN-AGER2 table. First, the employment date for Sammie Jones is incorrect and needs to be changed to August 1, 2005. The command shown in Figure 5-19 can be issued to correct the employment date.

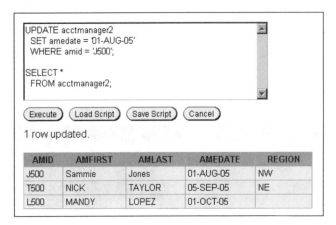

FIGURE 5-19 Command to correct the Amedate column entry

Notice the SET clause in Figure 5-19. The SET clause indicates only the new value that should be entered into the column indicated—it is not involved with the original value. The original value is simply overwritten.

The WHERE clause is used to identify exactly which record should be altered. In this case, the easiest way to specify that only the employment date for Sammie Jones should be changed is to include a WHERE clause with the condition that the Amid column must be equal to J500. Because the Amid column is the primary key for the ACCTMANAGER2 table, no two records can have the same Amid, and only the record for Sammie Jones will be affected by the update.

Next, the northern regions are being closed and the account managers assigned to these regions need to be reassigned to the western region. You can accomplish this by using the command shown in Figure 5-20.

Notice that in this case, two of the three existing rows in the table were modified.

What if you need to modify more than one column of a row? You can do this by adding multiple columns in the SET clause of the UPDATE command. Let's say that the employment date for Mandy Lopez needs to be changed to be October 10, 2005, and she needs to be assigned to the southern region. Execute the statement in Figure 5-21.

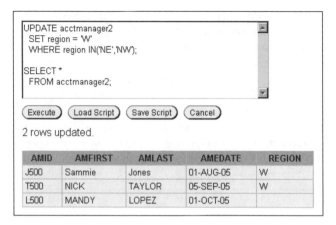

FIGURE 5-20 UPDATE command to reassign regions

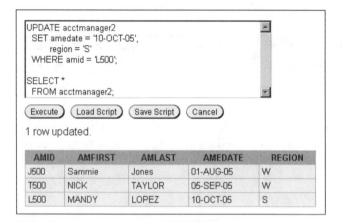

FIGURE 5-21 Updating multiple columns

SUBSTITUTION VARIABLES

In some cases, it might seem like a lot of effort just to add a record to a table—and even more effort to modify an existing record. This is especially true if you need to add or modify 10 or 20 different records. For example, the CUSTOMERS table will need to contain data that identifies the marketing region for each customer. After the Region column is added to the CUSTOMERS table using the ALTER TABLE command, every customer's record will need to be updated with the value for the new column. Depending on the strategy used, the

INSERT INTO command will have to be reissued several times—at least once for every identified region. Rather than type the same command again and again for the few values that are different, it is much simpler to use a substitution variable.

A **substitution variable** in a SQL command instructs Oracle 10g to use a substituted value in place of the variable at the time the command is actually executed. To include a substitution variable in a SQL command, simply enter an ampersand (&) followed by the name to be used for the variable in the necessary location.

First, let's look at the command to modify the records of all customers residing in California, so that the Region column contains the value of W, representing the western region. The command is shown in Figure 5-22.

```
UPDATE customers
SET region = 'W'
WHERE state = 'CA';
```

FIGURE 5-22 Command to update the Region column of the CUSTOMERS table

> **NOTE**
>
> The Region column was added to the CUSTOMERS table when the prech5.sql script was run at the beginning of the chapter.

Next, to alter the command shown in Figure 5-22 so you can reuse it for each state in which JustLee Books has customers, you need to enter a substitution variable in place of the value for the State column. Furthermore, the SET clause can contain a substitution variable, so the same command could be used for every region to be entered. The new command is shown in Figure 5-23.

```
UPDATE customers
SET region = '&Region'
WHERE state = '&State';
```

FIGURE 5-23 UPDATE command with substitution variables

When Oracle 10g executes the command shown in Figure 5-23, the user is first prompted to enter a value for the substitution variable named Region. The name of a substitution variable does not need to be the same as an existing column name; however, it should be an indicator of the data being requested from the user. Because the SET clause needs the value for the region to be entered by the user, the variable was named Region. After a user has entered a value for the Region column, the user is then asked for the value of the second substitution variable in the WHERE clause. The State substitution variable will be used to define exactly which rows will be updated. The output is shown in Figure 5-24.

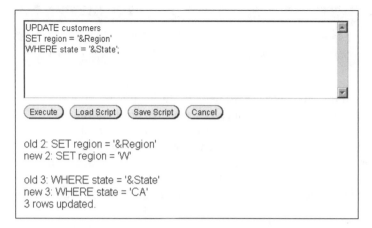

FIGURE 5-24 Output for UPDATE with substitution variables

NOTE

Regarding Figure 5-24, you should understand that if you are using iSQL*Plus, substitution variable prompts appear on a separate screen. Only the final screen is displayed in the figure.

NOTE

If the display of the old and new values becomes annoying, you can issue the command SET VERIFY OFF in SQL*Plus, and Oracle 10*g* suppresses the messages. The command SET VERIFY ON instructs Oracle 10*g* to start displaying the messages again.

As shown in the output displayed in Figure 5-24, when the UPDATE command is executed, Oracle 10*g* asks the user to enter the data value for the Region column. After the user responds with the requested data, Oracle 10*g* replaces the variable in the command with the entered data value. After the first substitution is complete, the user is then asked to enter a value for the second substitution variable—the state. To allow the user to verify the process that is occurring, the software displays the old and new values for the variables on the screen.

The statement can easily be reexecuted: In iSQL*Plus the statement remains in the work area, so you would need only to click Execute to reexecute the UPDATE. In the client SQL*Plus tool, when the command is first executed, it is placed in the SQL buffer. The buffer always contains the last SQL command processed. The contents of the buffer can be executed by typing **RUN** or the forward slash (/) at the SQL> prompt, and pressing the **Enter** key. Through the use of substitution variables, the statement becomes interactive, and the user can continue to update the Region column for as many states as necessary.

NOTE

The buffer contains only the last SQL command entered. However, it does not store SQL *PLUS commands, such as DESCRIBE.

If a user can't complete all the customer record updates during one session, the command can be permanently stored in a script file that can be executed at a later time. To create the script file, type the command in a simple word-processing program, such as Notepad or WordPad. For the client SQL*Plus tool, however, the file must be saved with **sql** as the file extension. After the command is saved in a script file, it can be executed at a later time by issuing the command START d:*filename* at the SQL> prompt, where d: indicates the location of the file. In iSQL*Plus, the load feature can be used to bring in statements from a file.

NOTE

Application developers embed SQL commands into their programs to support events that will occur via screens or Web pages. Variables are used to write SQL statements that will be used to support application screens to allow the user entries on the screen to be pushed into the SQL statement.

DELETING ROWS

When you need to remove rows from database tables, use the **DELETE** command. Compared to some of the other commands covered in this chapter, the DELETE command is incredibly simple—perhaps even too simple! The syntax for the DELETE command is shown in Figure 5-25.

```
DELETE FROM tablename
[WHERE condition];
```

FIGURE 5-25 Syntax for the DELETE command

The syntax for the DELETE command does not allow the user to specify any column names in the DELETE clause. This is logical because *DELETE applies to an entire row and cannot be applied to specific columns within a row.* The WHERE clause is optional and is used to identify the row(s) to be deleted from the specified table. A word of caution is appropriate here.

CAUTION

If you omit the WHERE clause, all rows are deleted from the specified table.

Suppose you are notified that Sammie Jones moved to the Customer Service Department and should no longer be listed in the ACCTMANAGER2 table. Because Jones' information has already been inserted, the DELETE command is needed to remove the row from the table. That account manager can be removed from the ACCTMANAGER2 table by using the DELETE command, as shown in Figure 5-26.

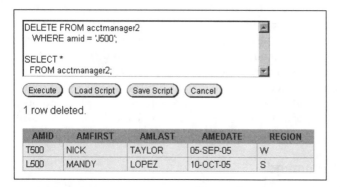

FIGURE 5-26 DELETE command used to remove a row from the ACCTMANAGER2 table

The WHERE clause of the DELETE command is used to identify the exact record to be removed from the ACCTMANAGER2 table. After the record is deleted, the table will contain only two records, as shown in Figure 5-26.

The WHERE clause in this DELETE statement caused only a single row to be deleted. Just as with an UPDATE statement, however, a group of rows could be deleted as well.

Be aware that a DELETE statement without a WHERE clause deletes all the rows in the table, as shown in Figure 5-27.

```
DELETE FROM acctmanager2;
```

FIGURE 5-27 DELETE command without the WHERE clause

Because the command in Figure 5-27 does not contain a WHERE clause, no specific record is identified for deletion. Therefore, the command deletes all the records in the ACCTMANAGER2 table.

CAUTION

Before completing the next section, review the data contents of the ACCTMANAGER2 table by using a SELECT statement. The table needs to contain the three rows inserted at the beginning of this chapter, as shown in Figure 5-10. Execute INSERT and/or UPDATE statements, if needed, to ensure that the table contains all three rows with the same data as shown in Figure 5-10. Make sure to COMMIT your changes before proceeding.

TRANSACTION CONTROL STATEMENTS

Changes to data made by DML commands are not permanently saved to the table when you execute the SQL statement. This allows you the flexibility of issuing **transaction control** statements either to save the modified data or to undo the changes if they were made in error. Until the data has been permanently saved to the table, no other users will be able to view any of the changes you have made. A **transaction** is a term used to describe a group of DML statements representing data actions that logically should be performed together. A common example is a bank transaction. For example, assume you withdraw $500 from a savings account and want to put half of this amount in your checking account and the other half in a mutual fund account. If the system crashed after the withdrawal but before the deposits to the other two accounts were accomplished, would you lose $500? Not if transaction control is in effect. The transaction control statements determine at which points the DML activity is permanently saved. In the bank example, the save will not occur until all three actions are entered.

COMMIT and ROLLBACK Commands

A **COMMIT** command, either implicitly or explicitly issued, permanently saves the DML statements previously issued. An **explicit** COMMIT occurs when the command is explicitly issued by entering a **COMMIT**; statement. The COMMIT command **implicitly** occurs when the user exits SQL*Plus. An implicit COMMIT also occurs if a DDL command, such as CREATE or ALTER TABLE, is issued. In other words, if a user adds several records to a table and then creates a new table, the records that were added before the DDL command was issued are automatically committed. In Oracle 10g, a **transaction** consists of a series of statements that have been issued and not committed. A transaction could consist of one SQL statement or 2000 SQL statements issued over an extended period of time. The duration of a transaction is defined by when a commit implicitly or explicitly occurs.

A program can manage the statements that should be grouped together into a transaction by determining when to issue a commit based on an event. For example, if a purchase order has been entered containing the purchase of several items, this would involve an insert into the ORDERS table and multiple inserts into the ORDERITEMS table. A commit covering all these inserts may not be issued until the user clicks a Finalize Order button and the credit card approval code is returned. At this point, the commit saves all the inserts permanently as a group.

NOTE

You can verify that data has not been permanently saved by issuing an UPDATE statement in one session and then logging in to a second session to view the table and note that the changes are not visible. Then in the first session, issue a COMMIT statement. Now check the data in the second session to see the changes.

Unless a DML operation is committed, it can be undone by issuing the **ROLLBACK** command. For example, if you have not exited Oracle 10g since beginning to work through the examples in this chapter, executing a **ROLLBACK**; statement would reverse all the rows that you entered or altered during your work in this chapter. Or, in the purchase order example that was described previously, a Cancel button would allow the data entry user to undo all the inserts issued during the entry of the order. Thus, the ROLLBACK command reverses all DML operations performed since the last commit was performed. By contrast, **commands such as CREATE TABLE, TRUNCATE TABLE, and ALTER TABLE cannot be rolled back because they are DDL commands, and a COMMIT occurs automatically when they are executed**. Note, however, that if the system crashes, a rollback automatically occurs after Oracle 10g restarts, and any operations not previously committed are undone.

To ensure that all the operations performed thus far in this chapter are safe from being accidentally reversed, issue the command shown in Figure 5-28 before continuing with the remaining examples in this chapter.

```
COMMIT;
```

FIGURE 5-28   Command to permanently save data changes

When designing programs for the Oracle 10g database, developers will sometimes use the SAVEPOINT command to create a type of bookmark within a transaction. A common use of this command is in the banking industry. For example, suppose that a customer is making both a deposit and a withdrawal through an ATM. If the customer first makes a deposit and then requests a withdrawal, but canceled the withdrawal before the money is dispensed, is the entire customer transaction canceled? To address this issue, a program can be designed to commit the deposit as one transaction and then begin the withdrawal process as a separate transaction. However, some designers have the program issue the command syntax SAVEPOINT *name*; after the deposit is completed to identify a particular "point" within a transaction. If a subsequent portion of the transaction is canceled, the program simply issues the command syntax ROLLBACK TO SAVEPOINT *name*;—and any SQL statements issued after the SAVEPOINT command are not permanently updated to the database. A COMMIT command would still need to be executed to update the database with any data that was added or changed by the first part of the transaction.

To see how this works, let's make several changes to the ACCTMANAGER2 table using transaction control statements. First, execute two UPDATE statements with a SAVEPOINT established between them, as shown in Figure 5-29.

The query shows that the region has been changed on two of the account manager records. What would happen if a ROLLBACK statement were issued at this point? Both UPDATES would be undone and the region value would return to the original value prior to the UPDATE statements. What if you want to undo only the last UPDATE? Issue a ROLLBACK TO SAVEPOINT statement, as shown in Figure 5-30.

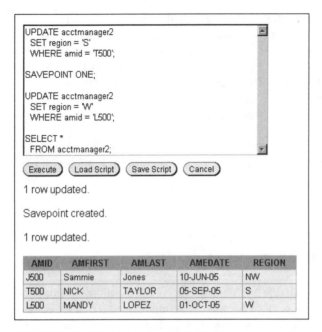

FIGURE 5-29 Establishing a SAVEPOINT

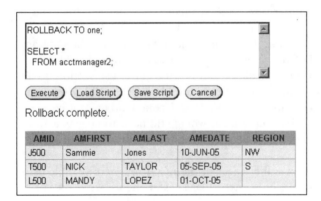

FIGURE 5-30 Undo changes to SAVEPOINT

Figure 5-30 shows that the second UPDATE was undone; however, the first UPDATE statement (before the SAVEPOINT) is intact. If you want to save the first UPDATE, you could now issue a COMMIT statement. But what if you need to undo the first UPDATE as well? Issue a ROLLBACK to undo the first UPDATE, as shown in Figure 5-31.

FIGURE 5-31 Undo all changes

TABLE LOCKS

The development of DML and transaction control statements needs to include a recognition of the fact that most database systems involve numerous concurrent users. What would happen if two users tried to change the same record at the same time? Which change would be saved to the table? When DML commands are issued, Oracle 10g, by default, performs a row level lock, which implicitly "locks" the row or rows being affected, so no other user can change the same row(s). In addition, a lock is indicated on the table so other users cannot attempt to lock the whole table while the row lock is active. The lock is a **shared lock**, in that other users can view the data stored in the table, but they cannot alter the structure of the table or perform other types of DDL operations, in addition to not being able to change the specific rows that are locked.

LOCK TABLE command

Although rarely used outside a program, a user can explicitly lock a table in **SHARE** mode by issuing the **LOCK TABLE** command. The syntax for this command is shown in Figure 5-32.

```
LOCK TABLE tablename IN SHARE MODE;
```

FIGURE 5-32 Syntax for LOCK TABLE in SHARE mode

When DDL operations are performed, Oracle 10g places an **exclusive lock** on the table so that no other user can alter the table or try to add or update the contents of the table. If an exclusive lock exists on a table, no other user can obtain an exclusive lock or a shared lock on the same table. In addition, if a user has a shared lock on a table, no other user can place an exclusive lock on the same table. If necessary, the user can instruct Oracle 10g to lock a table in **EXCLUSIVE** mode, using the command syntax shown in Figure 5-33.

```
LOCK TABLE tablename IN EXCLUSIVE MODE;
```

FIGURE 5-33   Syntax for LOCK TABLE in EXCLUSIVE mode

C A U T I O N

Always be careful when explicitly locking a table. If one user locks a portion of a table in SHARE mode, and another user locks a different portion of a table, and the completion of one of the commands depends on a portion of a table locked by the other user, a deadlock occurs. Usually, Oracle 10g detects deadlocks automatically and returns an error message to one of the users. When an error message is returned, the lock is also released, so the command issued by the other user is completed. *Locks (including exclusive locks) are automatically released if the user issues a transaction control statement, such as ROLLBACK or COMMIT, or when the user exits the system.*

SELECT...FOR UPDATE Command

A data consistency issue can occur when a user looks at the contents of a record, makes a decision based upon those contents, and then updates the record—only to find out that between the SELECT command and the UPDATE command, the contents of the record have changed. For example, suppose that you are assigned the task of increasing the retail price of certain books, and you are told to base the new retail price on a percentage of the cost of the book. As you begin to update retail prices, you realize someone has updated the cost of the books. Ugh! Now what should you do?

As previously mentioned, DML operations are not permanently stored in a table until a COMMIT command is issued. To provide a consistent view for all users accessing the table in a multiuser environment, no changes can be seen by other users until the changes have been committed. This can create major headaches when working with transaction-type tables that are constantly being changed to reflect new orders, account balances, and so on.

To avoid this type of problem, you can use the **SELECT...FOR UPDATE** command to view the contents of a record when it is anticipated that the record will need to be modified. The SELECT...FOR UPDATE command places a shared lock on the record(s) to be changed and prevents any other user from acquiring a lock on the same record(s). The syntax is the same as a regular SELECT statement, except the FOR UPDATE clause is added at the end of the command, as shown in Figure 5-34.

```
SELECT columnnames,…
FROM tablename, …
[WHERE condition]
FOR UPDATE;
```

FIGURE 5-34   Syntax for the SELECT...FOR UPDATE command

If a user decides to update a record, a regular UPDATE command is used to perform the change. However, if the user does not change any of the data selected by the SELECT... FOR UPDATE command, a COMMIT or ROLLBACK command must still be issued, or the rows selected will remain locked, and no other users will be able to make changes to those rows.

Chapter Summary

- Data manipulation language (DML) includes the INSERT, UPDATE, DELETE, COMMIT, and ROLLBACK commands.

- The INSERT INTO command is used to add new rows to an existing table.

- The column list specified in the INSERT INTO clause must match the order of the data provided in the VALUES clause.

- You can use a NULL value in an INSERT INTO command by including the keyword NULL, omitting the column from the column list of the INSERT INTO clause, or entering two single quotation marks in the position of the NULL value.

- To assign a DEFAULT option value, a column must be excluded from the column list in an INSERT.

- If rows are copied from a table and entered into an existing table through the use of a sub-query in the INSERT INTO command, the VALUES clause must be omitted because it is irrelevant.

- You can change the contents of a row or group of rows with the UPDATE command.

- You can use substitution variables to allow you to execute the same command several times with different data values.

- DML operations are not permanently stored in a table until a commit command is issued either implicitly or explicitly.

- A set of DML operations that are committed as a block is considered a transaction.

- Uncommitted DML operations can be undone by issuing the ROLLBACK command.

- A SAVEPOINT serves as a marker for a point in a transaction and allows only a portion of the transaction to be rolled back.

- Use the DELETE command to remove records from a table. If the WHERE clause is omitted, all rows in the table are deleted.

- Table locks can be used to prevent users from mistakenly overwriting changes made by other users.

- Table locks can be in SHARE mode or EXCLUSIVE mode.

- EXCLUSIVE MODE is the most restrictive table lock and prevents any other user from obtaining any locks on the same table.

- A lock is released when a transaction control statement is issued, a DDL statement is executed, or the user exits the system using the EXIT command.

- SHARE mode allows other users to obtain shared locks on other portions of the table, but it prevents users from obtaining an exclusive lock on the table.

- The SELECT...FOR UPDATE command can be used to obtain a shared lock for a specific row or rows. The lock is not released unless a DDL command is issued or the user exits the system.

Chapter 5 Syntax Summary

The following table presents a summary of the syntax that you have learned in this chapter. You can use the table as a study guide and reference.

SYNTAX GUIDE		
Command	Description	Example
Optional SELECT clauses		

INSERT	Adds new row(s) to a table. The user can include a subquery to copy row(s) from existing table.	`INSERT INTO acctmanager`
		`VALUES ('T500', 'NICK TAYLOR',`
		`'05-SEP-05', 'NE');`
		or
		`INSERT INTO acctmanager2`
		`SELECT amid, amname, amedate, region`
		`FROM acctmanager`
		`WHERE amedate <='01-OCT-05';`

UPDATE	Adds data to, or modifies data within, an existing row	`UPDATE acctmanager`
		`SET amedate = '05-SEP-05'`
		`WHERE amid = 'J500';`

COMMIT	Permanently saves changed data in a table	`COMMIT;`
ROLLBACK	Allows the user to "undo" uncommitted changes to data	`ROLLBACK;`
DELETE	Removes row(s) from a table	`DELETE FROM acctmanager` `WHERE amid = 'D500';`

LOCK TABLE	Prevents other users from making changes to a table	`LOCK TABLE customers IN SHARE MODE;`
		or
		`LOCK TABLE customers IN EXCLUSIVE`
		`MODE;`

SELECT...FOR UPDATE	Creates a shared lock on a table to prevent another user from making changes to data in specified columns	`SELECT cost`
		`FROM books`
		`WHERE category = 'COMPUTER'`
		`FOR UPDATE;`

Interactive Operator		

&	Identifies a substitution variable; allows the user to be prompted to enter a specific value for the substitution variable	`UPDATE customers`
		`SET region = '&Region'`
		`WHERE state = '&State';`

Review Questions

1. Which command should you use to copy data from one table and have it added to an existing table?

2. Which command can you use to change the existing data in a table?

3. When do the changes generated by DML operations become permanently stored in the database tables?

4. Explain the difference between explicit and implicit locks.

5. If you add a record to the wrong table, what is the simplest way to remove the record from the table?

6. How does Oracle 10*g* identify a substitution variable in a SQL command?

7. How are NULL values included in a new record being added to a table?

8. When should the VALUES clause be omitted from the INSERT INTO command?

9. What happens if a user attempts to add data to a table, and the addition would cause the record to violate an enabled constraint?

10. Explain the usefulness of a constraint naming convention.

Multiple Choice

1. Which of the following is a correct statement?
   a. A commit is implicitly issued when a user exits the system using the EXIT command.
   b. A commit is implicitly issued when a DDL command is executed.
   c. A commit is automatically issued when a DML command is executed.
   d. All of the above are correct.
   e. Both a and b are correct.
   f. Both a and c are correct.

2. Which of the following is a valid SQL statement?
   a. SELECT * WHERE amid = 'J100' FOR UPDATE;
   b. INSERT INTO homework10VALUES (SELECT * FROM acctmanager);
   c. DELETE amid FROM acctmanager;
   d. rollback;
   e. all of the above

3. Which of the following commands can be used to add rows to a table?
   a. INSERT INTO
   b. ALTER TABLE ADD
   c. UPDATE
   d. SELECT...FOR UPDATE

4. Which of the following will delete all the rows in the HOMEWORK10 table?
   a. DELETE * FROM homework10;
   b. DELETE *.* FROM homework10;
   c. DELETE FROM homework10;
   d. DELETE FROM homework10 WHERE 9amid = '*';
   e. Both c and d will delete all the rows in the HOMEWORK10 table.

5. Which of the following statements will obtain a shared lock on at least a portion of a table named HOMEWORK10?

   a. SELECT * FROM homework10 WHERE col2 IS NULL FOR UPDATE;

   b. INSERT INTO homework10 (col1, col2, col3)VALUES ('A', 'B', 'C');

   c. UPDATE homework10 SET col3 = NULLWHERE col1 = 'A';

   d. UPDATE homework10 SET col3 = LOWER(col3)WHERE col1 = 'A';

   e. all of the above

6. Assuming the HOMEWORK10 table has three columns (Col1, Col2, and Col3 in the order listed), which of the following commands will store a NULL value in Col3 of the HOME-WORK10 table?

   a. INSERT INTO homework10 VALUES ('A', 'B', 'C');

   b. INSERT INTO homework10 (col3, col1, col2)VALUES (NULL, 'A', 'B');

   c. INSERT INTO homework10 VALUES (NULL, 'A', 'B');

   d. UPDATE homework10SET col1 = col3;

7. Which of the following symbols designates a substitution variable?

   a. &

   b. $

   c. #

   d. _

8. Which of the following commands can you use to suppress the display of old and new values when executing a command containing substitution variables?

   a. SUPPRESS ON

   b. SET VERIFY ON

   c. SET VERIFY OFF

   d. SET DISPLAY OFF

   e. SET MESSAGE OFF

9. Which of the following commands locks the HOMEWORK10 table in EXCLUSIVE mode?

   a. LOCK TABLE homework10 EXCLUSIVELY;

   b. LOCK TABLE homework10 IN EXCLUSIVE mode;

   c. LOCK TABLE homework10 TO OTHER USERS;

   d. LOCK homework10 IN EXCLUSIVE mode;

   e. Both b and d lock the table in EXCLUSIVE mode.

10. You issue the following command: INSERT INTO homework10 (col1, col2, col3) VALUES ('A', NULL, 'C'). The command will fail if which of the following statements is true?

   a. Col1 has a PRIMARY KEY constraint enabled.

   b. Col2 has a UNIQUE constraint enabled.

   c. Col3 is defined as a DATE column.

   d. None of the above would cause the command to fail.

11. Which of the following will release a lock currently held by a user on the HOMEWORK10 table?

    a. A commit command is issued.

    b. A DCL command is issued to end a transaction.

    c. The user exits the system.

    d. A rollback command is issued.

    e. all of the above

    f. none of the above

12. Assume you have added eight new orders to the ORDERS table. Which of the following is true?

    a. Other users can view the new orders as soon as you execute the INSERT INTO command.

    b. Other users can view the new orders as soon as you issue a ROLLBACK command.

    c. Other users can view the new orders as soon as you exit the system or execute a COMMIT command.

    d. Other users can view only the new orders if they obtain an exclusive lock on the table.

13. Which of the following commands will remove all orders placed before April 1, 2006?

    a. DELETE FROM orders WHERE orderdate < '01-APR-06';

    b. DROP FROM orders WHERE orderdate < '01-APR-06';

    c. REMOVE FROM orders WHERE orderdate < '01-APR-06';

    d. DELETE FROM orders WHERE orderdate > '01-APR-06';

14. How many rows can be added to a table per execution of the INSERT INTO...VALUES command?

    a. 1

    b. 2

    c. 3

    d. unlimited

15. You accidentally deleted all the orders in the ORDERS table. How can the error be corrected after a COMMIT command has been issued?

    a. ROLLBACK;

    b. ROLLBACK COMMIT;

    c. REGENERATE RECORDS orders;

    d. None of the above will restore the deleted orders.

16. Which of the following is the standard extension used for a script file?

    a. SPT

    b. SRT

    c. SCRIPT

    d. SQL

17. A rollback will not automatically occur when:

    a. A DDL command is executed.

    b. A DML command is executed.

    c. The user exits the system.

    d. The system crashes.

18. What is the maximum number of rows that can be deleted from a table at one time?

    a. 1

    b. 2

    c. 3

    d. unlimited

19. Which of the following is a correct statement?

    a. If you attempt to add a record that violates a constraint for one of the table's columns, the valid columns for the row will be added.

    b. A subquery nested in the VALUES clause of an INSERT INTO command can return only one value without generating an Oracle 10*g* error message.

    c. If you attempt to add a record that violates a NOT NULL constraint, a blank space will automatically be inserted in the appropriate column so Oracle 10*g* can complete the DML operation.

    d. None of the above statements is correct.

20. What is the maximum number of records that can be modified with the UPDATE command?

    a. 1

    b. 2

    c. 3

    d. unlimited

Hands-On Assignments

To perform the following assignments, refer to the tables created in the prech05.sql script at the beginning of the chapter.

1. Add a new row in the ORDERS table with the following data: Order# = 1021, Customer# = 1009, and Order date = July 20, 2005.

2. Modify the zip code on order 1017 to 33222.

3. Save the changes permanently to the database.

4. Add a new row in the ORDERS table with the following data: Order# = 1022, Customer# = 2000, and Order date = August 6, 2005. Describe the error raised and what caused the error.

5. Add a new row in the ORDERS table with the following data: Order# = 1023 and Customer# = 1009. Describe the error raised and what caused the error.

6. Create a script using substitution variables that will allow a user to set a new cost amount based on an ISBN for a book.

Data Manipulation and Transaction Control

7. Execute the script and set the following: isbn = 1059831198 and cost = $20.00.

8. Execute a command that will undo the change in the previous step.

9. Delete Order# 1005.

10. Execute a command that will undo the previous deletion.

Advanced Challenge

To perform the following activity, refer to the tables in Appendix A.

Currently, the contents of the Category column in the BOOKS table are the actual name for each category. This presents a problem if one user enters 'COMPUTER' for the Computer Category and another user enters 'COMPUTERS'. To avoid this and other problems that may occur, the database designers have decided to create a CATEGORY table that will contain a code and description for each category. The structure for the CATEGORY table should be as follows:

COLUMN NAME	DATATYPE	WIDTH	CONSTRAINTS
CATCODE	VARCHAR2	3	Primary Key
CATDESC	VARCHAR2	11	Not Null

The data for the CATEGORY table is as follows:

CATCODE	CATDESC
BUS	BUSINESS
CHN	CHILDREN
COK	COOKING
COM	COMPUTER
FAL	FAMILY LIFE
FIT	FITNESS
SEH	SELF HELP
LIT	LITERATURE

Required:

- Create the CATEGORY table and populate it with the given data.
- Add a column to the BOOKS table called Catcode.
- Store the correct category code in the BOOKS table, based upon each book's current category.
- Verify that the correct categories have been assigned in the BOOKS table.

- Delete the Category column from the BOOKS table.
- Add a foreign key constraint that requires all category codes entered into the BOOKS table to already exist in the CATEGORY table.
- Commit all changes after you have verified their accuracy.

Case Study: *City Jail*

Execute the CityJail_5.sql script to rebuild the CRIMINALS and CRIMES tables of the City Jail database. The statements at the beginning of this script will drop existing tables in your schema with the same table names.

Review the script so that you are familiar with the table structure and constraints.

1. Create and execute statements to perform the following DML activities. Save the changes permanently to the database.

    a. Create a script to allow a user to add new criminals (providing prompts to the user) to the CRIMINALS table.

    b. Add the following criminals using the script created in the previous step. No value needs to be entered at the prompt if it should be set to the DEFAULT column value. Query the CRIMINALS table to confirm new rows have been added.

CRIMINAL_ID	LAST	FIRST	STREET	CITY	STATE	ZIP	PHONE	V_STATUS	P_STATUS
1015	Fenter	Jim		Chesapeake	VA	23320		N	N
1016	Saunder	Bill	11 Apple Rd	Virginia Beach	VA	23455	7678217443	N	N
1017	Painter	Troy	77 Ship Lane	Norfolk	VA	22093	7677655454	N	N

    c. Add a column named Mail_flag to the CRIMINALS table. The column should be assigned a datatype of CHAR(1).

    d. Set the Mail_flag column to a value of 'Y' for all criminals.

    e. Set the Mail_flag column to 'N' for all the criminals who do not have a street address recorded in the database.

    f. Change the phone number for criminal 1016 to 7675659032.

    g. Remove criminal 1017 from the database.

2. Execute a DML statement to accomplish each of the following actions. Each statement will produce a constraint error. Document the error number and message and briefly explain the cause of the error. If your DML statement generates a syntax error rather than a constraint violation error, revise your statement to correct any syntax errors. Recall that you can review the table creation script file to identify table constraints.

    a. Add a crime record using the following data: Crime_id = 100, Criminal_id = 1010, Classification = M, Date_charged = July 15, 2005, Status = PD.

    b. Add a crime record using the following data: Crime_id = 130, Criminal_id = 1016, Classification = M, Date_charged = July 15, 2005, Status = PD.

    c. Add a crime record using the following data: Crime_id = 130, Criminal_id = 1016, Classification = P, Date_charged = July 15, 2005, Status = CL.

CHAPTER **6**

ADDITIONAL DATABASE OBJECTS

LEARNING OBJECTIVES

After completing this chapter, you should be able to do the following:

*   Define the purpose of a sequence and state how it can be used in a database

*   Explain why gaps might appear in the integers generated by a sequence

*   Use the CREATE SEQUENCE command to create a sequence

*   Identify which options cannot be changed by the ALTER SEQUENCE command

*   Use NEXTVAL and CURRVAL in an INSERT command

*   Explain the main index structures: B-tree and Bitmap

*   Introduce variations on conventional indexes, including a function-based index and an Index Organized Table

*   Create indexes using the CREATE INDEX command

*   Remove an index using the DELETE INDEX command

*   Verify index existence via the data dictionary

*   Create and remove a public synonym

INTRODUCTION

The tables and constraints created in previous chapters are considered database objects. A **database object** is anything that has a name and a defined structure. Three other database objects commonly used in Oracle 10*g* are sequences, indexes, and synonyms. This chapter examines each of these objects. The following list identifies the role of each of these objects:

- A **sequence** generates sequential integers that can be used by organizations to assist with internal controls or simply to serve as a primary key for a table.

- A database **index** serves the same basic purpose as an index in a book, allowing users to quickly locate specific records.

- A **synonym** is a simpler name, like a nickname, given to an object with a complex name. Some organizations' naming conventions create complex object names that are difficult for users to remember. A synonym can be created to provide an alternative name to identify database objects. A synonym can be either a **private synonym**, used by an individual to reference objects owned by that person, or it can be a **public synonym** to be used by others to access an individual's database objects.

This chapter demonstrates how to create, maintain, and delete sequences and how to create and delete indexes and synonyms. Figure 6-1 provides an overview of this chapter's contents.

DESCRIPTION	COMMAND SYNTAX
Create a sequence to generate a series of integers	```CREATE SEQUENCE sequencename [INCREMENT BY value] [START WITH value] [{MAXVALUE value	NOMAXVALUE}] [{MINVALUE value	NOMINVALUE}] [{CYCLE	NOCYCLE}] [{ORDER	NOORDER}] [{CACHE value	NOCACHE}];```
Alter a sequence	```ALTER SEQUENCE sequencename [INCREMENT BY value] [{MAXVALUE value	NOMAXVALUE}] [{MINVALUE value	NOMINVALUE}] [{CYCLE	NOCYCLE}] [{ORDER	NOORDER}] [{CACHE value	NOCACHE}];```

FIGURE 6-1   Overview of chapter contents

DESCRIPTION	COMMAND SYNTAX
Drop a sequence	DROP SEQUENCE *sequencename*;
Create a B-tree index	CREATE INDEX *indexname* ON *tablename* (*columnname*, ...);
Create a Bitmap index	CREATE BITMAP INDEX *indexname* ON tablename (columnname, ...);
Create a function-based index	CREATE INDEX *indexname* ON *tablename* (expression);
Create an Index Organized Table	CREATE TABLE tablename (columnname datatype, columnname datatype) ORGANIZATION INDEX;
Drop an index	DROP INDEX *indexname*;
Create a synonym	CREATE [PUBLIC] SYNONYM *synonymname* FOR *objectname*;
Drop a synonym	DROP [PUBLIC] SYNONYM synonymname;

FIGURE 6-1 Overview of chapter contents (continued)

NOTE

Before attempting to work through the examples provided in this chapter, you should open the Chapter 6 folder in your Data Files. Run the prech06.sql file to ensure that all necessary tables and constraints are available.

SEQUENCES

A sequence is a database object that you can use to generate a series of integers. These integers are most commonly used either to generate a unique primary key for each record or for internal control purposes. A brief overview of these two concepts follows.

When you use values generated by a sequence as a primary key, there is no true correlation between the number assigned to a record and the entity it represents. However, depending on the parameters used to create the sequence, database users can be assured that no two records will receive the same primary key value. This is especially important if different users are assigned the task of entering records into a database table, because the possibility exists that more than one user may attempt to assign the same primary key value to different records. For example, if several customer service representatives are entering new customers at the same time, how are customer numbers assigned?

Are all the customer service representatives in the same room and asking one another, "What number did your last customer receive?" Not likely. Chances are, they are using a sequence, and, thus, each customer service representative can be certain that each customer's number is unique.

NOTE

When a database-generated value is used as a primary key value, it is often referred to as a surrogate key. If existing data, such as a book ISBN, is used to provide the table's primary key value, it is often referred to as a natural key.

A sequence can also be used to provide business and auditing controls. Every organization should have some control mechanisms to avoid problems with transaction auditing, embezzlement, and accounting errors. Most organizations use sequential numbers to track checks, purchase orders, invoices, or anything else that can be used to record economic events. Through the use of sequential numbers, an auditor can determine whether items such as checks or invoices are missing, which, in turn, can reveal accounting problems, such as unrecorded transactions, or whether employees have obtained blank checks or invoices for their own use.

Creating a Sequence

You create a sequence with the CREATE SEQUENCE command using the syntax shown in Figure 6-2. Optional commands are shown in square brackets. Curly brackets indicate that one of the two options shown can be used, but not both.

```
CREATE SEQUENCE sequencename
[INCREMENT BY value]
[START WITH value]
[{MAXVALUE value | NOMAXVALUE}]
[{MINVALUE value | NOMINVALUE}]
[{CYCLE | NOCYCLE}]
[{ORDER | NOORDER}]
[{CACHE value | NOCACHE}];
```

FIGURE 6-2 Syntax of the CREATE SEQUENCE command

Notice that most of the items in a CREATE SEQUENCE command are optional; not much is required to create a basic sequence. The statement shown in Figure 6-3 creates a sequence named CUSTOMERS_CUSTOMER#_SEQ.

```
CREATE SEQUENCE customers_customer#_seq;
```

FIGURE 6-3 Create a sequence

You may have some questions regarding this sequence. By what amount will the numbers generated be incremented? When will the sequence run out of numbers to generate? What number will be the first number generated? Because the statement in Figure 6-3 does not explicitly set any of the options, the default values for the options will be in effect. This is why it is important to understand all the option settings before creating a sequence. Let's look at each of the syntax elements in Figure 6-2. Note that some elements, such as CACHE, must indicate a value (how many numbers to generate and store in memory), whereas other options, such as NOCACHE, do not need a value indicated.

The **CREATE SEQUENCE** keywords are followed by the name used to identify the sequence. A standard naming convention for sequences is to include _seq at the end of the sequence name, which makes it easier to identify as a sequence.

The **INCREMENT BY** clause is used to specify the intervals that should exist between two sequential values. For checks and invoices, this interval is usually one. However, for sequences that represent credit card or bank account numbers, for example, the interval might be much larger, so that no two account numbers are very similar (for example 13,519 might be a more appropriate interval). If the sequence is incremented by a positive value, the values generated by the sequence are in an ascending order. However, if a negative value is specified, the values generated by the sequence are in a descending order. If the user needs the values to be in a descending order, the keywords INCREMENT BY must be used. *If the INCREMENT BY clause is not included when the sequence is created, the default setting is used, in which the sequence is increased by one for each integer generated.*

The **START WITH** clause is used to establish the starting value for the sequence. Oracle 10g begins each sequence with the value of *one* unless another value is specified in the START WITH clause. For example, if you want all customer numbers to consist of four-digit numbers, the START WITH clause can be assigned the value of 1000 to avoid having the first 999 customers assigned account numbers with fewer than four digits.

TIP

The START WITH value is a critical consideration if you already have existing data. For example, if you are importing existing data into a new database, you need to consider what values already exist for the column(s) that the sequence will populate. For example, if you already have existing customer numbers ranging from 1000 to 2500, you need to set a START WITH value of 2501 for the sequence used to populate the Customer Number column.

Continuing with the syntax listed in Figure 6-2, the **MINVALUE** and **MAXVALUE** clauses are used to establish a minimum or maximum value, respectively, for the sequence. If the sequence were incremented with a positive value, the MINVALUE clause would not make sense. By the same logic, you would assume that if the sequence were incremented with a negative value (that is, the values in the sequence decrease), a MAXVALUE clause would not be necessary, either. However, this is not the case. If a negative increment is used and you set a MINVALUE, the MAXVALUE must also be set. Typically in this case, the MAXVALUE is set to the same value as the START value.

By default, if you do not specify the minimum and maximum values, Oracle 10g assumes **NOMINVALUE** and **NOMAXVALUE**. When the NOMINVALUE option is

Additional Database Objects

assumed—or assigned—the lowest possible value for an increasing sequence is 1, and the lowest possible value for a decreasing sequence is -10^{26}, or -100,000,000,000,000,000,000,000,000. For the NOMAXVALUE option, 10^{27}, or 1,000,000,000,000,000,000,000,000,000 is the highest possible value for an ascending sequence, and -1 is the highest possible value for a descending sequence.

The **CYCLE** and **NOCYCLE** options are used to determine whether Oracle 10g should begin reissuing values from the sequence after the minimum or maximum value has been reached. If the CYCLE option is specified and Oracle 10g reaches the maximum value for an ascending sequence or the minimum value for a descending sequence, the CYCLE option instructs Oracle 10g to begin the cycle of numbers over again. If the sequence is being used to generate values for a primary key, this can cause problems if the sequence tries to assign a value that already exists in the table. However, some organizations reuse check numbers, order numbers, and so forth after an extended period of time rather than let the numbers become astronomically large. Therefore, the sequence must be allowed to reuse the same sequence of numbers. If a user does not specify a cycle option, Oracle 10g applies the default NOCYCLE option to the sequence. If the NOCYCLE option is in effect, Oracle 10g does not generate any numbers after the minimum or maximum value has been reached, and an error message is returned when the user requests another value from the sequence.

The **ORDER** and **NOORDER** options are used in **application cluster environments** in which multiple users might be requesting sequence values at the same time during large transactions (such as when printing a large quantity of checks or invoices). The ORDER option instructs Oracle 10g to return the sequence values in the same order in which the requests are received. If the option is not specified in the CREATE SEQUENCE command, the NOORDER option is assumed by default. In cases in which the sequence is being used to generate a primary key, the order of the sequence value is not a problem because each value is still unique.

Generating sequence values can slow down the processing of requests by other users, especially if large volumes of those values are requested in a short period of time. If the **NOCACHE** option is specified when the sequence is created, each number is generated when the request for a sequence number is received. However, if the nature of an organization's transactions requires large amounts of sequential numbers throughout a session, the **CACHE** option can be used to have Oracle 10g pregenerate a set of values and store those values in the server's memory. Then, when a user requests a sequence value, the next available value is assigned—without Oracle 10g having to generate the number. On the other hand, if the CACHE option is not specified when the sequence is created, Oracle 10g assumes a default option of CACHE 20, and it automatically stores 20 sequential values in memory for users to access.

Keep this in mind when working with sequences and cached values: When a value is generated, that value has been assigned and cannot be regenerated until the sequence begins a new cycle. Therefore, if Oracle 10g caches 20 values, the values have been generated regardless of whether they are actually used. If the system crashes, or if the users do not use the values, the values generated are lost. If the sequence is being used for internal control purposes, some of the gaps that appear in the sequence could be due to nonusage after a system crash. Because these cannot be documented, this can be a cause for concern. For example, suppose that 50 sequential numbers to be assigned as order numbers are cached. After a few orders have been received, the Oracle 10g server crashes and has to be restarted. All unassigned numbers are now lost, and a gap exists in the order number sequence. Gaps may also appear if transactions are rolled back because the sequence value is already considered used and cannot be returned for "reuse."

Sequences are not assigned to a specific column or table. Sequences are independent objects and, therefore, the same sequence can be used by different users to generate values that are inserted into a number of different tables. In other words, the same sequence could be used to generate order numbers and customer numbers. This would result in gaps in the sequence values that appear in each table. Although this would not cause concern if the values were being used for a primary key column that only requires unique values, it would be a problem if the sequence values were used for internal control purposes. If the objective of the sequence is to provide a means for auditing control and make certain no check, invoices, and so on are missing, a sequence should be used only for one table, and the values generated should not be cached.

Let's look at an example of how sequences can work. All orders placed by JustLee Books' customers are assigned a four-digit number to uniquely identify each order. As the sales volume increases, multiple data entry clerks will be entering orders into the ORDERS table. Thus, it is possible that two clerks could try to enter the same order number if they were identifying order numbers manually. Although the PRIMARY KEY constraint for the ORDERS table would prevent two orders from having the same number, it still slows down the data entry process because one of the clerks must choose a different order number and then reenter the order. To avoid this problem, you need to create a sequence to generate the order numbers used in the ORDERS table.

The sequence should be named ORDERS_ORDER#_SEQ to identify that this is a sequence object that was created to generate order numbers for orders in the ORDERS table. However, using this rationale is not an absolute requirement. You should note that although the name was assigned to indicate its purpose, the value can still be used in any table. The INCREMENT BY clause needs to instruct Oracle 10g that each number generated should be increased by the value of one. Because the last order number stored in the ORDERS table is 1020, the sequence needs to start at 1021, so there is no gap in the order numbers and no previously assigned value is duplicated. Oracle 10g should also be instructed not to cache any values in memory, which means each value will be generated only when it is requested. The CYCLE option needs to indicate that the sequence values are not to be reused after the maximum value has been reached. For any clause that is not explicitly included, the default values are in effect. The command to create this sequence is shown in Figure 6-4.

```
CREATE SEQUENCE orders_order#_seq
   INCREMENT BY 1
   START WITH 1021
   NOCACHE
   NOCYCLE;
```

FIGURE 6-4   CREATE SEQUENCE command to generate a sequence for order numbers

If the CREATE SEQUENCE command is executed successfully, you should see the results displayed in Figure 6-5.

FIGURE 6-5   Results of the CREATE SEQUENCE statement

After Oracle 10g has returned the message "Sequence created," you can verify that the sequence exists by querying the **USER_OBJECTS** table in the data dictionary using a SELECT statement, as shown in Figure 6-6. This statement identifies all the existing sequences.

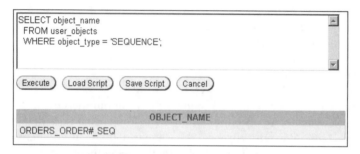

FIGURE 6-6   Verifying sequence creation with USER_OBJECTS

NOTE

Results will vary depending on the Oracle account being used. If you are using an Oracle administrator account such as SYSTEM, you will receive more rows of output displaying sequences automatically generated during the database installation.

TIP

The USER_OBJECTS table contains information on all the existing objects, including tables, sequences, indexes, and views. You can use the WHERE clause of this query to list only the type of objects you need to see at a given time.

To verify the individual settings for various options of a sequence, you can query the **USER_SEQUENCES** table of the data dictionary, as shown in Figure 6-7. This is also a quick way to identify which value is the next value to be assigned in the sequence—without accidentally generating a number. The next value to be assigned in a sequence created with the NOCACHE option is indicated in the last number column in the results of the SELECT query.

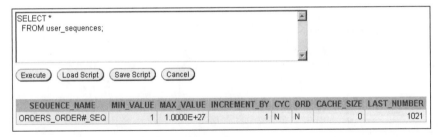

```
SELECT *
  FROM user_sequences;
```

(Execute) (Load Script) (Save Script) (Cancel)

SEQUENCE_NAME	MIN_VALUE	MAX_VALUE	INCREMENT_BY	CYC	ORD	CACHE_SIZE	LAST_NUMBER
ORDERS_ORDER#_SEQ	1	1.0000E+27	1	N	N	0	1021

FIGURE 6-7   Verifying sequence option settings

> **NOTE**
>
> Results will vary depending on the Oracle account being used. If you are using an Oracle administrator account such as SYSTEM, you will receive more rows of output displaying sequences automatically generated during the database installation.

Using Sequence Values

You can access sequence values through the two pseudocolumns NEXTVAL and CURRVAL. **Pseudocolumns** are data associated with table data, much like columns, but are not actually physical columns stored in the database. The pseudocolumn NEXTVAL (NEXT VALUE) is used to actually generate the sequence value. In other words, this calls the sequence object and requests the value of the next number in the sequence. After a value is generated, it is stored in the **CURRVAL** (CURRENT VALUE) pseudocolumn so a user can reference it again.

Let's look at an example. After the ORDERS_ORDER#_SEQ sequence is created, an order is received from customer 1010 on April 6, 2005, for one copy of *Big Bear and Little Dove*, to be shipped to 123 West Main, Atlanta, GA 30418. To process the order, the first step is to place the order information in the ORDERS table. The command to add the order to the ORDERS table is shown in Figure 6-8.

```
INSERT INTO ORDERS (order#, customer#, orderdate, shipdate,
                    shipstreet, shipcity, shipstate, shipzip)
VALUES (orders_order#_seq.nextval, 1010, '06-APR-05',
        NULL, '123 WEST MAIN', 'ATLANTA', 'GA', 30418);
```

FIGURE 6-8   Inserting a row using a sequence value

In Figure 6-8, the orders_order#_seq.nextval reference in the VALUES clause of the INSERT INTO command instructs Oracle 10g to generate the next sequential value from the ORDERS_ORDER#_SEQ sequence. Because the reference is listed as the first column of the VALUES clause, the value generated is stored in the first column of the column list, which is identified in the INSERT INTO clause of the command as order#. The NEXTVAL pseudocolumn is preceded by the name of the sequence, which identifies the sequence that should generate the value. Figure 6-9 shows the order added to the ORDERS table.

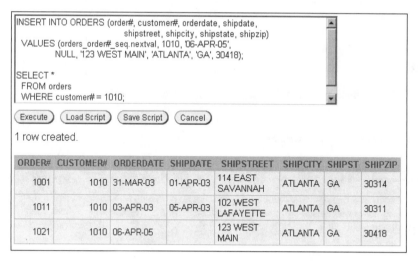

FIGURE 6-9 Order added using a sequence value for the order#

After the row has been added to the ORDERS table, the SELECT command can be used to view the order number assigned to the new order. The next step is to add the ordered item, *Big Bear and Little Dove*, to the ORDERITEMS table.

NOTE

If an error message occurs when you try to insert the new order, make certain the order number does not already exist. If it does, run the prech06.sql script as instructed at the beginning of this chapter to reset your tables to match the contents of this chapter.

When adding a book to the ORDERITEMS table, the order number must be entered. One approach is to query the ORDERS table to determine the number assigned to the order. However, because the user did not generate another number by referencing the NEXTVAL pseudocolumn again, the assigned order number is still stored as the value in the CURRVAL pseudocolumn. The CURRVAL value of a sequence is the last value generated by the sequence in the user session. Therefore, the contents of CURRVAL can be used to add the order number to the ORDERITEMS table, and the user does not have to recall the assigned order number. You can use the INSERT INTO command shown in Figure 6-10 to insert the order number, item number, ISBN, and quantity ordered into the ORDERITEMS table.

```
INSERT INTO orderitems (order#, item#, isbn, quantity)
  VALUES (orders_order#_seq.currval, 1, 8117949391, 1);
```

FIGURE 6-10 INSERT INTO command referencing CURRVAL

The command shown in Figure 6-10 instructs Oracle 10*g* to place the value stored in the CURRVAL pseudocolumn for the ORDERS_ORDER#_SEQ sequence into the first column of the ORDERITEMS table. *Any reference to CURRVAL does not cause Oracle 10g to generate a new order number.* However, if the example had referenced NEXTVAL in Figure 6-10, a new sequence number would have been generated, and the order number entered in the ORDERITEMS table would not have been the same as the order number already stored in the ORDERS table. Figure 6-11 shows CURRVAL inserted into the ORDERITEMS table.

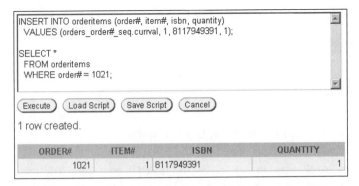

FIGURE 6-11 Using CURRVAL to insert an order detail row

> **NOTE**
>
> When a user logs in to Oracle 10*g*, no value is initially stored in the CURRVAL pseudocolumn and the current value is NULL. After a NEXTVAL call has been issued to generate a sequence value, CURRVAL stores that value until the next value is generated. *CURRVAL contains only the last value generated.*

Altering Sequence Definitions

The settings for a sequence can be changed by using the **ALTER SEQUENCE** command. However, any changes made are applied only to values generated *after* the modifications are made. The only restrictions that apply to changing the sequence settings are as follows:

- The START WITH clause cannot be changed—the sequence would have to be dropped and re-created to make this change.
- The changes cannot make previously issued sequence values invalid (for example they cannot change the defined MAXVALUE to a number less than a sequence number that has already been generated).

As shown in Figure 6-12, the ALTER SEQUENCE command follows the same syntax as the CREATE SEQUENCE command. Only the options that need to be added or modified need to be included in the ALTER SEQUENCE command. Any options you do not specify in the ALTER SEQUENCE command remain at their current setting.

```
ALTER SEQUENCE sequencename
  [INCREMENT BY value]
  [{MAXVALUE value | NOMAXVALUE}]
  [{MINVALUE value | NOMINVALUE}]
  [{CYCLE | NOCYCLE}]
  [{ORDER | NOORDER}]
  [{CACHE value | NOCACHE}];
```

FIGURE 6-12   Syntax of the ALTER SEQUENCE command

Suppose that management decides that all order numbers should increase by a value of 10, rather than by 1. Figure 6-13 shows the command to change the INCREMENT BY setting for the ORDERS_ORDER#_SEQ sequence.

```
ALTER SEQUENCE orders_order#_seq
  INCREMENT BY 10;
```

FIGURE 6-13   Command to change the INCREMENT BY setting for a sequence

Because no other settings were changed in the ALTER SEQUENCE command, their values are unaffected. The new setting for the sequence can be viewed from the USER_SEQUENCES data dictionary table, as shown in Figure 6-14. The last number now indicates that the next number to be issued from the sequence is 1031, which is 10 more than the start value of 1021.

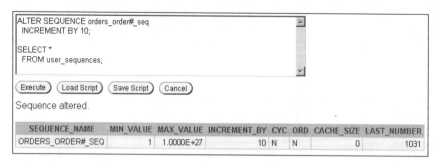

FIGURE 6-14   New settings for the ORDERS_ORDER#_SEQ

Removing a Sequence

A sequence can be deleted using the **DROP SEQUENCE** command. The syntax for the command is shown in Figure 6-15.

```
DROP SEQUENCE sequencename;
```

FIGURE 6-15   Syntax of the DROP SEQUENCE command

When a sequence is dropped, it does not affect any values previously generated and stored in a database table. When the DROP SEQUENCE command is successfully executed, the user receives the message "Sequence dropped," as shown in Figure 6-16.

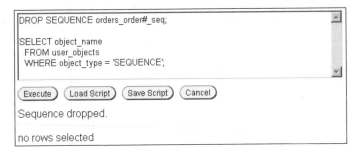

FIGURE 6-16   Dropping the ORDERS_ORDER#_SEQ sequence

INDEXES

An Oracle 10g **index** is a database object that stores a map of column values and the ROWIDs of matching table rows. A **ROWID** is the physical address of a table row. A database index is much like the index at the end of this textbook. If you look up a particular topic such as "primary key," you can quickly scan the handful of alphabetically sorted index pages to determine the page location of this topic. If no index existed, you would need to scan through the entire book to find the references to this topic, and this could be very time consuming. In a similar way, database indexes make data retrieval more efficient.

A common challenge in managing databases is improving data retrieval speed. As tables become populated with many rows, the processing of operations involved in a query search (WHERE conditions) and sorting (ORDER BY and joins) take increasing amounts of time. Much of this increase in execution time is due to disk I/O or disk reads (reading data from the physical disk drives). A CREATE TABLE statement in Oracle 10g by default creates a **heap-organized table,** which is an unordered collection of data. As rows of data are inserted, they are physically added to the table in no particular order. As rows are deleted, the space can be reused by new rows. Therefore, if a search condition such as "WHERE zip = 90404" is included in a SELECT statement, a full table scan is performed. In the **full table scan,** each row of the table is read and the zip value is checked to determine whether it satisfies the condition. If the table contains 10 million rows, each of the 10 million rows is read into memory and reviewed.

This scenario raises several core issues regarding query performance and the need for indexes:

- First, disk I/O is typically the largest contributor in the total execution time of a query. Indexes are the primary means of reducing disk I/O.
- Second, if the data of the column used in a search condition has high selectivity or a high number of distinct values, the full table scan results in reading many more rows than needed. For example, if only 100 rows contained a zip value of 90404, over nine million rows would be read that did not match the condition.
- Third, the amount of buffer pool space used to hold the full table data affects other system users.

The database **buffer pool** serves as the shared cache memory area of the database server. As users perform SQL queries, the data is placed in memory and can be reused from memory for quicker retrieval versus performing disk I/O operations again. As the buffer space fills, previous query data is cleared to make space for the more recent query data. A full table scan on a large table could use a significant amount of buffer space and, therefore, clear a large amount of cached data and cancel the benefits of data caching.

NOTE

Much of the Oracle documentation uses the term "cardinality" instead of "selectivity" to describe the level of distinct values within a table column. If a column contains many distinct values, it is described as having high cardinality.

Many of the issues can be resolved by appropriately applying indexes to table columns. Oracle 10g provides a number of different indexes. To elaborate on the functionality and physical structure of an index, we'll discuss in detail the B-tree index, which is the default index structure, and the Bitmap index. These two indexes represent the two main physical index structures used in an Oracle database. Following this discussion, two other index variations based on usage goals are introduced: function-based indexes and index organized tables.

B-tree Index

The B-tree or Balanced-tree index is the most common index used in Oracle. You can create this type of index with a basic CREATE INDEX statement. For example, many queries on customer data search for specific zip code values or a range of zip code values.
Figure 6-17 displays the SQL statement that is used to create an index on the Zip column of the CUSTOMERS table.

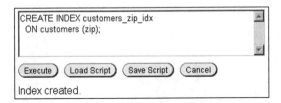

FIGURE 6-17 Creating an index on the Zip column

Now that you have an index, examine the organization of a B-tree index. Figure 6-18 displays a diagram depicting the organization of a B-tree index using the zip code example.

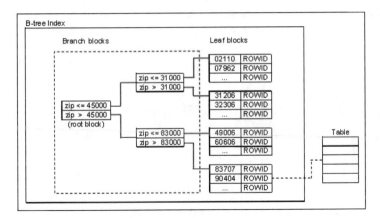

FIGURE 6-18 B-tree index organization

Data in the Oracle database is stored in basic structures called **data blocks.** The amount of data that a block can contain is determined by the size setting, which is defined in the database configuration. Both indexes and table data are stored in data blocks. The index begins with the root node block, which provides the initial breakdown or ranges of column values. The upper blocks (branch blocks) of a B-tree index contain index data that points to lower-level index blocks. The branch blocks continue to provide value breakdowns until the ranges are narrow enough to be divided into block-wide ranges (leaf blocks). The lowest–level index blocks (leaf blocks) contain every indexed data value and a corresponding ROWID that is used to locate the actual table row. The leaf blocks are doubly linked in that they are linked not only to a branch block, but also to the previous and next leaf nodes to support range value searches as well as specific (equality) value searches.

Additional Database Objects

So, for the example, a query that searches for a specific zip code value of 90404, branch 1, branch 2, and leaf 4 blocks are read. At this point, the ROWID (pointer to a table row) is used to read the matching data block and the identified row from the CUSTOMERS table. In total, four data blocks are read to resolve the query (assuming a single row matches the specified zip code value). What if the search found two matching rows? Two ROW-IDs would be identified in the leaf 4 block read and, if the two table rows were stored on two different table data blocks, the read would then involve one additional data block or a total of five data blocks. In contrast, if the CUSTOMERS table consists of 100,000 data blocks, a full table scan would require reading 100,000 rather than four or five data blocks! As you can see, the use of indexes significantly reduces disk I/O or reads.

NOTE

A B-tree index is referred to as a Balanced index because it attempts to make all leaf blocks have equal depth so that every value search should result in an equal number of index block reads.

An index can be created either implicitly by Oracle 10g or explicitly by a user (as you did with the previous CREATE INDEX statement). Oracle 10g automatically creates an index whenever a primary key or unique constraint is created for a column. Because both of these constraints enforce uniqueness, the purpose of adding the index is to allow Oracle 10g to determine whether a value exists in a table without having to perform a full table scan. Oracle 10g accesses the index whenever a value is inserted into or changed in a column that has been designed as a primary key column or as a column that can contain only unique values.

TIP

If you are adding a primary or unique key constraint on a column that already is involved in another constraint, Oracle 10g might not automatically generate an index for the column. Be careful to confirm the creation of the index. You learn how to check the data dictionary for index information later in this chapter.

So, if the purpose of an index is to improve data retrieval efficiency, why not just create an index for every column? Although indexes can speed up row retrieval, their use is not always appropriate. Consider the following issues:

- Because an index is a database object based on table values, Oracle 10g must update it *every* time a DML operation is performed on an underlying table. Thus, if you have a table that is frequently modified (update, insert, delete), the speed for processing the update slows down because Oracle 10g must now update both the table *and* the index. Furthermore, if a table has 10 indexes and a row is added, all 10 indexes must be individually updated.
- B-tree indexes are typically beneficial only if a small percentage of the table is expected to be returned in query results. Because having an index requires that Oracle 10g first examine the index to identify records that meet the criteria and then to retrieve the rows from the actual table, large result sets could require Oracle 10g to do additional work. If a majority of table blocks need

to read to retrieve the requested data, the index reads might actually add more block reads than a full table scan. In fact, having an index in this situation might slow down data retrieval. Various users and publications have guidelines on indexes. For example, some guidelines state that indexes should be used if the query condition will return less than 10% of the table rows. These guidelines do not apply to all situations; the best way to prove whether an index is beneficial is to test it. Chapter 15 introduces the testing of query performance and determining when an index is being used for query execution.

NOTE

Understanding the statement execution plan and the database optimizer is critical to analyzing index usage and applicability. The execution or explain plan identifies the steps the database system will use to resolve the query, including whether an index scan or full table scan will be used. The optimizer provides the logic used by the database system in determining the best path of execution based on the information available. Oracle supports two optimizers: rule and cost-based. Chapter 15 introduces these performance-tuning topics.

- Typically, small tables will not benefit from indexes. In this case, a full table scan may possibly be accomplished as fast as the combined effort of reading index blocks and then the appropriate table blocks.
- More storage space is required for the database because the index objects are additional database objects in the database.

Given these considerations, you should first weigh the benefits of improving query performance against the decreased performance for data manipulation actions. Determine the volume of queries and DML statements regularly executed on the database and which type of statement has performance priority. In some circumstances, the database query activity is the priority (that is, the database is used primarily to support queries and little DML activity occurs or is of less importance in terms of performance). Under those circumstances, indexes should be considered and tested for columns that are frequently used in WHERE conditions or for sorting operations including table joins. On the other hand, if an operational database experiences a high percentage of DML actions rather than query actions, and if DML performance is the priority, index creation should be minimized. However, consider using unique indexes on appropriate columns because this type of index assists performance in verifying that duplicate values are not entered into a column during DML operations.

As index candidate columns are identified, perform tests to determine if the index actually improves data retrieval. Testing would involve measuring query execution time with and without the index and comparing the results. This can be accomplished using such SQL*Plus tools as the TIMING feature and AUTOTRACE to review the explain plan. Chapter 15 covers these tools.

Sorting operations include the use of an Order By clause and table joins. Sorting a large number of rows for output can be both memory and process intensive as data rows are manipulated into order within memory. This is particularly true if the rows contain large amounts of selected data. Indexes can improve sorting operations by allowing rows to be retrieved in sorted order. Join operations, which link data between two tables, involve sorting operations. These operations assist in matching row values of the two tables by ordering the rows by the column(s) that will be used to join the tables.

A unique index is typically created automatically when a primary key or unique key constraint is defined on a column(s). Unique indexes can also be explicitly created by including the UNIQUE keyword in the CREATE TABLE statement, as shown in Figure 6-19. This statement creates a unique index on the Title column of the BOOKS table.

```
CREATE UNIQUE INDEX books_title_idx
   ON books(title);
```

FIGURE 6-19 Explicitly creating a unique index

If an index is created to improve performance on large sorting operations, consider whether the sort order required is ascending or descending. By default, an index is created using an ascending order on the index column value. However, a descending sort can be used in an index by indicating a DESC sort option upon index creation, as shown in Figure 6-20.

```
CREATE INDEX customers_zip_desc_idx
   ON customers(zip DESC);
```

FIGURE 6-20 Indicating a descending sort for index values

The existence of NULL values in a column also must be considered in creating indexes for a column that is frequently used in search conditions. A B-tree index will not include any rows with NULL values in the indexed column. Therefore, if the column contains a high volume of NULL values, an index might be useful, because it will eliminate searching through all the NULL value rows with a full table scan. However, if your search condition were hunting for NULL values (that is, WHERE zip IS NULL), a full table scan would be performed regardless of whether a B-tree index exists on the column. In this case, a function-based index may be of use. We cover this type of index later in this chapter.

The indexes presented so far consist of only a single column. However, indexes can include multiple columns. Indexes that include multiple columns of a table are called **composite** or **concatenated indexes.** For example, an index on customer name could be created using the statement shown in Figure 6-21.

```
CREATE INDEX customer_name_idx
    ON customers(lastname, firstname);
```

FIGURE 6-21 Creating a composite index

This index could improve the performance of queries that include a search condition on both the last and first name columns. This is generally more efficient than creating two separate single-column indexes, as less I/O is required to read a single index rather than two separate indexes. In addition, the selection results of two separate indexes do not have to be combined. In this case, the indexed key value would contain both the first and last name of the customer. Putting the most selective column first typically produces the most efficient result. This index can also be used on search conditions involving only the leading column (Last name).

N O T E

The maximum number of columns that can be included in a B-tree index is 32. A Bitmap index can contain up to 30 columns.

Bitmap Indexes

A Bitmap index is very different in structure and use compared to a B-tree index. A Bitmap index is useful for improving queries on columns that have low selectivity (low cardinality, or a small number of distinct values). The index is a two-dimensional array that contains one column for each distinct value in the column being indexed. Each row is linked to a ROWID and contains a bit (0 or 1) that indicates whether the column value matches this index value. For example, the CUSTOMERS table contains a Region column that may have one of eight possible values: 'N', 'NW', 'NE', 'S', 'SE', 'SW', 'W', or 'E'. A Bitmap index on the Region column can be created with the command shown in Figure 6-22, which adds the keyword BITMAP to the CREATE INDEX command.

```
CREATE BITMAP INDEX customers_region_idx
  ON customers(region);
```

FIGURE 6-22 Creating a Bitmap index on the Region column

Figure 6-23 displays a diagram depicting the organization of a Bitmap index using the Region column example.

Notice in Figure 6-23 that each row has only one bit turned on (that is, set to 1), which indicates the region value for the corresponding row of the CUSTOMERS table. The Bitmap index structure can be particularly beneficial in queries involving compound conditions (AND and OR operator usage). For example, a WHERE clause might attempt to identify all male customers in the NW region (this assumes we have a gender column in the

Bitmap index

Region

N	NW	NE	S	SE	SW	W	E
0	1	0	0	0	0	0	0
0	0	0	0	1	0	0	0
1	0	0	0	0	0	0	0
0	1	0	0	0	0	0	0
0	0	0	0	0	1	0	0

FIGURE 6-23 Organization of a Bitmap index

CUSTOMERS table). In this type of query, if a Bitmap index exists on both columns, the Bitmap index information for the two column values (Gender = 'M', Region = 'NW') can be combined to identify the rows to be queried from the table.

When you use Bitmap indexes, query performance improvements are offset by less efficient DML statement execution—just as with B-tree indexes. Bitmap indexes tend to cause poor DML performance when new values are added to the indexed column because a new Bitmap column must be added to the index.

Function-based Indexes

A Function-based index can be useful if a query search is based on an expression (calculated value) or a function. After an expression or function is used on the column in the search condition, a Non-function-based index is ignored. For JustLee Books, one commonly used search criterion is profit. Management might be looking for values that fall above or below a certain dollar value for profit. To speed up the retrieval of rows that meet a given condition, you can create an index based on the calculated profit for each book. The only difference in the CREATE INDEX command is that the expression or function on which the index is based is provided in the ON clause rather than in just the column name. For example, to create an index on the BOOKS table for the dollar profit returned by each book, you could use the command shown in Figure 6-24.

```
CREATE INDEX books_profit_idx
  ON books(retail-cost);
```

FIGURE 6-24 Creating a function-based index

NOTE

You can create a function-based index in a B-tree type structure (default) or a bitmap structure. Add the keyword "bitmap" to create the function in a bitmap structure.

A function-based index can also assist in improving query performance for search conditions on NULL values. By default, NULL values in an indexed column cause that row not to be indexed. To work around this issue, you could use a function-based index using the NVL function. For example, JustLee routinely executes queries to identify orders that have not been shipped. The WHERE clause would check for a NULL value in the Shipdate column. A basic B-tree index on this column would not include NULL value rows and, therefore, would result in a full table scan. However, creating a function-based index, as shown in Figure 6-25, allows NULL values to be indexed.

```
CREATE INDEX orders_shipdate_idx
  ON orders(NVL(shipdate,'null'));
```

FIGURE 6-25   Creating a function-based index for NULL values

NOTE

The function-based index is used under two circumstances: 1) If the condition in the WHERE clause is "WHERE shipdate IS NULL", an optimizer hint needs to be added in the applicable queries (Chapter 15 covers hints), or 2) if the WHERE clause uses the same function as used to create the index "WHERE NVL-(shipdate,'null') = 'null'.

Index Organized Tables

An **Index Organized Table** (IOT) is a variant of the B-tree index structure and is used as an alternative to the conventional heap-organized table. This structure stores the contents of the entire table in a B-tree index with rows sorted in the primary key value order. This combines the index and table into a single structure. Search and sort operations involving the primary key column can be improved with this index. This has an advantage over other types of indexes in that only one physical object is needed to house both the index and the data values.

The leaf blocks of an IOT contain the primary key value and the entire row of data. A primary key is required to create an IOT because it is used as the row identifier. A search by ROWID is not required in this type of index. As the B-tree index becomes the table structure, creating an IOT is accomplished at the time of table creation. To create an IOT, add the keywords ORGANIZATION INDEX to the CREATE TABLE statement. For example, if you anticipate performing many searches and sort operations on the ISBN column of the BOOKS table, you could create an IOT, as shown in Figure 6-26.

```
CREATE TABLE books2
(ISBN VARCHAR2(10),
 title VARCHAR2(30),
 pubdate DATE,
 pubID NUMBER (2),
 cost NUMBER (5,2),
 retail NUMBER (5,2),
 category VARCHAR2(12),
   CONSTRAINT books2_isbn_pk PRIMARY KEY(isbn))
 ORGANIZATION INDEX;
```

FIGURE 6-26   Creating an IOT for the BOOKS table

Notice that the table is named BOOKS2 to avoid conflict with the existing BOOKS table. As with other types of indexes, an IOT can affect DML operations because the primary key order must be maintained in this structure.

Verifying an Index

After an index has been either implicitly or explicitly created, you can use the USER_INDEXES data dictionary view to determine that the index exists. Figure 6-27 shows the command to verify an index's existence for a specific table.

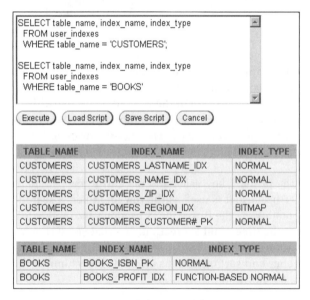

FIGURE 6-27 Query USER_INDEXES to identify indexes

Querying USER_INDEXES verifies what indexes exist; however, it does not identify which columns each index includes. The USER_IND_COLUMNS data dictionary view includes information regarding the columns included in each index. Figure 6-28 shows the command to list the index column information for the CUSTOMERS table.

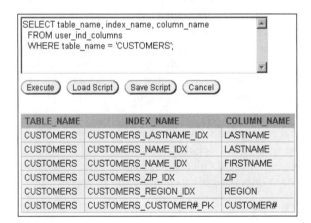

FIGURE 6-28 More index details from USER_IND_COLUMNS

Removing an Index

If an inappropriate index exists (for example, queries on an indexed column tend to return a large number of rows or updates are slow), you can delete an index using the **DROP INDEX** command. An index cannot be modified; if you need to change an existing index, you have to delete and then re-create it. The syntax for the DROP INDEX command is shown in Figure 6-29.

```
DROP INDEX indexname;
```

FIGURE 6-29 Syntax for the DROP INDEX command

As with other DROP and DDL commands, after the DROP INDEX command is executed, the statement cannot be rolled back, and the index is no longer available to Oracle 10g. Figure 6-30 shows the command to drop the BOOKS_PROFIT_IDX index.

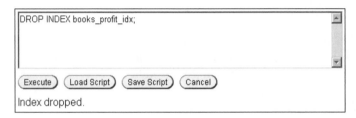

FIGURE 6-30 Dropping BOOKS_PROFIT_IDX

Additional Database Objects

SYNONYMS

A synonym is an alternative name or alias for a database object such as a table or a sequence. A synonym is used for several reasons:

- A synonym provides the user with the convenience of not having to use a full object name, including the schema name, if required.
- A synonym hides the actual object name, thus contributing to greater security.
- Synonyms assist in minimizing application modification by allowing the alias to be used in the application code. When object changes are necessary (object name changes, object schema changes), this does not require a change in application code, but instead a modification of the synonym to identify the correct object.

Simplifying schema references can be an important benefit of using synonyms. When a user creates an object, unless otherwise specified, it belongs to the schema of that user. By grouping objects according to the owner, multiple objects can exist in the same database that have the same object name. *This is only possible if each object belongs to a different schema.*

For example, suppose that a user named Jeff creates a table called PROFITTABLE. Unless Jeff indicates otherwise, the table is an object in the schema called Jeff. If any other user who has permission wants to access Jeff's table, the user would have to identify the table using the correct schema name in the FROM clause of the SELECT statement (for example Jeff.PROFITTABLE). If the name of the table is not prefixed by a schema name, Oracle 10g searches for the table only in the schema of the user who issued the SELECT statement. If a table with the same name does not exist in the user's schema, Oracle 10g returns an error message indicating that the table does not exist.

This type of scenario can cause problems if several users must frequently access a table. In the case of JustLee Books, different users (customer service representatives) enter orders into the ORDERS table. Before a user can enter an order into the table, the individual must remember who actually owns the table, and then prefix the table name with the correct schema. To simplify this process, Oracle 10g allows you to create synonyms that serve as a substitute for an object name.

The syntax for the CREATE SYNONYM command is shown in Figure 6-31.

```
CREATE [PUBLIC] SYNONYM synonymname
  FOR objectname;
```

FIGURE 6-31   Syntax for creating a synonym

Note the following elements in Figure 6-31:

- The synonym name given in the CREATE SYNONYM clause identifies the substitute name, or permanent alias, for the object listed in the FOR clause.

- The optional PUBLIC keyword can be used so any user in the database can use that synonym to refer to the object.
- The object listed in the FOR clause can be the name of a table, constraint, view, or any other Oracle 10g object.

For example, if the user who owns the ORDERS table wants other users to be able to access the table without having to reference the correct schema, the synonym shown in Figure 6-32 can be created.

```
CREATE PUBLIC SYNONYM orderentry
  FOR orders;
```

FIGURE 6-32   Command to create a public synonym

Because the PUBLIC keyword is included in the command in Figure 6-32, any user can reference the ORDERS table by using ORDERENTRY as the table name. Because the PUBLIC keyword allows any database user to use the synonym, only someone with DBA privileges is allowed to delete the synonym. Why? To make certain that one user doesn't delete the synonym, which could impact the work of other users.

NOTE

User privileges are discussed in Chapter 7.

Look at the SELECT statement in Figure 6-33. The statement uses a synonym.

As shown in Figure 6-33, after a synonym is created, you can substitute the synonym for the object name in a command. When Oracle 10g tries to find that object in the database, the software takes this path: First, it searches for an object that has the same name as the synonym; if no object is found, it searches for private synonyms with the synonym's name; if no private synonym is found, it searches for public synonyms with the synonym's name; if no public synonym is found, Oracle 10g returns an error message, indicating that the object does not exist.

Many users like to create synonyms to avoid typing long or complex object names. You can create synonyms for individual use by simply omitting the PUBLIC keyword from the CREATE SYNONYM clause. *If the PUBLIC keyword is not included, the individual who created the private synonym is the only one who can use it.* However, you can use a private synonym to reference objects contained in someone else's schema. For example, if user Jane frequently references the table named PROFITTABLE in user Jeff's schema, Jane can create a synonym for the object Jeff.PROFITTABLE, and she will not need to remember to include the schema name when she accesses his table.

Deleting a Synonym

The **DROP SYNONYM** command is used to drop both private and public synonyms. However, if the synonym being dropped is a public synonym, you *must* include the keyword PUBLIC. The syntax for the DROP SYNONYM command is shown in Figure 6-34.

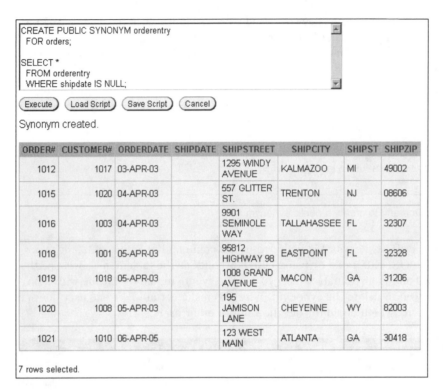

FIGURE 6-33 Creating and using a synonym

```
DROP [PUBLIC] SYNONYM synonymname;
```

FIGURE 6-34 Syntax of the DROP SYNONYM command

If you attempt to drop a public synonym and forget to include the PUBLIC keyword, Oracle 10g returns an error message, indicating that a private synonym by that name does not exist. To drop the public synonym ORDERENTRY that you just created, you could use the command shown in Figure 6-35.

```
DROP PUBLIC SYNONYM orderentry;
```

FIGURE 6-35 Command to delete the ORDERENTRY synonym

NOTE

If you have DBA privileges, the DROP SYNONYM command executes successfully, and the synonym is no longer available. However, if your account does not have the correct privileges, you cannot delete the synonym even if you are the user who created it.

After the DROP SYNONYM command has executed successfully, Oracle 10*g* returns the message "Synonym dropped." After the synonym has been deleted, any user referencing ORDERENTRY in a SQL statement receives an error message indicating that the object does not exist.

Chapter Summary

- A sequence can be created to generate a series of integers.
- The values generated by a particular sequence can be stored in any table.
- A sequence is created with the CREATE SEQUENCE command.
- Gaps in sequences might occur if the values are stored in various tables, if numbers are cached but not used, or if a rollback occurs.

- A value is generated by using the NEXTVAL pseudocolumn.
- The CURRVAL pseudocolumn will be NULL until a value is generated by NEXTVAL.
- The USER_OBJECTS data dictionary table can be used to confirm existence of all schema objects.
- The USER_SEQUENCES table of the data dictionary is used to view sequence settings.
- The ALTER SEQUENCE command is used to modify an existing sequence. The only settings that cannot be modified are the START WITH option and any option that would be invalid because of previously generated values.
- The DROP SEQUENCE command deletes an existing sequence.
- An index can be created to speed up the query process.
- DML operations are always slower when indexes exist.
- Oracle 10*g* automatically creates an index for PRIMARY KEY and UNIQUE constraints.
- An explicit index is created with the CREATE INDEX command.
- An index can automatically be used by Oracle 10*g* if a query criterion or sort operation is based on a column or expression used to create the index.
- The two main structures for indexes are B-tree and Bitmap.
- Function-based indexes are used to index an expression or the use of functions on a column or columns.
- An Index Organized Table is a table stored in a B-tree structure to combine the index and table into one object.
- Information about an index can be retrieved from the USER_INDEXES and USER_IND_COLUMNS views.
- An index can be dropped using the DROP INDEX command.
- An index cannot be modified. It must be deleted and then re-created.
- A synonym provides a permanent alias for a database object.
- A public synonym is available to any database user.
- A private synonym is available only to the user who created it.
- A synonym is created by using the CREATE SYNONYM command.
- A synonym is deleted by using the DROP SYNONYM command.
- Only a user with DBA privileges can drop a public synonym.

Chapter 6 Syntax Summary

The following table presents a summary of the syntax that you have learned in this chapter. You can use the table as a study guide and reference.

SYNTAX GUIDE		
Description	Command Syntax	Example

Create a sequence to generate a series of integers	CREATE SEQUENCE *sequencename*
	  [INCREMENT BY *value*]
	  [START WITH *value*]
	  [{MAXVALUE *value* |
	    NOMAXVALUE}]
	  [{MINVALUE *value* |
	    NOMINVALUE}]
	  [{CYCLE | NOCYCLE}]
	  [{ORDER | NOORDER}]
	  [{CACHE *value* | NOCACHE}];
		CREATE SEQUENCE
		  orders_ order#_seq
		  INCREMENT BY 1
		  START WITH 1021
		  NOCACHE
		  NOCYCLE;

Alter a sequence	ALTER SEQUENCE *sequencename*
	  [INCREMENT BY *value*]
	  [{MAXVALUE *value* |
	    NOMAXVALUE}]
	  [{MINVALUE *value* |
	    NOMINVALUE}]
	  [{CYCLE | NOCYCLE}]
	  [{ORDER | NOORDER}]
	  [{CACHE *value* | NOCACHE}];
		ALTER SEQUENCE
		  orders_order#_seq
		  INCREMENT BY 10;

Drop a sequence	DROP SEQUENCE *sequencename*;	**DROP SEQUENCE** **orders_order#_seq;**

Create a B-tree index	CREATE INDEX *indexname*
	ON *tablename*
	  (*columnname*, ...);
		CREATE INDEX
		  customers_lastname_idx
		ON customers(lastname);

Create a Bitmap index	CREATE BITMAP INDEX *indexname*
	ON *tablename*
	  (*columnname*, ...)
		CREATE BITMAP INDEX
		  customers_region_idx
		ON customers (region);

Create a function-based index	CREATE INDEX *indexname* ON *tablename*; (expression);	**CREATE INDEX** **books_profit_idx** **ON books(retail-cost);**
Create an Index Organized Table	CREATE TABLE tablename (columnname datatype, columnname datatype) ORGANIZATION INDEX;	**CREATE TABLE** **books(isbn NUMBER(10), title** **VARCHAR2(30))** **ORGANIZATION INDEX;**
Drop an index	DROP INDEX *indexname*;	**DROP INDEX** **books_profit_idx;**

Create a synonym	CREATE [PUBLIC] SYNONYM
	  synonymname FOR objectname;
		CREATE PUBLIC SYNONYM
		  orderentry
		FOR orders;

SYNTAX GUIDE		
Description	Command Syntax	Example
Drop a synonym	DROP [PUBLIC] SYNONYM synonymname;	DROP PUBLIC SYNONYM orderentry;

View Prefix	Description
ALL_	Displays objects accessible by the user
DBA_	Displays all objects contained in the database
USER_	Displays objects owned by the user
V$	Displays dynamic database statistics

Review Questions

1. How can a sequence be used in a database?
2. How can gaps appear in a sequence?
3. How can you indicate that the values generated by a sequence should be in descending order?
4. When is an index appropriate for a table?
5. What is the difference between the B-tree and Bitmap index organizations?
6. When does Oracle 10*g* automatically create an index for a table?
7. Under what circumstances should you not create an index for a table?
8. What is an IOT and under what circumstances might it be useful?
9. What command is used to modify an index?
10. What is the purpose of a synonym?

Multiple Choice

To answer the following questions, refer to the tables in Appendix A.

1. Which of the following generates a series of integers that can be stored in a database?

    a. a number generator

    b. a view

    c. a sequence

    d. an index

    e. a synonym

2. Which syntax is appropriate for removing a public synonym?

    a. DROP SYNONYM *synonymname*;

    b. DELETE PUBLIC SYNONYM *synonymname*;

c. DROP PUBLIC SYNONYM *synonymname*;

d. DELETE SYNONYM *synonymname*;

3. Which of the following commands can you use to modify an index?

   a. ALTER SESSION

   b. ALTER TABLE

   c. MODIFY INDEX

   d. ALTER INDEX

   e. none of the above

4. Which of the following generates an integer in a sequence?

   a. NEXTVAL

   b. CURVAL

   c. NEXT_VALUE

   d. CURR_VALUE

   e. NEXT_VAL

   f. CUR_VAL

5. Which of the following is a valid SQL statement?

   a. INSERT INTO publisher VALUES (pubsequence.nextvalue, 'HAPPY PRINTING', 'LAZY LARRY', NULL);

   b. CREATE INDEX a_new_index ON (firstcolumn*.02);

   c. CREATE SYNONYM pub FOR publisher;

   d. all of the above

   e. only a and c

   f. none of the above

6. Suppose that user Juan creates a table called MYTABLE that has four columns. The first column has a primary key constraint, the second column has a NOT NULL constraint, the third column has a CHECK constraint, and the fourth column has a FOREIGN KEY constraint. Given this information, how many indexes will Oracle 10*g* automatically create when the table and constraints are created?

   a. 0

   b. 1

   c. 2

   d. 3

   e. 4

7. Given the table created in Question 6, which of the following commands can Juan use to create a synonym that will allow anyone to access the table without having to identify his schema in the table reference?

   a. CREATE SYNONYM thetable
      FOR juan.mytable;

   b. CREATE PUBLIC SYNONYM thetable
      FOR mytable;

   c. CREATE SYNONYM juan
      FOR mytable;

   d. none of the above

8. Which of the following statements is true?

   a. A gap can appear in a sequence created with the NOCACHE option if the system crashes before a user can commit a transaction.

   b. Any unassigned sequence values will appear in the USER_SEQUENCE data dictionary table as unassigned.

   c. Only the user who creates a sequence is allowed to delete a sequence.

   d. Only the user who created a sequence is allowed to use the value generated by the sequence.

9. When is it inappropriate to manually create an index?

   a. when queries return a large percentage of the rows in the results

   b. when the table is small

   c. when the majority of the table processes are updates

   d. all of the above

   e. only a and c

10. If many queries on a table use compound conditions (AND, OR) on several columns, which index organization type would be most appropriate?

    a. IOT

    b. B-tree

    c. Bitmap

    d. function-based

11. If a column has low selectivity, this means:

    a. The column contains many distinct values.

    b. The column contains only a small number of distinct values.

    c. A WHERE clause is always used in a query on the column.

    d. The selectivity of a column cannot be determined.

12. Oracle 10g automatically creates an index for which type of constraint(s)?

    a. NOT NULL

    b. PRIMARY KEY

    c.   FOREIGN KEY

    d.   CHECK

    e.   none of the above

    f.   only a and b

    g.   only b and d

13. Which of the following settings cannot be modified with the ALTER SEQUENCE command?

    a.   INCREMENT BY

    b.   MAXVALUE

    c.   START WITH

    d.   MINVALUE

    e.   CACHE

14. Which node of the B-tree index contains ROWIDs?

    a.   branch blocks

    b.   root block

    c.   leaf blocks

    d.   none of the above, because the primary key is used to identify rows

15. If the CACHE or NOCACHE options are not included in the CREATE SEQUENCE command, which of the following statements is correct?

    a.   Oracle 10*g* automatically generates 20 integers and stores those in memory.

    b.   No integers are cached by default.

    c.   Only one integer is cached at a time.

    d.   The command will fail.

    e.   Oracle 10*g* automatically generates 20 three-digit decimal numbers and stores those in memory.

16. Which of the following is a valid command?

    a.   CREATE INDEX book_profit_idx
        ON (retail-cost)
        WHERE (retail-cost) > 10;

    b.   CREATE INDEX book_profit_idx
        ON (retail-cost);

    c.   CREATE FUNCTION INDEX book_profit_idx
        ON books
        WHERE (retail-cost) > 10;

    d.   both a and c

    e.   none of the above

17. Which of the following can be used to determine whether an index exists?

    a. DESCRIBE indexname;

    b. the USERS_INDEXES view

    c. the INDEXES table

    d. the USER_INDEX view

    e. all of the above

    f. none of the above

18. Which of the following is not a valid option for the CREATE SEQUENCE command?

    a. ORDER

    b. NOCYCLE

    c. MINIMUMVAL

    d. NOCACHE

    e. All of the above are valid options.

19. Which of the following commands can you use to insert a sequence value into a record?

    a. INSERT VALUE INTO

    b. INSERT SEQUENCE.NEXTVAL INTO

    c. INSERT INTO

    d. ADD NEXTVAL TO

20. Which of the following commands will create a private synonym?

    a. CREATE PRIVATE SYNONYM

    b. CREATE NONPUBLIC SYNONYM

    c. CREATE SYNONYM

    d. CREATE PUBLIC SYNONYM

Hands-On Assignments

To perform the following assignments, refer to the tables in Appendix A.

1. Create a sequence to use for populating the Customer# column of the CUSTOMERS table. When setting the start and the increment values, keep in mind that data already exists in this table. The options should be set to not cycle the values, not cache any values, and no minimum or maximum values should be declared.

2. Add a new customer row using the sequence created in Question 1. The only data currently available for the customer is as follows: last name = Shoulders, first name = Frank, and zip = 23567.

3. Create a sequence that will generate integers starting with the value 5. Each value should be three less than the previous value generated. The lowest possible value should be 0, and the sequence should not be allowed to cycle. Name the sequence MY_FIRST_SEQ.

4. Issue a SELECT statement that will display NEXTVAL for MY_FIRST_SEQ three times. Because the value is not being placed in a table, use the DUAL table in the FROM clause of the SELECT statement. What causes the error on the last SELECT?

5. Change the setting of MY_FIRST_SEQ so the minimum value that can be generated is −1000.

6. Create a private synonym that will allow you to reference the MY_FIRST_SEQ object as NUMGEN.

7. Use a SELECT statement to view the CURRVAL of NUMGEN. Delete the NUMGEN synonym and MY_FIRST_SEQ.

8. Create a Bitmap index on the CUSTOMERS table to speed up queries that search for customers based on their state of residence. Verify that the index exists, and then delete the index.

9. Create a B-tree index on the customer's Last Name column. Verify that the index exists by querying the data dictionary. Remove the index from the database.

10. Many queries search by the number of days to ship (number of days between the order and shipping dates). Create an appropriate index that might improve the performance of these queries.

Advanced Challenge

To perform the following activity, refer to the tables in Appendix A.

Management forecasts that the volume of orders to be processed by JustLee Books will more than triple in the next few months. In fact, the Human Resources Department has already begun recruiting new data entry clerks, account managers, and IT personnel to compensate for the anticipated growth. As the amount of work begins to increase, the database's performance could suffer, and queries and DML operations might take longer to complete. In addition, some of the new users will not be familiar with some of the names that have been assigned to various views, table names, and so on.

- Using the training you have received, determine appropriate uses for at least three each of sequences, indexes, and synonyms that can address needed functionality for the JustLee Books database. In a memo to management, you should identify each sequence, index, and synonym that is needed and the rationale supporting your suggestions. You should also state any drawbacks that might affect the performance of the database if the changes are implemented.

Case Study: *City Jail*

1. The head DBA has requested the creation of a sequence for the primary key columns of the CRIMINALS and CRIMES tables. After creating the sequences, add a new criminal named "Johnny Capps" to the CRIMINALS table using the appropriate sequence. (Use any values for the remainder of columns.) A crime needs to be added for the criminal also. Add a row to the CRIMES table referencing the sequence value already generated for the Criminal_id and using the appropriate sequence to generate the Crime_id value. (Use any values for the remainder of columns.)

2. The last name, street, and phone number columns of the CRIMINALS table are used quite often in the WHERE clause condition of a query. Create objects that might improve the retrieval of data for these queries.

3. Would a Bitmap index be appropriate for any columns in the City Jail database (assuming that the columns are used in search and/or sort operations)? If so, identify the columns and explain why a bitmap index is appropriate for these columns.

4. Would using the City Jail database be any easier with the creation of any synonyms? Explain why or why not.

CHAPTER 7

USER CREATION AND MANAGEMENT

LEARNING OBJECTIVES

After completing this chapter, you should be able to do the following:

- Explain the concept of data security
- Create a new user account
- Identify two types of privileges: system and object
- Grant privileges to a user
- Address password expiration requirements
- Change the password of an existing account
- Create a role
- Grant privileges to a role
- Assign a user to a role
- View privilege information
- Revoke privileges from a user and a role
- Remove a user and roles

INTRODUCTION

The preceding chapters of this text have introduced the creation of database objects. After you've created a database, proper access to database objects must be assigned to users. This chapter focuses on setting up database users, controlling their access to database objects, and defining what actions they can perform on those objects. For example, if an employee's main job for JustLee Books is inputting new orders, this employee needs access only to the customer and order tables, and the actions she can perform on those tables should be limited to selecting and inserting items on those tables. To understand how this can be accomplished, this chapter examines creating, maintaining, and dropping user accounts; granting and revoking privileges; and simplifying the administration of privileges through the use of roles.

These duties are typically performed by a database administrator (DBA) or security officer. However, everyone involved with a database should understand the basic user access principles covered in this chapter. For example, if you are involved in application development, you will most likely participate in determining the appropriate privileges needed by users.

Figure 7-1 provides an overview of the commands presented in this chapter. The commands are grouped by category, rather than by the order in which you will encounter them in this chapter.

COMMAND DESCRIPTION	COMMAND SYNTAX
Creating, Maintaining, and Dropping User Accounts	
Create a user	`CREATE USER username` `IDENTIFIED BY password;`

Change or expire a password	`ALTER USER username`
	`[IDENTIFIED BY newpassword]`
	`[PASSWORD EXPIRE];`

Drop a user	`DROP USER username;`
Granting and Revoking Privileges	

Grant object privileges to users or roles	`GRANT {objectprivilege|ALL} [(columnname),`
	`         objectprivilege (columnname)]`
	`ON objectname`
	`TO {username|rolename|PUBLIC}`
	`[WITH GRANT OPTION];`

Grant system privileges to users or roles	`GRANT systemprivilege [, systemprivilege, ...]`
	`TO username|rolename [, username|rolename, ...]`
	`[WITH ADMIN OPTION];`

Revoke object privileges	`REVOKE objectprivilege [,...objectprivilege]`
	`ON objectname`
	`FROM username|rolename;`

FIGURE 7-1 Overview of chapter contents

COMMAND DESCRIPTION	COMMAND SYNTAX
Granting and Revoking Roles	
Create a role	CREATE ROLE *rolename*;
Grant a role to a user	GRANT *rolename* [, *rolename*] TO *username* [, *username*];
Assign a default role to a user	ALTER USER *username* DEFAULT ROLE *rolename*;
Set or enable a role	SET ROLE *rolename*;
Add a password to a role	ALTER ROLE *rolename* IDENTIFIED BY *password*;
Revoke a role	REVOKE *rolename* FROM *username*|*rolename*;
Drop a role	DROP ROLE *rolename*;

FIGURE 7-1   Overview of chapter contents (continued)

> **NOTE**
>
> Before attempting to work through the examples provided in this chapter, you should open the Chapter 7 folder in your Data Files. Run the prech07.sql file to ensure that all necessary tables and constraints are available.

> **CAUTION**
>
> Most of the commands used in this chapter require database administrator privileges. If you are using an account accessing a server and receive an "insufficient privileges" error upon execution of a command, check with your instructor. If you are using your own installation of Oracle, use the SYSTEM user, which has DBA privileges. The SYSTEM user is automatically created during Oracle installation.

DATA SECURITY

Most organizations store data in some type of electronic database. For example, a bank typically manages all customer financial accounts by using a computer database. Imagine the problems that could arise from improper access to bank account data. Problems could range from significant financial loss to identity theft. Company data needs to be protected from a variety of threats. Some threats arise from natural disasters, such as floods, fires, and tornadoes. However, the biggest threat to an organization's data often comes in the form of people—computer criminals and the organization's own employees. Computer criminals, sometimes referred to as hackers or cyber criminals, attempt to illegally access a system and then might copy, manipulate, or delete the data it contains. By contrast, employees can also create havoc with an organization's data—often more easily because they are assigned legitimate system access. Disgruntled employees might justify their actions as a way of "getting even" for a promotion they didn't receive or as a way to retaliate if they feel their employer doesn't value them. However, even employees with good intentions can damage an organization's data by accidentally deleting records or inserting data into the wrong table.

Data security is a wide-ranging topic that covers many facets of information technology, including network access and authentication, operating system manipulation, application security, database account management, physical protection, and user monitoring. This chapter addresses only one area of data security—database account management—and more specifically, it addresses the creation of database user accounts with passwords and the management of the specific database privileges assigned to that user. The tasks described in this chapter are important for limiting what a user can do after being logged in to a database and for minimizing any intentional or unintentional damage that could be accomplished, while still granting sufficient rights for users to do their jobs.

Database security involves a two-stage process of 1) authentication to identify the user and 2) authorization to allow user access to objects. **Authentication** is the process of identifying an individual attempting to connect to a system, typically based on a user name and password. This process does not identify the objects that the user can access. We will address authentication by creating a user that can connect to the database via SQL*Plus. **Authorization** is the granting of object privileges to a user based on their identity. This is addressed by issuing GRANT commands for specific privileges.

Now, let's look at database user management in the context of a newly hired employee at JustLee Books. This chapter traces the steps for creating a database account for that employee and then granting the user privileges to access database data.

CREATING A USER

When an individual is hired at JustLee Books, the new employee's supervisor notifies the DBA of the new employee's name and requests an Oracle 10g database account. The supervisor also tells the DBA the kind of duties the new employee will need to perform. Based on the employee's job responsibilities, the DBA will determine what database objects he can access.

Creating User Names and Passwords

The first step in creating a new user account is to determine the user's name and password. By default, a user name can contain up to 30 characters, including numbers, letters, and the underscore, dollar sign, and number symbols (_, $, and #). A user account is created using the **CREATE USER** command with the syntax shown in Figure 7-2.

```
CREATE USER username
 IDENTIFIED BY password;
```

FIGURE 7-2 Syntax of the CREATE USER command

NOTE

For user creation commands to execute successfully, you must use an account with database administrator privileges.

Suppose that the DBA has been notified that a new data entry clerk, Ron Thomas, has been hired by JustLee Books and will need an Oracle 10*g* user account. In most cases, the DBA will create the account by using a coding scheme for the user's account name, such as the user's first initial followed by his last name. Because the new employee's name is Ron Thomas, we will assign him the account name of "rthomas." As shown in the syntax of the CREATE USER command in Figure 7-2, a password is entered in the IDENTIFIED BY clause when the account is created. Usually, a temporary password is assigned using the PASSWORD EXPIRE option, and the user is allowed to change the password after he has logged in to the database. This allows the user to create a password that is easier to remember than one that is randomly generated, but one that would still be difficult for other individuals to guess. However, the PASSWORD EXPIRE option is not required to create an account. To create the account for the new employee, use the CREATE USER command shown in Figure 7-3.


```
CREATE USER rthomas
 IDENTIFIED BY little25car
 PASSWORD EXPIRE;
```

FIGURE 7-3 Command to create an account for a new employee

NOTE

Many organizations use a coding scheme to assign account names to simplify user tracking and assist in producing unique names.

The CREATE USER command in Figure 7-3 creates a new user account with the user name of RTHOMAS and the password LITTLE25CAR. Note that Oracle 10*g* account names and passwords are not case sensitive. After the command is executed, the "User created" message is displayed.

Even though an account now exists for the new employee, he cannot log in to the database yet. If the employee attempted to log in at this point, the system would prompt him to change his password, but then it would issue a "login denied" message. A user requires some minimum privileges to connect to the database and establish a session; however, no privileges have been granted to this user's account thus far. The next section covers granting of privileges.

ASSIGNING USER PRIVILEGES

Privileges allow Oracle 10g users to execute certain SQL statements. Two types of privileges exist: system privileges and object privileges. **System privileges** allow access to the Oracle 10g database and let users perform DDL operations such as CREATE, ALTER, and DROP on database objects (for example, tables, views). **Object privileges** allow users to perform DML operations such as INSERT and UPDATE on the data contained within the database objects. Let's look at both types of privileges.

System Privileges

Almost 200 system privileges are availble in Oracle 10g. The ability to create, alter, and drop database objects such as tables and sequences are system privileges. Other system privileges apply to database access and user accounts. For example, to connect to Oracle 10g, a user must have the CREATE SESSION privilege. To create new user accounts, a user must have the CREATE USER privilege. You can use the optional ANY keyword in granting a system privilege to allow the user to perform the privilege systemwide. For example, the DROP ANY TABLE command allows a user to delete or truncate any table that exists in the database—not just tables in their own schema. You can view all the system privileges available in Oracle 10g through the data dictionary view SYSTEM_PRIVILEGE_MAP. Figure 7-4 displays just a portion of the output from querying this view.

Notice that DML operations such as INSERT and DELETE are included in the system privileges list with the ANY keyword (for example, INSERT ANY TABLE). DML operations are considered object rather than system privileges unless the ANY keyword is assigned to the privilege. For example, the INSERT ANY TABLE command gives the user the ability to add rows to any table, regardless of whether she owns the table or has explicit permission to access that particular table. Because the privilege is effective systemwide (that is, the privilege can be used on any schema), it is reclassified from an object privilege to a system privilege.

Granting System Privileges

System privileges are assigned or granted to users with the GRANT command. Figure 7-5 shows the syntax for the GRANT command for system privileges.

It is useful to look at the elements of the statement in Figure 7-5.

- The system privilege being assigned is identified in the GRANT clause. If more than one system privilege is being granted, commas separate the privileges listed. System privileges are not granted for a particular database object; therefore, no database objects are referenced.
- The user(s) or role(s) receiving the system privileges is identified in the TO clause. Roles define a group of users and are discussed later in this chapter.
- The WITH ADMIN OPTION is used to allow any user or role identified in the TO clause to grant the system privilege(s) to any other database users.

Now return to the account creation for the new employee, Ron Thomas. The employee needs to be granted the system privilege CREATE SESSION to connect to the Oracle 10g database. Figure 7-6 shows the command used to accomplish this.

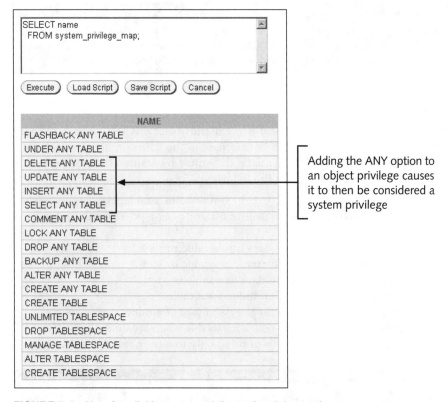

```
SELECT name
  FROM system_privilege_map;
```

(Execute) (Load Script) (Save Script) (Cancel)

NAME
FLASHBACK ANY TABLE
UNDER ANY TABLE
DELETE ANY TABLE
UPDATE ANY TABLE
INSERT ANY TABLE
SELECT ANY TABLE
COMMENT ANY TABLE
LOCK ANY TABLE
DROP ANY TABLE
BACKUP ANY TABLE
ALTER ANY TABLE
CREATE ANY TABLE
CREATE TABLE
UNLIMITED TABLESPACE
DROP TABLESPACE
MANAGE TABLESPACE
ALTER TABLESPACE
CREATE TABLESPACE

Adding the ANY option to an object privilege causes it to then be considered a system privilege

FIGURE 7-4 List of available system privileges (partial output)

```
GRANT systemprivilege [, systemprivilege, …]
TO username|rolename [, username|rolename, …]
[WITH ADMIN OPTION];
```

FIGURE 7-5 Syntax of the GRANT command for system privileges

```
GRANT CREATE SESSION
  TO rthomas;
```

FIGURE 7-6 Command to give CREATE SESSION privilege to new employee

The only privilege that has been assigned to the new user is the ability to log in to Oracle 10*g*. Log in as user RTHOMAS to test the account. Notice that because the account was created with the PASSWORD EXPIRE option, you are prompted to change the password. However, after being logged in to the database, the user cannot perform any actions, as no other privileges have been granted. For example, if the user attempts to execute a SELECT statement to view the contents of the BOOKS table, an error message is returned.

NOTE

The BOOKS table exists in another schema so the new user, RTHOMAS, would need to qualify the table name with the schema name. For example, if the BOOKS table was created in the SCOTT schema, the FROM clause of the query would need to reference the table as SCOTT.BOOKS.

Object Privileges

When a user creates an object, he automatically has all the object privileges associated with that object. However, if other users need to have access to the data contained in a database object, or the ability to manipulate the data, they must be granted the privilege to do so. Object privileges in Oracle 10g include the following:

- SELECT: allows users to display data contained in a table, view, or sequence. The SELECT privilege allows a user to generate the next sequence value from a sequence by using NEXTVAL.
- INSERT: allows users to insert data into a table or view.
- UPDATE: allows users to modify data in a table or view.
- DELETE: allows users to delete data in a table or view.
- INDEX: allows users to create an index for a table.
- ALTER: can be used only to alter the definition of a table or sequence.
- REFERENCES: allows users to reference a table when creating a FOREIGN KEY constraint. This privilege can be granted only to a user, not to a role.

NOTE

There are a total of 15 object privileges in Oracle 10*g*. The remaining object privileges are: DEBUG, which allows users to access code via a debugger; EXECUTE, which allows users to run a stored function or procedure; FLASHBACK, which allows users to issue flashback queries; READ, which allows users to view binary files (BFILEs) in a directory; ON COMMIT REFRESH, which allows users to refresh a view; QUERY REWRITE, which allows users to create a view for query rewrite; UNDER, which allows users to create a view based on an existing view; and WRITE, which allows users to write to a directory.

Granting Object Privileges

Object privileges are also granted to users using the GRANT command. The syntax for the GRANT command is shown in Figure 7-7.

```
GRANT {objectprivilege|ALL} [(columnname),
       objectprivilege (columnname)]
 ON objectname
 TO {username|rolename|PUBLIC}
 [WITH GRANT OPTION];
```

FIGURE 7-7 Syntax of the GRANT command for object privileges

Examine the individual clauses in Figure 7-7.

- The GRANT clause is used to identify the object privilege(s) being assigned. The INSERT, UPDATE, and REFERENCES privileges can also be assigned to specific columns within a table or view. If the object privilege is being assigned to a specific column, the column name should be included in the GRANT clause, within parentheses, after the privilege name. Rather than identifying individual object privileges, the ALL keyword can be substituted to indicate that all object privileges are to be granted. *Either an object privilege or the ALL keyword must be used after the GRANT keyword.* Be careful when granting users all available object privileges because this will provide them with the ability to perform any DML operation on the named object.

- The **ON** clause is used to identify the object (for example, table, view, sequence) to which the privilege(s) applies.
- The **TO** clause identifies the user or role (discussed in a later section) receiving the privilege. Multiple users or roles can receive privileges from the same GRANT command by providing the names in a list, separated by commas. If all database users should receive the privilege being granted, the PUBLIC keyword can be used in the TO clause, rather than a list of names.
- The **WITH GRANT OPTION** gives the user the ability to grant the same object privileges to other users.

Now consider several different sample scenarios and their associated commands for granting privileges to Ron Thomas, as shown in Figure 7-8.

SAMPLE SCENARIO	GRANT COMMAND
Ron Thomas needs the ability to select rows from and insert rows into the CUSTOMERS table, which is in the SCOTT schema.	```GRANT select, insert ON scott.customers TO rthomas;```
Ron Thomas needs the ability to select any data from the CUSTOMERS table but only be able to modify the following columns: LASTNAME and FIRSTNAME.	```GRANT select, update(lastname, firstname) ON scott.customers TO rthomas;```
Ron Thomas is assigned full responsibility for the CUSTOMERS table and, therefore, needs the ability to perform any activity on the CUSTOMERS table and the right to grant those privileges to other users.	```GRANT ALL ON scott.customers TO rthomas WITH GRANT OPTION;```

FIGURE 7-8 Examples of granting object privileges to a user

NOTE

The WITH GRANT OPTION cannot be used when granting object privileges to roles; it only applies to individual users.

Assume that the new employee needs only to view the contents of the BOOKS table, but doesn't need to add, delete, or change anything. In this case, the GRANT command needs to be issued to provide the necessary object privilege, as shown in Figure 7-9. *You will need to replace the scott schema name with the appropriate schema name that contains the JustLee Books database.*


```
GRANT SELECT
ON scott.books
TO rthomas;
```

FIGURE 7-9 Command used to provide the SELECT privilege

NOTE

Creating a synonym for the BOOKS table would eliminate the need to indicate the schema name in this command.

Because the command shown in Figure 7-9 does not include the WITH GRANT OPTION, the user rthomas is not allowed to assign the SELECT privilege for the BOOKS table to any other user. However, Ron Thomas can now view the contents of the BOOKS table by using a SELECT command, even though he cannot perform any DML or DDL operations. Execute the GRANT command from within a database administrator account. Then log in as RTHOMAS and test the privilege by executing a SELECT statement on the BOOKS table.

PASSWORD MANAGEMENT

Users frequently forget their passwords, and so the DBA routinely receives requests to reset account passwords. After an account has been created, the simplest approach for changing the password is by using the **ALTER USER** command. The DBA can reset the current password and mark the password as "expired," which forces the user to set up a new password as soon as the user attempts to connect to the database. The syntax for the ALTER USER command is shown in Figure 7-10.

```
ALTER USER username
[IDENTIFIED BY newpassword]
[PASSWORD EXPIRE];
```

FIGURE 7-10 Syntax of the ALTER USER command

The **IDENTIFIED BY** clause is used to specify the new password. The command to reset RTHOMAS is shown in Figure 7-11.

```
ALTER USER rthomas
 IDENTIFIED by rxy22b
  PASSWORD EXPIRE;
```

FIGURE 7-11   ALTER USER command used to reset password

Execute the command in Figure 7-11 from your DBA account. Now, to test the modification, log in as RTHOMAS with the new password of RXY22B. You should be prompted to change the password and then the login should complete successfully.

After a user is successfully logged in to an account, the user can also change the account's password, using the ALTER USER command with the IDENTIFIED BY clause. After Ron logs in to his account and decides that the new password he selected is too difficult to remember, he could issue the command shown in Figure 7-12.

```
ALTER USER rthomas
 IDENTIFIED BY monster42truck;
```

FIGURE 7-12   Command used to change a password

The command shown in Figure 7-12 can be issued by the account owner or by anyone with the ALTER USER system privilege.

CAUTION

When changing an account password, Oracle 10*g* does not require the person issuing the command to know the current password for the account. Therefore, this privilege should be assigned with caution. Also, users should be warned that if they log in to Oracle 10*g* and then leave their workstation while still connected to the database, anyone could sit down at their computer and change the password for their account. Because Oracle 10*g* allows a user multiple logins to the database for the same account at the same time, someone could be causing damage to the database while the "real" user is performing legitimate tasks. The illegal password change would go undetected until the user's next login attempt.

In addition, SQL*Plus provides mechanisms for the user to change her password after logging in to an account. The specific action would depend on which SQL*Plus interface is being used.

- SQL*Plus Client Tool: Issue a PASSWORD command at the SQL prompt. The user will be prompted for the old password and then the new password.
- iSQL*Plus Internet Tool: Click Preferences and Change Password links, which will provide a screen prompting for the old and new password.

From a DBA's perspective, the previous discussion regarding logins and passwords just touches the surface of the ways you can control user logins. For example, Oracle 10g supports a variety of authentication methods beyond the database authentication addressed here. Methods include externally processed authentication, authentication that is processed globally by (SSL), and authentication processed by proxy. Creating user profiles is another feature available to provide more complex password management, as well as including password aging rules and verification complexity (for example, setting requirements on the content and length of passwords).

In addition, authentication also raises the topic of encryption. **Encryption** refers to the scrambling of data to make it unreadable to anyone other than the sender and receiver. Encryption plays a role in both protecting data as it is transmitted via network communications and by storing data in an encrypted form. Oracle provides encryption for passwords during login transmission and stores users' passwords in the data dictionary in an encrypted format. However, encryption of other data should be considered. For example, if a customer is providing a credit card number via the Internet, how can this information be protected? This calls for securing transmissions using a protocol such as SSL and, if the credit card number is to be stored, encrypting data columns using Oracle encryption tools.

NOTE

Beware of using the ALTER USER command to change user passwords. This command is transmitted in clear text, including the new password. It is recommended that users should utilize the mechanisms within SQL*Plus to change passwords.

These topics are beyond the scope of this text. However, the basic concepts introduced regarding user account creation and passwords are fundamental to understanding user access.

UTILIZING ROLES

In most cases, a user needs more privileges than just the CREATE SESSION system privilege and the SELECT object privilege for one table. For example, as a data entry clerk, Ron Thomas will probably need to enter the ship date for orders, update information in the CUSTOMERS table, and so on. In fact, all JustLee Books' data entry clerks need to perform these same tasks. Assigning each of the necessary individual privileges to each of the appropriate employees could become quite cumbersome. Rather than assign the same privileges again and again to users who need identical privileges, a simpler approach is to assign a group of privileges to a role, and then assign the role to appropriate users.

A **role** is a group, or collection, of privileges. In most organizations, roles correlate to users' job duties. For example, customer service representatives might need to view all the data in each database table. However, they probably would not need to have the privilege of updating the data in the BOOKS table (ISBN, cost, retail price, etc.). By grouping individuals based on the tasks they need to perform, you can create roles that have been assigned the privileges required by each group. Thus, rather than assign each user a series of individual privileges, you can just assign a collection of privileges—a role—to the users. In addition, if you have some employees with job duties covering multiple areas (for example, a branch supervisor), you can assign multiple roles to this employee, as depicted in Figure 7-13.

204

Billing

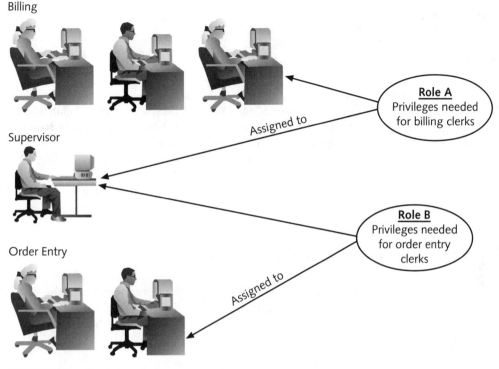

Supervisor

Order Entry

FIGURE 7-13   Depiction of role assignment

Creating and Assigning Roles

Before you can assign privileges to a role, the role object must be created by using the **CREATE ROLE** command. The syntax for the command is shown in Figure 7-14.

```
CREATE ROLE rolename;
```

FIGURE 7-14   Syntax of the CREATE ROLE command

After the role has been created, you can grant system and/or object privileges to the role, using the same syntax as when granting the privileges directly to the user. The *only exception* is that an object privilege cannot be granted to a role with the WITH GRANT OPTION. After all the privileges have been assigned to a role, the role can then be assigned to all relevant users, using the GRANT command. The syntax for using the GRANT command to grant a role to a user is shown in Figure 7-15.

```
GRANT rolename [, rolename]
TO username [, username];
```

FIGURE 7-15   Syntax used to grant a role to a user

For example, suppose that each order entry clerk should be allowed to issue the SELECT, INSERT, and UPDATE commands for the CUSTOMERS, ORDERS, and ORDERITEMS tables in the JustLee Books database. Rather than remember exactly which privileges are required every time a new order entry clerk is hired, a DBA could create a role called ORDERENTRY that could be assigned in lieu of the individual privileges. To create the role, connect as the DBA user and issue the command shown in Figure 7-16.

```
CREATE ROLE orderentry;
```

FIGURE 7-16   Command used to create the ORDERENTRY role

Now that you've created the role, you can assign the necessary privileges to the newly created ORDERENTRY role, using the commands shown in Figure 7-17. Keep in mind that you should enter the correct schema in place of SCOTT—this should be the schema you are using that contains the JustLee Books database tables.

```
GRANT SELECT, INSERT, UPDATE
  ON scott.customers
  TO orderentry;
GRANT SELECT, INSERT, UPDATE
  ON scott.orders
  TO orderentry;
GRANT SELECT, INSERT, UPDATE
  ON scott.orderitems
  TO orderentry;
```

FIGURE 7-17   Commands used to grant privileges to the ORDERENTRY role

After assigning privileges to the ORDEREntry role, you can assign the role to any new order entry clerk by granting the role to the employee. Grant the role to Ron Thomas by issuing the command shown in Figure 7-18.

```
GRANT orderentry
  TO rthomas;
```

FIGURE 7-18   Command used to grant the ORDERENTRY role to RTHOMAS

After you've assigned Ron Thomas the ORDERENTRY role, he can execute SELECT, UPDATE, and INSERT commands on any table in the JustLee Books database.

Users can be assigned several roles based on the different types of tasks they usually perform. For example, suppose you create an ORDERENTRY role and a BILLING role to address both of the employee groups presented at the beginning of this section in Figure 7-13. Then, you create a user name of SDAVIS for the supervisor whose name is Scott Davis. How can you grant both of the roles to the supervisor? The command shown in Figure 7-19 accomplishes this task.

```
GRANT orderentry, billing
  TO sdavis;
```

FIGURE 7-19 Command used to assign multiple roles to a user

You can also define a role by including a group of previously defined roles. For example, JustLee Books might have a number of supervisors who would need the same set of roles assigned. To simplify assignment, you could create a role that includes both the ORDERENTRY and BILLING roles and then assign it to each supervisor needing both these roles. The commands in Figure 7-20 create a role named SUPERVISOR, which includes two other roles.

```
CREATE ROLE supervisor;
GRANT orderentry, billing
  TO supervisor;
```

FIGURE 7-20 Command used to create a role that includes two roles

Predefined Roles

You don't necessarily have to create roles from scratch. Oracle has a set of predefined roles available for assigning user privileges. Figure 7-21 lists several of the predefined roles available in Oracle 10g.

You can grant these roles to users in the same way you grant roles that you explicitly create. However, the predefined roles are not typically applicable to assigning privileges to application users. These roles are more closely associated with privileges that developers and database administrators would require. In addition, Oracle recommends that organizations should create their own roles rather than depending on the predefined roles. If any modifications to predefined roles are enacted as the database changes, organizations using the predefined roles need to apply necessary changes.

ROLE NAME	PRIVILEGES INCLUDED
CONNECT	ALTER SESSION, CREATE CLUSTER, CREATE DATABASE LINK, CREATE SEQUENCE, CREATE SESSION, CREATE SYNONYM, CREATE TABLE, CREATE VIEW
RESOURCE	CREATE CLUSTER, CREATE INDEXTYPE, CREATE OPERATOR, CREATE PROCEDURE, CREATE SEQUENCE, CREATE TABLE, CREATE TRIGGER, CREATE TYPE
DBA	All system privileges WITH ADMIN OPTION
EXP_FULL_DATABASE	Provides the privileges required to perform full and incremental database exports
IMP_FULL_DATABASE	Provides the privileges required to perform full database imports

FIGURE 7-21   Sample of the predefined roles available in Oracle 10*g*

Default Roles

A user who has been assigned several different roles does not have to have all the roles activated upon login. A user can be assigned a default role that is automatically enabled whenever the user logs in to the database. The default role should consist of only those privileges the user will frequently need. Privileges that are rarely needed (and that could cause problems in the database if incorrectly used) should be assigned to other roles the user can assume when necessary. For example, one role a user is assigned might contain ALTER TABLE privileges, which are used infrequently by the user. In this case, it might be more desirable to have this role activated only when needed to avoid any unintended table modifications.

You use the ALTER USER and SET ROLE commands to control how roles are activated for a user. After the user account has been created, an ALTER USER command can be issued to assign a default role to a user by using the syntax shown in Figure 7-22.

```
ALTER USER username
DEFAULT ROLE
(rolename|ALL [EXCEPT role1, role2, …]|NONE);
```

FIGURE 7-22   Syntax used to assign a default role to a user

As shown in the ALTER USER statement syntax, users can have none, one, or many roles set as the default role. Including a specific role name causes only this specific role to be enabled when the user logs in. The ALL option enables all the roles assigned to the user. If the EXCEPT clause is used, all roles with the exception of those listed in this clause are enabled. The NONE option disables all assigned roles, requiring the user to enable a role after logging in. Figure 7-23 demonstrates ALTER USER commands for several scenarios.

SAMPLE SCENARIO	GRANT COMMAND
Ron Thomas needs only the ORDERENTRY role enabled upon login.	`ALTER USER rthomas` `    DEFAULT ROLE orderentry;`
Scott Davis, the supervisor, has been assigned three roles: ORDERENTRY, BILLING, and MOD_TABLES. All privileges are needed routinely so all roles should be enabled upon login.	`ALTER USER sdavis` `    DEFAULT ROLE ALL;`

Scott Davis, the supervisor, has been assigned three roles: ORDERENTRY, BILLING, and MOD_TABLES. The MOD_TABLES role privileges are needed only periodically, so all roles except this one should be enabled upon login.	`ALTER USER sdavis`
	`    DEFAULT ROLE`
	`    ALL EXCEPT mod_tables;`

FIGURE 7-23   Examples of setting DEFAULT roles

Enabling Roles After Login

After connecting to the database, a user might need to assume or enable a role or set of privileges other than those assigned as his default role(s). The user can issue a **SET ROLE** command to do this. The syntax for the SET ROLE command is provided in Figure 7-24.

```
SET ROLE rolename;
```

FIGURE 7-24   Syntax of the SET ROLE command

Note that users cannot set their role to any role that has not already been assigned to them. For example, a user could not issue the command **SET ROLE DBA;** unless the individual had already been assigned that role by an authorized user.

As a safety precaution, some database administrators add a password to a role. For example, suppose that employee Scott has been assigned certain privileges through the DBA role. However, he uses the DBA role only when he needs to perform certain types of operations. Suppose that one day, he is logged in to the Oracle 10g database, performing day-to-day activities that require only his normal privileges. He is summoned away for a quick meeting and doesn't remember to log out of the server. Because no special privileges are currently available, this might not present a problem. However, if a disgruntled employee knows that Scott has access to the DBA role, all that employee needs to do is sit down at Scott's computer and set the role to DBA with the SET ROLE command. With those types of privileges available, the disgruntled employee could do a lot of damage in a very short period of time.

To avoid this type of problem, administrators add passwords to roles that have important and potentially dangerous privileges. To add a password to a role, simply use the **ALTER ROLE** command. The syntax for the ALTER ROLE command is provided in Figure 7-25.

```
ALTER ROLE rolename
 IDENTIFIED BY password;
```

FIGURE 7-25 Syntax of the ALTER ROLE command

After the ALTER ROLE command has been used to add a password to a role, any user attempting to use the role will be required to enter the password, or it will not be enabled.

NOTE

A password will not be required for roles that have been assigned as a default role for a user. It is required only when the SET ROLE command is issued to enable a role.

VIEWING PRIVILEGE INFORMATION

Various data dictionary views can be queried to determine the privileges currently assigned to a user or role. Some of the more commonly used data dictionary views are described in Figure 7-26. A "Y" in the DBA column indicates that DBA privileges are required to view the listed data dictionary view.

DBA	DATA DICTIONARY VIEW	INFORMATION
Y	dba_roles	All roles defined in the database
Y	role_tab_privs	Table privileges assigned to roles
Y	Role_sys_privs	System privileges assigned to roles
Y	dba_role_privs	Roles assigned to users
Y	dba_tab_privs	All table privileges assigned
N	USER_SYS_PRIVS	System privileges granted to current user
N	USER_TAB_PRIVS	Privileges on objects for the current user
N	user_role_privs	Roles assigned to the current user
N	role_tab_privs	Table privileges assigned to current user via roles
N	session_roles	Roles currently enabled for current user
N	Session_privs	Privileges active for current user that have not been assigned via a role

FIGURE 7-26 Views for privilege information

The DBA at JustLee Books might need to confirm which table privileges are assigned via the ORDERENTRY role. The statement in Figure 7-27 accomplishes this task.

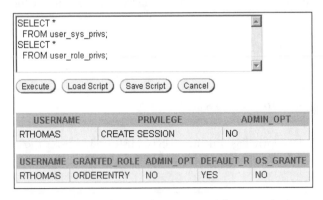

FIGURE 7-27   Verifying privileges assigned by a role

After being logged in, the user RTHOMAS might want to check which system privileges and which roles are currently enabled. The statements in Figure 7-28 accomplish this task.

```
SELECT *
 FROM user_sys_privs;
SELECT *
 FROM user_role_privs;
```

(Execute)   (Load Script)   (Save Script)   (Cancel)

USERNAME	PRIVILEGE	ADMIN_OPT
RTHOMAS	CREATE SESSION	NO

USERNAME	GRANTED_ROLE	ADMIN_OPT	DEFAULT_R	OS_GRANTE
RTHOMAS	ORDERENTRY	NO	YES	NO

FIGURE 7-28   Verifying active system privileges and roles

REMOVING PRIVILEGES AND ROLES

Just as easily as privileges and roles can be assigned, they can also be removed or dropped.

Revoking Privileges and Roles

Privileges granted to a user or role can be removed by using the **REVOKE** command. The syntax for the REVOKE command to remove a system privilege from a user or role is shown in Figure 7-29.

```
REVOKE systemprivilege [,...systemprivilege]
 FROM username|rolename;
```

FIGURE 7-29 Syntax used to revoke a system privilege

The REVOKE command can also be used to revoke object privileges, using the syntax shown in Figure 7-30.

```
REVOKE objectprivilege [,...objectprivilege]
 ON objectname
 FROM username|rolename;
```

FIGURE 7-30 Syntax used to revoke an object privilege

In addition, the REVOKE command can be used to revoke a role from an account, using the syntax shown in Figure 7-31.

```
REVOKE rolename
 FROM username|rolename;
```

FIGURE 7-31 Command used to revoke a role from an account

For example, to remove the DELETE privilege on the CUSTOMERS table from the ORDERENTRY role, the DBA could issue the command shown in Figure 7-32.

```
REVOKE delete
 ON customers
 FROM orderentry;
```

FIGURE 7-32 Command used to revoke an object privilege from a role

To revoke the ORDERENTRY role from Ron Thomas, the DBA can issue the command shown in Figure 7-33.

```
REVOKE orderentry
FROM rthomas;
```

FIGURE 7-33 Command used to revoke the ORDERENTRY role from a user

CAUTION

When revoking an object privilege that was originally granted using the WITH GRANT OPTION, the privilege is revoked not only from the specified user, but also from any other users to she might have subsequently granted the privilege. Revoking a system privilege that was originally granted with the WITH ADMIN OPTION has no cascading effect on other users.

Dropping a Role

A role can be deleted from the Oracle 10*g* database through the **DROP ROLE** command. The syntax for the DROP ROLE command is shown in Figure 7-34.

```
DROP ROLE rolename;
```

FIGURE 7-34 Syntax of the DROP ROLE command

When a role is removed from the database, users lose all privileges derived from that role. The only way the users will be able to use the privileges previously assigned by the role is to receive those privileges again, either via direct grants of the same privileges or by creating another role.

For example, suppose that you decide that the ORDERENTRY role should be more restrictive and specify exactly which columns can be updated in certain tables. The simplest solution is to drop the existing role and re-create it with a new DATAENTRY role. You can drop the role using the command shown in Figure 7-35.

```
DROP ROLE orderentry;
```

FIGURE 7-35 Command used to drop the ORDERENTRY role

NOTE

Now that the ORDERENTRY role is not available to Ron Thomas, he no longer has access to any of the tables.

Dropping a User

At times, accounts need to be removed from the system for various reasons, the most common being employee departure due to events such as retirement. Or, suppose, for example, that

Ron's supervisor has just informed you that the correct spelling of the new employee's last name is Tomas, not Thomas. There is no ALTER USER option available to change the user name for an account. Instead, you must delete the existing account and re-create Ron's account with the correct spelling. The DROP USER command is used to remove a user account from an Oracle 10g database. The syntax for the DROP USER command is shown in Figure 7-36.

```
DROP USER username;
```

FIGURE 7-36 Syntax of the DROP USER command

The command **DROP USER rthomas;** drops the existing account named "rthomas." You would then need to create a new account with the correct "rtomas" spelling.

Chapter Summary

- Database account management is only one facet of data security.
- A new user account is created with the CREATE USER command. The IDENTIFIED BY clause contains the password for the account.
- System privileges are used to grant access to the database and to create, alter, and drop database objects.
- The CREATE SESSION system privilege is required before a user can access his account on the Oracle server.
- The system privileges available in Oracle 10g can be viewed through the SYSTEM_PRIVILEGE_MAP.
- Object privileges allow users to manipulate data in database objects.
- Privileges are given through the GRANT command.
- The ALTER USER command, combined with the PASSWORD EXPIRE clause, can be used to force a user to change her password upon the next attempted login to the database.
- The ALTER USER command, combined with the IDENTIFIED BY clause, can be used to change a user's password.
- Privileges can be assigned to roles to make the administration of privileges easier.
- Roles are collections of privileges.
- The ALTER USER command, combined with the DEFAULT ROLE keywords, can be used to assign a default role(s) to a user.
- A role can be enabled in a session using the SET ROLE command.
- Privileges can be revoked from users and roles using the REVOKE command.
- Roles can be revoked from users using the REVOKE command.
- A role can be deleted using the DROP ROLE command.
- A user account can be deleted using the DROP USER command.

Chapter 7 Syntax Summary

The following table presents a summary of the syntax that you have learned in this chapter. You can use the table as a study guide and reference.

SYNTAX GUIDE		
Command Description	Command Syntax	Example
Creating, Maintaining, and Dropping User Account		
Create a user	`CREATE USER username` `IDENTIFIED BY password;`	`CREATE USER rthomas` `IDENTIFIED BY little25car;`

Change or expire a password	`ALTER USER username`
	 `[IDENTIFIED BY newpassword]`
	 `[PASSWORD EXPIRE];`
		`ALTER USER rthomas`
		 `IDENTIFIED BY`
		 `   monstertruck42;`

Drop a user	`DROP USER username;`	`DROP USER rthomas;`
Granting and Revoking Privileges		

Grant object privileges to users or roles	`GRANT {objectprivilege|ALL}`
	 `   [(columnname),`
	 `    objectprivilege`
	 `    (columnname)]`
	 `ON objectname`
	 `TO {username|rolename|PUBLIC}`
	 `[WITH GRANT OPTION];`
		`GRANT select, insert`
		 `ON customers`
		 `TO rthomas`
		 `WITH GRANT OPTION;`

Grant system privileges to users or roles	`GRANT systemprivilege`
	 `   [, systemprivilege, ...]`
	 `TO username|rolename`
	 `   [, username|rolename,`
	 `   ...]`
	 `[WITH ADMIN OPTION];`
		`GRANT CREATE SESSION`
		 `TO rthomas;`

Revoke object privileges	`REVOKE objectprivilege`
	 `   [,...objectprivilege]`
	 `ON objectname`
	 `FROM username|rolename;`
		`REVOKE INSERT`
		 `ON customers`
		 `   FROM rthomas;`

Granting and Revoking Roles		
Create a role	`CREATE ROLE rolename;`	`CREATE ROLE orderentry;`
Grant a role to a user	`GRANT rolename [, rolename]` `TO username [, username];`	`GRANT orderentry TO rthomas;`
Assign a default role to a user	`ALTER USER username` `   DEFAULT ROLE rolename;`	`ALTER USER rthomas DEFAULT` `   ROLE orderentry;`
Set or enable a role	`SET ROLE rolename;`	`SET ROLE DBA;`
Add a password to a role	`ALTER ROLE rolename` `IDENTIFIED BY password;`	`ALTER ROLE orderentry` `IDENTIFIED BY apassword;`

SYNTAX GUIDE		
Command Description	Command Syntax	Example
Granting and Revoking Roles		
Revoke a role	`REVOKE rolename` `FROM username|rolename;`	`REVOKE orderentry` `FROM rthomas;`
Drop a role	`DROP ROLE rolename;`	`DROP ROLE orderentry;`

Review Questions

To answer the following questions, refer to the tables in Appendix A.

1. What is the purpose of data security?
2. What does a database account with the CREATE SESSION privilege allow the user to do?
3. How is a user password assigned in Oracle 10*g*?
4. What is a privilege?
5. If you are logged in to Oracle 10*g*, how can you determine which privileges are currently available to your account?
6. What types of privileges are available in Oracle 10*g*? Define each type.
7. What is the purpose of a role in Oracle 10*g*?
8. How can you assign a password to a role?
9. What happens if you revoke an object privilege that was granted with the WITH GRANT OPTION? What if the privilege is revoked from a user who had granted the same object privilege to three other users?
10. How can you remove a user account from Oracle 10*g*?

Multiple Choice

To answer the following questions, refer to the tables in Appendix A.

1. Which of the following commands can be used to change a password for a user account?
   a. ALTER PASSWORD
   b. CHANGE PASSWORD
   c. MODIFY USER PASSWORD
   d. ALTER USER...PASSWORD
   e. none of the above

2. Which of the following statements assigns the role of Customerrep as the default role for Maurice Cain?

   a. ALTER ROLE mcain DEFAULT ROLE customerrep;
   b. ALTER USER mcain TO customerrep;
   c. SET DEFAULT ROLE customerrep FOR mcain;
   d. ALTER USER mcain DEFAULT ROLE customerrep;
   e. SET ROLE customerrep FOR mcain;

3. Which of the following statements is most accurate?

   a. Authentication procedures will prevent any data stored in the Oracle 10g database from becoming stolen or damaged.
   b. Authentication procedures are used to limit unauthorized access to the Oracle 10g database.
   c. Oracle 10g authentication will not prevent anyone from accessing data in the database if the individual has a valid operating system account.
   d. Authentication procedures restrict the type of data manipulation operations that can be executed by a user.

4. Which of the following statements will create a user account named DeptHead?

   a. CREATE ROLE depthead IDENTIFIED BY apassword;
   b. CREATE USER depthead IDENTIFIED BY apassword;
   c. CREATE ACCOUNT depthead;
   d. GRANT ACCOUNT depthead;

5. Which of the following privileges must be granted to a user's account before the user can connect to the Oracle 10g database?

   a. CONNECT
   b. CREATE SESSION
   c. CONNECT ANY DATABASE
   d. CREATE ANY TABLE

6. Which of the following privileges allows a user to truncate tables in a database?

   a. DROP ANY TABLE
   b. TRUNCATE ANY TABLE
   c. CREATE TABLE
   d. TRUNC TABLE

7. Which of the following tables or views will display the current enabled privileges for a user?

   a. SESSION_PRIVS
   b. SYSTEM_PRIVILEGE_MAP
   c. USER_ASSIGNED_PRIVS
   d. V$ENABLED_PRIVILEGES

8. Which of the following commands will only eliminate the user ELOPEZ's ability to enter new books into the BOOKS table?

    a. REVOKE insert
       ON books
       FROM elopez;

    b. REVOKE insert
       FROM elopez;

    c. REVOKE INSERT INTO
       FROM elopez;

    d. DROP insert
       INTO books
       FROM elopez;

9. Which of the following commands is used to assign a privilege to a role?

    a. CREATE ROLE

    b. CREATE PRIVILEGE

    c. GRANT

    d. ALTER PRIVILEGE

10. Which of the following options will require a user to change his password at the time of the next login?

    a. CREATE USER

    b. ALTER USER

    c. IDENTIFIED BY

    d. PASSWORD EXPIRE

11. Which of the following options allows a user to grant system privileges to other users?

    a. WITH ADMIN OPTION

    b. WITH GRANT OPTION

    c. DBA

    d. ASSIGN ROLES

    e. SET ROLE

12. Which of the following is an object privilege?

    a. CREATE SESSION

    b. DROP USER

    c. INSERT ANY TABLE

    d. UPDATE

13. Which of the following privileges can be granted only to a user, and not to a role?
   a. SELECT
   b. CREATE ANY
   c. REFERENCES
   d. READ
   e. WRITE

14. Which of the following is used to grant all the object privileges for an object to a specified user?
   a. ALL
   b. PUBLIC
   c. ANY
   d. OBJECT

15. Which of the following identifies a collection of privileges?
   a. an object privilege
   b. a system privilege
   c. DEFAULT privilege
   d. a role

16. Which of the following is true?
   a. If the DBA changes the password for a user while the user is connected to the database, the connection automatically terminates.
   b. If the DBA revokes the CREATE SESSION privilege of a user account, the user cannot connect to the database.
   c. If a user is granted the privilege to create a table and the privilege is revoked after the user creates a table, the table is automatically dropped from the system.
   d. all of the above

17. Which of the following commands can be used to eliminate the receptionist role?
   a. DELETE ROLE receptionist;
   b. DROP receptionist;
   c. DROP ANY ROLE;
   d. none of the above

18. Which of the following will display a list of all system privileges available in Oracle 10*g*?
   a. SESSION_PRIVS
   b. SYS_PRIVILEGE_MAP
   c. V$SYSTEM_PRIVILEGES
   d. SYSTEM_PRIVILEGE_MAP

19. Which of the following can be used to change the role that is currently enabled for a user?
    a. SET DEFAULT ROLE
    b. ALTER ROLE
    c. ALTER SESSION
    d. SET ROLE

20. Which of the following is an object privilege?
    a. DELETE ANY
    b. INSERT ANY
    c. UPDATE ANY
    d. REFERENCES

Hands-On Assignments

To perform the following assignments, refer to the tables in Appendix A.

1. Create a new user account. The name of the account should be a combination of your first initial and your last name.

2. Attempt to log in to Oracle 10*g* using the newly created account.

3. Assign privileges to the new account that allow the new user to connect to the database, create new tables, and alter an existing table.

4. Using a properly privileged account, create a role named customerrep that allows new rows to be inserted into the ORDERS and ORDERITEMS tables and allows rows to be deleted from those tables.

5. Assign the account created in Assignment 1 the customerrep role.

6. Log in to Oracle 10*g* using the new account created in Assignment 1. Determine the privileges currently available to the account.

7. Revoke the privilege to delete rows in the ORDERS and ORDERITEMS tables from the customerrep role.

8. Revoke the customerrep role from the account created in Assignment 1.

9. Delete the customerrep role from the Oracle 10*g* database.

10. Delete the user account created in Assignment 1.

Advanced Challenge

To perform the following activity, refer to the tables in Appendix A.

There are three major classifications for employees who do not work for the Information Systems Department of JustLee Books: account managers, who are responsible for the marketing activities of the company (for example, promotions based on customers' previous purchases or for specific books); data entry clerks, who enter inventory updates (for example, add new books and publishers, change prices); and customer service representatives, who are responsible for adding new customers and entering orders into the database. Each employee group has different tasks to perform; therefore, they will need different privileges for the various tables in the database. To simplify the administration of system and object privileges, a role should be created for each employee group.

Create a document for your supervisor that contains the following information:

1. List the tables that each group of employees needs to access from these tables: BOOKS, CUSTOMERS, ORDERS, ORDERITEMS, AUTHOR, BOOKAUTHOR, PUBLISHER, and PROMOTION.

2. Name the privileges needed by each group of employees.

3. For each group of employees, name a role that contains the appropriate privileges for that group.

4. For each group of employees, list the exact command(s) necessary to create and assign specific privileges to their role.

5. Explain your rationale for the privileges granted to each role.

Case Study: *City Jail*

The City Jail organization is preparing to deploy the new database to four departments. The departments and associated duties regarding the database are described in the following table.

DEPARTMENT	# EMPLOYEES	DUTIES
Criminal Records	8	1. Add new criminals and crime charges 2. Make changes to criminal and crime charge data as needed for corrections or updates 3. Keep the police officer information up to date 4. Maintain the crime codes list
Court Recording	7	1. Enter and modify all court appeals information 2. Enter and maintain all probation information 3. Maintain the probation officer list
Crimes Analysis	4	1. Analyze all criminal and court data to identify trends 2. Query all crimes data as needed to prepare federal and state reports
Data Officer	1	1. Remove crimes, court, and probation data based on approved requests from any of the other departments

Based on the department duties outlined in the table, develop a plan to assign privileges to employees in all four departments. The plan should include the following:

- A description of what types of objects will be required

- A listing of commands that would be required to address user creation for each department

RESTRICTING ROWS AND SORTING DATA

LEARNING OBJECTIVES

After completing this chapter, you should be able to do the following:

- Use a WHERE clause to restrict the rows returned by a query
- Create a search condition using mathematical comparison operators
- Use the BETWEEN...AND comparison operator to identify records within a range of values
- Specify a list of values for a search condition using the IN comparison operator
- Search for patterns using the LIKE comparison operator
- Identify the purpose of the % and _ wildcard characters
- Join multiple search conditions using the appropriate logical operator
- Perform searches for NULL values
- Specify the order for the presentation of query results using an ORDER BY clause

INTRODUCTION

Chapters 3 through 7 covered the construction of database objects, including tables, constraints, indexes, sequences, and users. This chapter shifts the focus back to querying a database. In Chapter 2, you learned how to retrieve specific fields from a table. However, unless you used the DISTINCT or UNIQUE keyword, your results included every record. In some instances, you will only want to see the records that meet a certain condition or conditions—a process referred to as **selection**. Because selection reduces the number of records retrieved by a query, it might be easier to locate a particular record in the output. In addition, it might be easier to identify trends if the data is presented in a sorted order. This chapter explains how to perform queries using search conditions and methods for sorting results. In particular, you'll see how the WHERE clause of the SELECT statement can be used as a search condition and how the ORDER BY clause can be used to present results in a specific sequence. You have already seen the WHERE clause used in the UPDATE and DELETE DML statements to limit the number of rows affected by the modification. Figure 8-1 provides an overview of this chapter's topics.

ELEMENT	DESCRIPTION
WHERE clause	Used to specify condition(s) that must be true for a record to be included in the query results
ORDER BY clause	Used to specify the sorted order for presenting the results of a query
Mathematical comparison operators (=, <, >, <=, >=, <>, !=, ^=)	Used to indicate how a record should relate to a specific search value
Other comparison operators (BETWEEN...AND, IN, LIKE, IS NULL)	Used in conditions with search values that include patterns, ranges, or null values
Logical operators (AND, OR, NOT)	Used to join multiple search conditions (AND, OR) or reverse the meaning of a given search condition (NOT)

FIGURE 8-1 Keywords and operators used to restrict and sort rows

NOTE

Before attempting to work through the examples provided in this chapter, you should open the Chapter 8 folder in your Data Files. Run the prech08.sql file to ensure that all necessary tables and constraints are available. This script will rebuild the JustLee Books database. Do not be concerned with errors from the DROP TABLE commands, which delete any existing tables of the same names.

WHERE CLAUSE SYNTAX

To retrieve records based upon a given condition in an Oracle 10g database, add the WHERE clause to the SELECT statement. As indicated by square brackets ([ ]), the WHERE clause is optional. When used, it should be listed beneath the FROM clause in the SELECT statement, as shown in Figure 8-2.

```
SELECT      [DISTINCT|UNIQUE] (*, column [ AS alias], …)
  FROM      table
  [WHERE    condition]
  [GROUP BY group_by_expression]
  [HAVING   group_condition]
  [ORDER BY column];
```

FIGURE 8-2   Syntax of the SELECT command

A **condition** identifies what must exist or a requirement that must be met for a record to be included in the results. Oracle 10*g* searches through each record to determine whether the condition is TRUE. If a record meets the given condition, it is returned in the results of the query. For a simple search of a table, the condition portion of the WHERE clause uses the following format:

```
<column name> <comparison operator> <another named column or a value>
```

For example, suppose that you need a list containing the last name of every customer living in the state of Florida. You would use the SQL statement shown in Figure 8-3.

```
SELECT lastname, state
  FROM customers
  WHERE state = 'FL';
```

FIGURE 8-3   Query to perform a simple search based upon a given condition

As shown in Figure 8-3, the query specifies the Lastname and State columns stored in the CUSTOMERS table as the data to list in the output. However, you want to see only the records of the customers who have the letters FL stored in the State field. Thus, in the WHERE clause, WHERE is the keyword, State is the name of the column to be searched, the comparison operator "equal to" (=) means it must contain the exact value specified, and the specified value is FL. Notice the single quotation marks around FL, which designate FL as a string literal. Also note that the value of FL is in uppercase to match the format in which the data was entered into the State field. Figure 8-4 shows the output of this SELECT statement.

NOTE

If you receive an error message, verify that the FL value was entered with single and not double quotation marks. (Double quotation marks, used in Chapter 2, are for column aliases.) If no rows are returned in the results, make certain the letters FL are capitalized. The data for JustLee Books was originally entered in uppercase characters and is, therefore, stored in the database tables in uppercase characters. Any value entered in a string literal (that is, within single quotation marks) is evaluated *exactly* as entered—both in spacing and case of the characters. Therefore, if a string literal is entered for a search condition, it must be in the same case as the data being searched, or no rows will be returned in the results. You can verify the case of data in a table by querying the table and reviewing the output. The default output reflects the case in which the field is stored. Although Oracle 10*g* is not case sensitive when evaluating keywords, table names, and column names, the *evaluation of data contained within a record is case sensitive*.

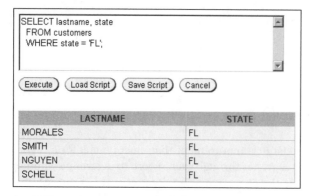

FIGURE 8-4   Results of a state-based search

The results returned from the query list the last name and state for each customer living in Florida. Notice that only four rows were returned, even though our table contained 20 customers. The WHERE clause restricts the number of records returned in the results to only those meeting the given condition of **state = 'FL'**.

Rules for Character Strings

As shown in the SQL command in Figure 8-3, the value of FL is shown within single quotation marks. Whenever you use a string literal as part of a search condition, the value must be enclosed within single quotation marks and, as a result, will be interpreted exactly as listed within the single quotation marks. By contrast, if the target field had consisted only of *numbers*, single quotation marks would not have been required. To demonstrate, suppose that you want to see all the data stored in the CUSTOMERS table for customer 1010. The Customer# field has a numeric datatype. Issue the SQL statement and compare it to the output shown in Figure 8-5.

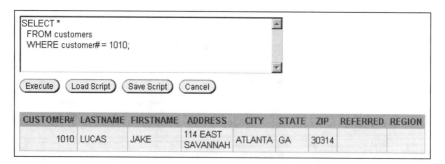

FIGURE 8-5   Search results for Customer 1010

In this example, the value of 1010 for the Customer# column is not enclosed in single quotation marks because the Customer# column has been defined to store only numbers. Thus, single quotation marks are not necessary.

Next, let's use the WHERE statement to search for a book with the ISBN of 1915762492. Figure 8-6 shows the input and the output of that query.

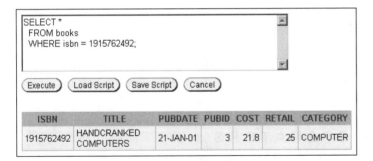

FIGURE 8-6 Search results for ISBN 1915762492

The ISBN column of the BOOKS table is defined as a character field, or text field, rather than a numeric field because some ISBNs could contain letters. In this particular instance, however, none of the values stored within the ISBN column contain any letters. Therefore, you were able to search the field using a search condition specified as a numeric value without any quotation marks. However, if the table had contained even *one* record that had a letter in the ISBN column, Oracle 10*g* would have returned an error message. In other words, it might have worked in this one case, but it might not always work. Using single quotation marks ultimately depends upon whether the field is defined to hold text or only numeric data. Therefore, *always use single quotation marks if the column is defined with anything other than a numeric datatype.*

N O T E

If you don't know if a column is defined to hold only numeric values, issue the DESCRIBE *tablename* command to see how table columns have been defined.

Rules for Dates

Sometimes, you might need to use a date as a search condition. Oracle 10*g* displays dates in the default format of DD-MON-YY, with MON being the standard three-letter abbreviation for the month. Because the Pubdate field contains letters and hyphens, it is not considered a numeric value when Oracle 10*g* performs searches. Therefore, the date value must be enclosed in single quotation marks. Figure 8-7 shows a query for books published on January 21, 2001.

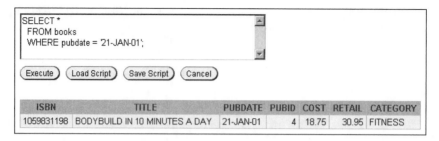

ISBN	TITLE	PUBDATE	PUBID	COST	RETAIL	CATEGORY
1059831198	BODYBUILD IN 10 MINUTES A DAY	21-JAN-01	4	18.75	30.95	FITNESS

FIGURE 8-7 Simple query using a date condition

COMPARISON OPERATORS

Thus far in this chapter, you have used an equal sign, or **equality operator**, to evaluate search conditions; basically, you instructed Oracle 10g to return only results containing the *exact* value you provided. However, there are many situations that are not based on an "equal to" condition. For example, suppose management needs a list of books for a proposed marketing campaign. The Marketing Department wants to include a gift with the purchase of any book that has a retail price of more than $55.00. Management wants to know which books can be specifically mentioned in an advertisement of this marketing campaign. The equality operator would not be appropriate in this situation; you would need a different comparison operator. A **comparison operator** indicates how the data should relate to the given search value (such as greater than or less than). In this case, you would need to use a comparison operator that means "greater than" (>) to determine which books meet the "more than $55.00" requirement. Execute the statement in Figure 8-8 to accomplish this task.

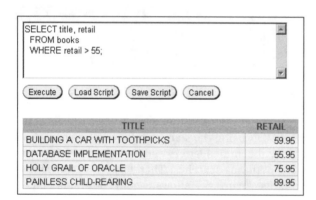

TITLE	RETAIL
BUILDING A CAR WITH TOOTHPICKS	59.95
DATABASE IMPLEMENTATION	55.95
HOLY GRAIL OF ORACLE	75.95
PAINLESS CHILD-REARING	89.95

FIGURE 8-8 Searching for books with a retail price greater than $55

Based on these results, you know that four books meet the condition for this sales promotion. Notice that you entered **55** as the value for the Retail price condition. This value could have also been entered as **55.00**. Oracle 10g accepts a period to indicate decimal positions without considering the entry to be a character value rather than a numeric value. However, if you had entered the dollar sign ($) or a comma (to indicate a thousands position), you would have received an error message indicating that the Retail field is numeric and the value entered was an "invalid character" (the dollar symbol ($) is treated as a formatting character, so $55.00 is not equivalent to 55.00). Unlike some other database management systems, Oracle 10g does not have a currency datatype, and it regards the comma and the dollar sign as characters.

The "greater than" (>) comparison operator can also be used with text and date fields. Suppose you are about to take a physical inventory of all books in stock. The procedure JustLee Books uses for taking a physical inventory is to give each employee a list of books, and then have the person record the quantity on hand. When creating the list of books, each person is responsible for a portion of the alphabet. For example, one person might be responsible for all books with titles falling in the A through D range.

Figure 8-9 shows how to create the list of books for the person who has been assigned to take an inventory of all books with a title that alphabetically occurs after the letters HO. All book titles with additional characters following HO or beginning with the letters HP and the letters that follow are listed.

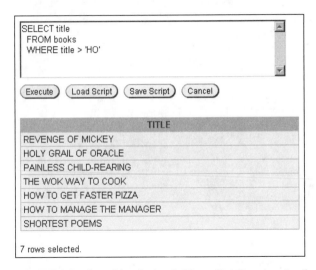

FIGURE 8-9 Searching for book titles with letters greater than HO

Now that you've examined "equal to" and "greater than" comparison operators, let's examine some others. Figure 8-10 shows comparison operators commonly used in Oracle 10g. Even though the first set of comparison operators shown are considered to be mathematical operators, these can be used for a variety of datatypes, including characters and dates.

COMPARISON OPERATORS	
Mathematical Comparison Operators	
=	Equality or "equal to"—for example, cost = 55.95
>	Greater than—for example, cost > 20
<	Less than—for example, cost < 20
<>, !=, or ^=	Not equal to—for example, cost <> 55.95 or cost != 55.95 or cost ^=55.95
<=	Less than or equal to—for example, cost <= 20
>=	Greater than or equal to—for example, cost >= 20
Other Comparison Operators	
[NOT] BETWEEN x AND y	Used to express a range—for example, searching for numbers BETWEEN 5 and 10. The optional NOT is used when searching for numbers that are NOT BETWEEN 5 AND 10.
[NOT] IN(x,y,...)	Similar to the OR logical operator. Can search for records which meet at least one condition contained within the parentheses— for example, Pubid IN (1, 4, 5) will return only books with a publisher id of 1, 4, or 5. The optional NOT keyword instructs Oracle to return books not published by Publisher 1, 4, or 5.
[NOT] LIKE	Used when searching for patterns if you are not certain how something is spelled—for example, title LIKE 'TH%'. Using the optional NOT indicates that records that do contain the specified pattern should not be included in the results.
IS [NOT] NULL	Used to search for records that do not have an entry in the specified field—for example, Shipdate IS NULL. Include the optional NOT to find records that do have an entry in the field—for example, Shipdate IS NOT NULL.

FIGURE 8-10 Comparison operators

In contrast to the "greater than" (>) operator that returns only rows with a value higher than the value in the stated condition, the "less than" (<) operator returns only values that are less than the stated condition. For example, the management of JustLee Books would like a list of all books that have a profit of less than 20% of the book's cost. Because the profit for a book is determined by subtracting the cost from the retail price of the book, this calculated value can be compared against the cost of the book multiplied by .20, or the decimal version of 20%. As shown in Figure 8-11, only one book generates less than a 20% profit margin.

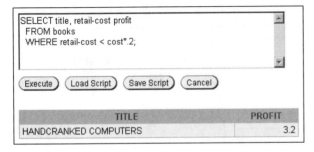

```
SELECT title, retail-cost profit
  FROM books
  WHERE retail-cost < cost*.2;
```

Execute   Load Script   Save Script   Cancel

TITLE	PROFIT
HANDCRANKED COMPUTERS	3.2

FIGURE 8-11   Searching Profit using the "less than" operator

The "greater than" and "less than" operators do not include values that exactly match the given condition. For example, if you want the results in Figure 8-11 to also include books that return a 20% profit, then the comparison operator must be changed to the "less than or equal to" operator (<=). Using the <= operator, any book that returns exactly 20% profit is also included in the results.

For example, suppose the Marketing Department is sorting paper files and requests a list of all customers who live in Georgia or in a state that alphabetically appears before the state of Georgia (that is, A through GA). The simplest way to identify those customers is to search for all customers using the condition **state <= 'GA'**, as shown in Figure 8-12.

```
SELECT firstname, lastname, state
  FROM customers
  WHERE state <= 'GA';
```

Execute   Load Script   Save Script   Cancel

FIRSTNAME	LASTNAME	STATE
BONITA	MORALES	FL
RYAN	THOMPSON	CA
LEILA	SMITH	FL
JORGE	PEREZ	CA
JAKE	LUCAS	GA
NICHOLAS	NGUYEN	FL
STEVE	SCHELL	FL
MICHELL	DAUM	CA
GREG	MONTIASA	GA

9 rows selected.

FIGURE 8-12   Searching State using the "less than or equal to" operator

Later, the Marketing Department requests that you identify all customers who live in Georgia or in a state that has a state abbreviation "greater than" Georgia's abbreviation of GA. Although you might think this is a little unusual because the previous list already included customers living in Georgia, you nevertheless create the list using the condition **state >= 'GA'** and obtain the names of 13 customers who meet that condition, as shown in Figure 8-13.

```
SELECT firstname, lastname, state
  FROM customers
  WHERE state >= 'GA';
```

(Execute)   (Load Script)   (Save Script)   (Cancel)

FIRSTNAME	LASTNAME	STATE
THOMAS	PIERSON	ID
CINDY	GIRARD	WA
MESHIA	CRUZ	NY
TAMMY	GIANA	TX
KENNETH	JONES	WY
JAKE	LUCAS	GA
REESE	MCGOVERN	IL
WILLIAM	MCKENZIE	MA
JASMINE	LEE	WY
BECCA	NELSON	MI
GREG	MONTIASA	GA
JENNIFER	SMITH	NJ
KENNETH	FALAH	NJ

13 rows selected.

FIGURE 8-13   Searching State using the "greater than or equal to" operator

Suppose, for another marketing analysis, the Marketing Department requests a list of all customers who *do not* live in the state of Georgia. You can perform this task by using the condition **state <> 'GA'** to generate the requested list, as shown in Figure 8-14.

NOTE

Using != or ^= in the query shown in Figure 8-14 returns the same results as using <> for the "not equal to" operator.

To demonstrate the use of comparison operators with dates, suppose you need to produce a list of all orders that were placed prior to April 2005. You could use the WHERE clause with a condition on the Orderdate column, as shown in Figure 8-15.

```
SELECT firstname, lastname, state
  FROM customers
  WHERE state <> 'GA';
```

Execute   Load Script   Save Script   Cancel

FIRSTNAME	LASTNAME	STATE
BONITA	MORALES	FL
RYAN	THOMPSON	CA
LEILA	SMITH	FL
THOMAS	PIERSON	ID
CINDY	GIRARD	WA
MESHIA	CRUZ	NY
TAMMY	GIANA	TX
KENNETH	JONES	WY
JORGE	PEREZ	CA
REESE	MCGOVERN	IL
WILLIAM	MCKENZIE	MA
NICHOLAS	NGUYEN	FL
JASMINE	LEE	WY
STEVE	SCHELL	FL
MICHELL	DAUM	CA
BECCA	NELSON	MI
JENNIFER	SMITH	NJ
KENNETH	FALAH	NJ

18 rows selected.

FIGURE 8-14   Searching State using the "not equal to" operator

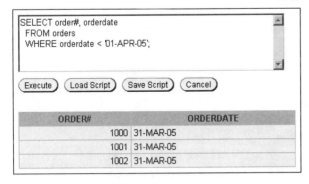

```
SELECT order#, orderdate
  FROM orders
  WHERE orderdate < '01-APR-05';
```

Execute   Load Script   Save Script   Cancel

ORDER#	ORDERDATE
1000	31-MAR-05
1001	31-MAR-05
1002	31-MAR-05

FIGURE 8-15   Searching a date value

BETWEEN...AND Operator

The BETWEEN...AND comparison operator is used when searching a field for values that fall within a specific range. Figure 8-16 shows a query to find any book whose publisher has an assigned ID between 1 and 3. Notice in the results that the range is inclusive and includes any publisher with the ID of 1, 2, or 3.

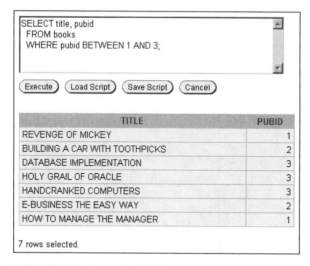

FIGURE 8-16   Searching Pubid using the BETWEEN...AND operator

Returning to the book inventory issue raised earlier, an alphabetical range of titles could be queried using the BETWEEN...AND comparison operator. The statement shown in Figure 8-17 identifies all the book titles that fall in the A through D range.

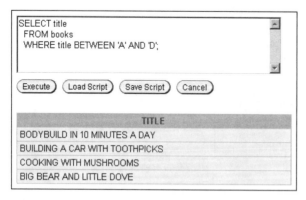

FIGURE 8-17   Searching for a character range using the BETWEEN...AND operator

IN Operator

The IN operator returns records that match one of the values given in the listed values. *Oracle 10g syntax requires the items listed to be separated by commas, and the entire list must be enclosed in parentheses.* The output of the query in Figure 8-18 shows that there are seven books currently in inventory that were published by Publisher 1, 2, or 5.

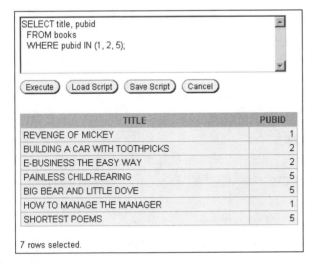

FIGURE 8-18   Searching Pubid using the IN operator

If a list of all customers who reside in either California or Texas is needed, you should include the state abbreviations in single quotation marks because they are string literals. The query in Figure 8-19 demonstrates how the IN operator can produce the desired condition for this task.

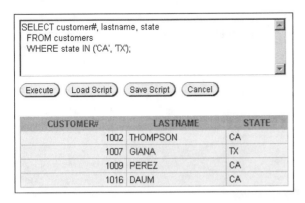

FIGURE 8-19   Searching State using the IN operator

LIKE Operator

The LIKE operator is unique in that it is used with wildcard characters to search for patterns. **Wildcard characters** are used to represent one or more alphanumeric characters. The wildcard characters available for pattern searches in Oracle 10g are the percent sign (%) and the underscore symbol (_). The percent sign is used to represent *any number of characters (zero, one, or more)*, whereas the underscore symbol represents exactly *one character*. For example, if you were trying to find any customer whose last name started with P and did not care about the remaining letters of the last name, you could enter the SQL statement shown in Figure 8-20.

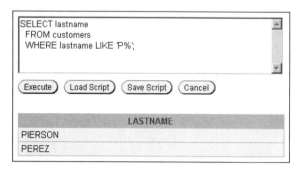

FIGURE 8-20   Searching using the LIKE operator with the % wildcard character

The results include two customers whose last names begin with a P. If, however, you are searching for customers whose last names contain a P in any position, you change the search pattern to '%P%'. The software interprets the pattern to mean, "It doesn't matter what is before the letter P or what comes after, but a P must be somewhere in the Last-name column."

Suppose that you are having difficulty reading the printout of an order for a customer because someone spilled coffee on the printed order form. You can tell that the first two digits of the Customer# are a "1" and a "0," and the last digit is "9." However, you can't read the third number. In this case, you could use an underscore character to represent the missing digit, as shown in Figure 8-21.

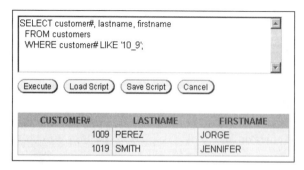

FIGURE 8-21   Searching using the LIKE operator with the _ wildcard character

Oracle 10g interprets the search condition in Figure 8-21 as, "Look for any customer number that begins with a 10, is followed by any character, and ends with a 9." The results return two customers, 1009 and 1019.

The percent sign and underscore symbol can also be combined in the same search condition to create more complex search patterns. Suppose that you need to identify every book ISBN that has the numeral 4 as its second numeral and ends with a 0. The actual pattern you are trying to identify can be stated as '_4%0' because you know just one number comes before the number 4 and that there will be additional numbers after the 4; you don't care what the numbers are or how many there are—as long as the last digit is 0. As shown in Figure 8-22, this search pattern identifies two books from the BOOKS table.

```
SELECT isbn, title
  FROM books
  WHERE isbn LIKE '_4%0';
```

(Execute) (Load Script) (Save Script) (Cancel)

ISBN	TITLE
3437212490	COOKING WITH MUSHROOMS
2491748320	PAINLESS CHILD-REARING

FIGURE 8-22   Searching using the LIKE operator with a mix of wildcard characters

NOTE

The regular expression REGEXP_LIKE extends the pattern matching capabilities of the LIKE operator. Regular expressions are covered in Chapter 10.

Although IS NULL and IS NOT NULL are comparison operators, they specifically address searches based on a column having or not having a NULL value, respectively. Therefore, the discussion of these two operators is presented in a later section of this chapter.

LOGICAL OPERATORS

There might be times when you need to search for records based on two or more conditions. In these situations, you can use **logical operators** to combine search conditions. The logical operators AND and OR are commonly used to combine search conditions. (The NOT operator, mentioned in Figure 8-10, is also a logical operator available in Oracle 10g, but it is used to reverse the meaning of search conditions, rather than to combine them.) When queried with a WHERE clause, each record in the table is compared to the stated condition. If the condition is TRUE when compared to a record, the record is included in the results. When the AND operator is used in the WHERE clause, both conditions combined by the AND operator must be evaluated as being TRUE, or the record is not included in the results.

Figure 8-23 shows a query for titles of books published by Publisher 3 *and* that are in the computer category. Because the search is for books meeting both conditions, the conditions are combined with the AND operator.

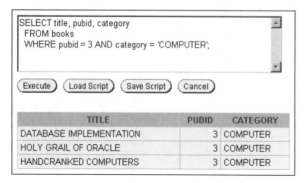

FIGURE 8-23   Searching with multiple conditions using the AND logical operator

NOTE

Recall that all data for JustLee Books are stored in the tables in uppercase letters because the data was entered in uppercase letters when the tables were originally created. If no rows are returned, make certain you typed COMPUTER in all capital letters.

On the other hand, if you want a list of books that either were published by Publisher 3 *or* are in the Computer Category, you can use the OR operator, as shown in Figure 8-24.

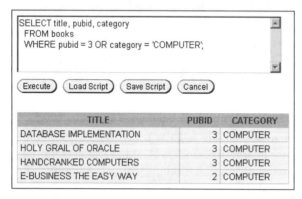

FIGURE 8-24   Searching with multiple conditions using the OR logical operator

With the OR operator, only one of the conditions must be TRUE to have the record included in the results. In Figure 8-24, the first three records met both conditions. However, the last record met only the **category='COMPUTER'** condition; the **pubid=3** condition was evaluated as FALSE for the last record. Because the OR operator was used, only one condition had to be TRUE, so even though that book was not published by Publisher 3, it was included in the output.

Let's also consider the order of logical operators. Because the WHERE clause can contain multiple types of operators, you need to understand the order in which they are resolved.

- Arithmetic operations are solved first.
- Comparison operators (<, >, =, LIKE, etc.) are solved next.
- Logical operators have a lower precedence and are evaluated last—in the order of NOT, AND, and, finally, OR.

If you need to change the order of evaluation, simply use parentheses to indicate the operators to be resolved first.

Look at the results of the query shown in Figure 8-25. The list includes books that are published by Publisher 4 *and* that cost more than $15.00. The list also includes any book from the Family Life Category. Although the OR operator was actually listed first in the WHERE clause, Oracle 10g first evaluated the Pubid and Cost conditions that are combined with the AND logical operator. After that was solved, the Category condition that preceded the OR logical operator was then considered.

Suppose that after examining the results of the previous query, you realize the order in which the logical operators were evaluated did not yield the output you wanted—to find any book that cost more than $15.00 and is either published by Publisher 4 or is in the Family Life Category. To have Oracle 10g evaluate the conditions in the desired order, you must use parentheses first to identify any book that is published by Publisher 4 or is categorized as Family Life. After the books meeting either the Category or Publisher condition are found, the Cost condition is then evaluated, and only those records that cost more than $15.00 are displayed. Notice how the query in Figure 8-26 returns results different from those previously shown in Figure 8-25.

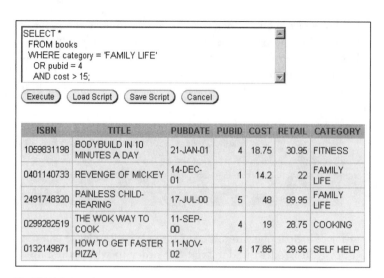

FIGURE 8-25 Searching using both AND and OR operators

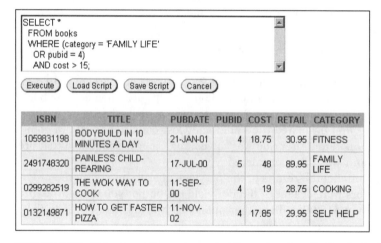

FIGURE 8-26 Using parentheses to control the order of evaluation for logical operators

TREATMENT OF NULL VALUES

When performing arithmetic operations or search conditions, NULL values can return unexpected results. A **NULL value** means that no value has been stored in that particular field. Do not confuse a NULL value with a blank space: *A NULL is the absence of data in a field; a field containing a blank space does contain a value—a blank space—and is, therefore, not NULL.* When searching for NULL values, you cannot use the equal sign (=)

because there is not a value available to use for comparison in the search condition. If you need to identify records containing a NULL value, you must use the IS NULL comparison operator.

For example, when an order is shipped to a customer, the date on which the order is shipped is entered into the ORDERS table. If a date does not appear in the Shipdate field, then the order has not yet shipped. To find any order that has not yet been sent to the customer, use the query shown in Figure 8-27.

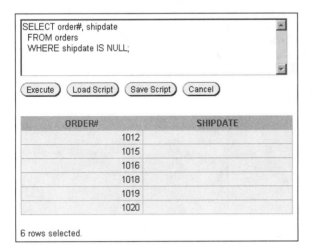

FIGURE 8-27   Searching for NULL values with the IS NULL operator

As shown in Figure 8-27, there are currently six orders outstanding. Notice that when searching for a NULL value, you simply state the field to be searched followed by the words IS NULL in the WHERE clause. If you want a list of all orders that have shipped (that is, the Shipdate column contains an entry), simply add the logical operator NOT. When searching for a field that is not NULL, you instruct Oracle 10g to return any records with data available in the named field, as shown in Figure 8-28.

Be aware that using "= NULL" search condition does not raise an error. This condition always returns a "no rows selected" message. Try the previous example replacing "IS NULL" with "= NULL" to see the results, as shown in Figure 8-29.

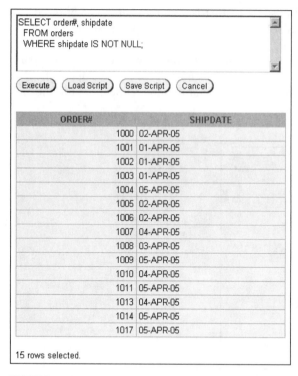

```
SELECT order#, shipdate
 FROM orders
 WHERE shipdate IS NOT NULL;
```

Execute Load Script Save Script Cancel

ORDER#	SHIPDATE
1000	02-APR-05
1001	01-APR-05
1002	01-APR-05
1003	01-APR-05
1004	05-APR-05
1005	02-APR-05
1006	02-APR-05
1007	04-APR-05
1008	03-APR-05
1009	05-APR-05
1010	04-APR-05
1011	05-APR-05
1013	04-APR-05
1014	05-APR-05
1017	05-APR-05

15 rows selected.

FIGURE 8-28   Searching for non-NULL values with the IS NOT NULL operator

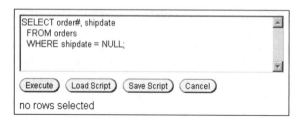

```
SELECT order#, shipdate
 FROM orders
 WHERE shipdate = NULL;
```

Execute Load Script Save Script Cancel

no rows selected

FIGURE 8-29   Erroneously using the " = NULL" operator

ORDER BY CLAUSE SYNTAX

Use the ORDER BY clause for displaying the results of a query in a sorted order. The ORDER BY clause is listed at the end of the SELECT statement, as shown in Figure 8-30.

```
SELECT     [DISTINCT|UNIQUE] (*, column [ AS alias], …)
  FROM     table
  [WHERE   condition]
  [GROUP BY group_by_expression]
  [HAVING   group_condition]
  [ORDER BY column];
```

FIGURE 8-30 Syntax of the SELECT statement

For example, to see a list of all publishers sorted by Name, enter the SQL statement shown in Figure 8-31.

FIGURE 8-31 Sorting results by the publisher name in ascending order

In the results of the query, the second column (NAME) is listed in ascending order. Note these important points:

- When sorting in ascending order, values are listed in this order:
  1. Numeric values
  2. Character values
  3. NULL values
- Unless you specify "desc" for descending, the ORDER BY clause sorts in ascending order by default.

To view the publishers in descending order by name, simply enter **DESC** after the name of the column. After changing the sort order to descending, you obtain the results shown in Figure 8-32.

If you want to make clear that a column is to be sorted in ascending order, you can specify **ASC** after the column name.

```
SELECT *
FROM publisher
ORDER BY name DESC;
```

Execute   Load Script   Save Script   Cancel

PUBID	NAME	CONTACT	PHONE
5	REED-N-RITE	SEBASTIAN JONES	800-555-8284
4	READING MATERIALS INC.	RENEE SMITH	800-555-9743
2	PUBLISH OUR WAY	JANE TOMLIN	010-410-0010
1	PRINTING IS US	TOMMIE SEYMOUR	000-714-8321
3	AMERICAN PUBLISHING	DAVID DAVIDSON	800-555-1211

FIGURE 8-32   Sorting results by the publisher name in descending order

If a column alias is given to a field in the SELECT clause, you can reference that field in the ORDER BY clause by using the column alias—although this is not required. Note the example in Figure 8-33.

```
SELECT pubid, name "Publisher Name", phone
  FROM publisher
  ORDER BY "Publisher Name";
```

Execute   Load Script   Save Script   Cancel

PUBID	Publisher Name	PHONE
3	AMERICAN PUBLISHING	800-555-1211
1	PRINTING IS US	000-714-8321
2	PUBLISH OUR WAY	010-410-0010
4	READING MATERIALS INC.	800-555-9743
5	REED-N-RITE	800-555-8284

FIGURE 8-33   Referencing a column alias in the ORDER BY clause

Remember, if the column alias contains a space, it must be enclosed in double quotation marks. This remains true even if the column alias is being used in the ORDER BY clause.

You can also use the ORDER BY clause with optional NULLS FIRST or NULLS LAST keywords to change the order for the listing of null values. By default, nulls are listed *last* when the results are sorted in ascending order, and *first* when they are sorted in descending order. The query in Figure 8-34 lists the last and first name of each customer and the customer number of the person who referred that customer to JustLee Books. The results are sorted in ascending order by the Referred column.

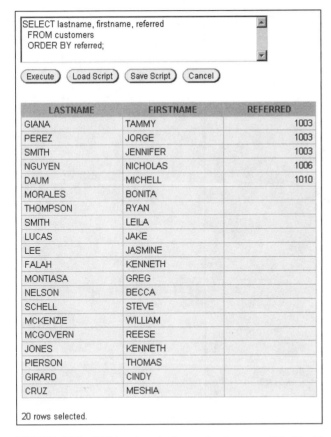

```
SELECT lastname, firstname, referred
  FROM customers
  ORDER BY referred;
```

(Execute) (Load Script) (Save Script) (Cancel)

LASTNAME	FIRSTNAME	REFERRED
GIANA	TAMMY	1003
PEREZ	JORGE	1003
SMITH	JENNIFER	1003
NGUYEN	NICHOLAS	1006
DAUM	MICHELL	1010
MORALES	BONITA	
THOMPSON	RYAN	
SMITH	LEILA	
LUCAS	JAKE	
LEE	JASMINE	
FALAH	KENNETH	
MONTIASA	GREG	
NELSON	BECCA	
SCHELL	STEVE	
MCKENZIE	WILLIAM	
MCGOVERN	REESE	
JONES	KENNETH	
PIERSON	THOMAS	
GIRARD	CINDY	
CRUZ	MESHIA	

20 rows selected.

FIGURE 8-34   NULL values in the sort column are listed last by default

Suppose, however, that you want the results sorted in ascending order, but you need to have the NULL values listed first. To override the placement of the null values, you can add NULLS FIRST in the ORDER BY clause; this instructs Oracle 10g to place the NULL values at the beginning of the list, and sort the remaining records in ascending order, as shown in Figure 8-35. (If you had used the descending sequence, you would use NULLS LAST to override the default order for the NULL values.)

Secondary Sort

In the previous examples, we specified only one column in the ORDER BY clause; we refer to this as a **primary sort**. In some cases, you might also want to include a secondary sort. A **secondary sort** provides a second field to sort by if an exact match occurs between two or more rows in the primary sort. For example, telephone books list residential customers alphabetically by last name. However, when two or more customers have the same last name, the customers are listed in alphabetical order by their first name. In other words, a primary sort is performed on last name and then, when necessary, a secondary sort is performed on first name.

```
SELECT lastname, firstname, referred
  FROM customers
  ORDER BY referred NULLS FIRST;
```

(Execute) (Load Script) (Save Script) (Cancel)

LASTNAME	FIRSTNAME	REFERRED
MORALES	BONITA	
THOMPSON	RYAN	
SMITH	LEILA	
PIERSON	THOMAS	
JONES	KENNETH	
LUCAS	JAKE	
MCKENZIE	WILLIAM	
LEE	JASMINE	
SCHELL	STEVE	
FALAH	KENNETH	
MONTIASA	GREG	
NELSON	BECCA	
MCGOVERN	REESE	
GIRARD	CINDY	
CRUZ	MESHIA	
GIANA	TAMMY	1003
PEREZ	JORGE	1003
SMITH	JENNIFER	1003
NGUYEN	NICHOLAS	1006
DAUM	MICHELL	1010

20 rows selected.

FIGURE 8-35 Using the NULLS FIRST option in the ORDER BY clause

NOTE

The limit on the number of columns that can be used in the ORDER BY clause is 255.

For illustrative purposes, the query in Figure 8-36 shows that customers are to be listed in descending order by state. When there is more than one customer living in a particular state, customers are to be sorted by city—in ascending order.

When looking at the results of the query, you can see that several states have multiple residents. Within those states, notice the customers are sorted in ascending order, according to the city in which they live, as instructed by the SQL statement. The descending sort order applied only to the column after which it was listed. Because City did not reference a sort order, the default value of ascending was assumed.

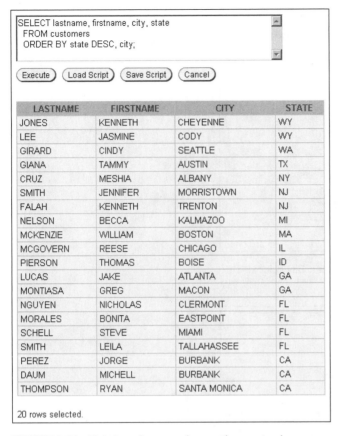

```
SELECT lastname, firstname, city, state
   FROM customers
   ORDER BY state DESC, city;
```

(Execute) (Load Script) (Save Script) (Cancel)

LASTNAME	FIRSTNAME	CITY	STATE
JONES	KENNETH	CHEYENNE	WY
LEE	JASMINE	CODY	WY
GIRARD	CINDY	SEATTLE	WA
GIANA	TAMMY	AUSTIN	TX
CRUZ	MESHIA	ALBANY	NY
SMITH	JENNIFER	MORRISTOWN	NJ
FALAH	KENNETH	TRENTON	NJ
NELSON	BECCA	KALMAZOO	MI
MCKENZIE	WILLIAM	BOSTON	MA
MCGOVERN	REESE	CHICAGO	IL
PIERSON	THOMAS	BOISE	ID
LUCAS	JAKE	ATLANTA	GA
MONTIASA	GREG	MACON	GA
NGUYEN	NICHOLAS	CLERMONT	FL
MORALES	BONITA	EASTPOINT	FL
SCHELL	STEVE	MIAMI	FL
SMITH	LEILA	TALLAHASSEE	FL
PEREZ	JORGE	BURBANK	CA
DAUM	MICHELL	BURBANK	CA
THOMPSON	RYAN	SANTA MONICA	CA

20 rows selected.

FIGURE 8-36   Using a primary and secondary sort column

Sorting by SELECT Order

The query statement in Figure 8-37 requests a list of customers who live in states CA and FL. The statement also specifies that the output should be listed with a primary descending sort on State and a secondary sort on City.

Oracle 10g also provides an abbreviated method for referencing the column to use for sorting if the field name is previously used in the SELECT clause. In the previous example, State and City were used in both the SELECT and ORDER BY clauses. Rather than relisting the field name in the ORDER BY clause, you can reference a field by its position in the column list of the SELECT clause. Because State is listed fourth in the SELECT clause and City is third, you can modify the SQL statement, as shown in Figure 8-38, and receive the same results.

FIGURE 8-37 Sorting on the State and City columns

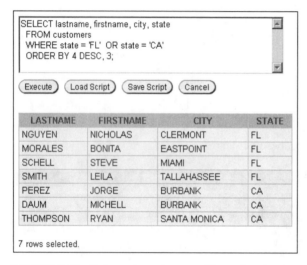

FIGURE 8-38 Referencing column positions of sort columns in the ORDER BY clause

Chapter Summary

- The WHERE clause can be included in a SELECT statement to restrict the rows returned by a query to only those meeting a specified condition.

- When searching a nonnumeric field, the search values must be enclosed in single quotation marks.

- Comparison operators are used to indicate how the record should relate to the search value.

- Mathematical comparison operators include =, >, <, >=, and <=, and the "not equal to" operators of <>, !=, and ^=.

- The BETWEEN...AND comparison operator is used to search for records that fall within a certain range of values.

- The IN comparison operator is used to identify a list of values that should be used for the search condition. A record must contain one of the values in the list to be included in the query results.

- The LIKE comparison operator is used with the percent and underscore symbols (% and _) to establish search patterns.

- Logical operators such as AND and OR can be used to combine several search conditions.

- Logical operators are always evaluated in the order of NOT, AND, and, finally, OR. Parentheses can be used to override the order of evaluation.

- When using the AND operator, all conditions must be TRUE for a record to be returned in the results. However, with the OR operator, only one condition must be TRUE.

- A NULL value is the absence of data, not a field with a blank space entered.

- Use the IS NULL comparison operator to match NULL values. The IS NOT NULL comparison operator finds records that do not contain NULL values in the indicated column.

- You can sort the results of queries by using an ORDER BY clause. When used, the ORDER BY clause should be listed last in the SELECT statement.

- By default, records are sorted in ascending order. Entering DESC directly after the column name sorts the records in descending order.

- Multiple columns can be used for sorting by listing each column name to be used in a single ORDER BY clause separated by commas. List first the column you want used for the primary sort. If an exact match occurs between two or more records, the next column listed is used to determine the correct order, and so on. You can specify columns by column name or by their position in the SELECT clause.

- A column does not have to be listed in the SELECT clause to serve as a basis for sorting.

Chapter 8 Syntax Summary

The following table presents a summary of the syntax that you have learned in this chapter. You can use the table as a study guide and reference.

SYNTAX GUIDE		
Element	Description	Example
Optional SELECT clauses		
WHERE clause	Specifies a search condition	SELECT * FROM customers WHERE state = 'GA';
ORDER BY clause	Specifies presentation order of the results	SELECT * FROM publisher ORDER BY name;
Mathematical Comparison Operators		
=	"Equality" operator—requires an exact match of the record data and the search value	WHERE cost = 55.95
>	"Greater than" operator—requires record to be greater than the search value	WHERE cost > 55.95
<	"Less than" operator—requires a record to be less than the search value	WHERE cost < 55.95
<>, !=, ^=	"Not equal to" operator—requires a record not to match the search value	WHERE cost <> 55.95 or WHERE cost != 55.95 or WHERE cost ^= 55.95
<=	"Less than or equal to" operator— requires a record to be less than or an exact match with the search value	WHERE cost <= 55.95
>=	"Greater than or equal to" operator—requires a record to be greater than or an exact match with the search value	WHERE cost >= 55.95
Other Comparison Operators		
[NOT] BETWEEN x AND y	Searches for records in a specified range of values	WHERE cost BETWEEN 40 AND 65
[NOT] IN(x,y,...)	Searches for records that match one of the items in the list	WHERE cost IN(22, 55.95, 13.50)
[NOT] LIKE	Searches for records that match a search pattern—used with wildcard characters	WHERE lastname LIKE '_A%'

SYNTAX GUIDE		
Element	Description	Example
Other Comparison Operators		
IS [NOT] NULL	Searches for records with a NULL value in the indicated column	`WHERE referred IS NULL`
Wildcard Characters		
%	Percent sign wildcard— represents any number of characters	`WHERE lastname LIKE '%R%'`
_	Underscore wildcard— represents exactly one character in the indicated position	`WHERE lastname LIKE '_A';`
Logical Operators		
AND	Combines two conditions together—record must match both conditions	`WHERE cost > 20 AND retail < 50`
OR	Requires a record to match only one of the search conditions	`WHERE cost > 20 OR retail < 50`

Review Questions

1. Which clause is used to restrict the number of rows returned from a query?
2. Which clause displays the results of a query in a specific sequence?
3. Which operator can you use to find any books with a retail price of at least $24.00?
4. Which operator should you use to find NULL values?
5. The IN comparison operator is similar to which logical operator?
6. When should single quotation marks be used in a WHERE clause?
7. What is the effect of using the NOT operator in a WHERE clause?
8. When should a percent sign (%) be used with the LIKE operator?
9. When should an underscore symbol (_) be used with the LIKE operator?

Multiple Choice

To answer the following questions, refer to the tables in Appendix A.

1. Which of the following SQL statements is not valid?

   a. SELECT address || city || state || zip "Address"
      FROM customers
      WHERE lastname = 'SMITH';

   b. SELECT * FROM publisher ORDER BY contact;

c. SELECT address, city, state, zip
   FROM customers
   WHERE lastname = "SMITH";

d. All of the above are valid and return the expected results.

2. Which clause is used to restrict rows?

   a. SELECT

   b. FROM

   c. WHERE

   d. ORDER BY

3. Which of the following SQL statements is valid?

   a. SELECT order# FROM orders WHERE shipdate = NULL;

   b. SELECT order# FROM orders WHERE shipdate = 'NULL';

   c. SELECT order# FROM orders WHERE shipdate = "NULL";

   d. None of the statements are valid.

4. Which of the following will return a list of all customers' names sorted in descending order by city within state?

   a. SELECT name FROM customers ORDER BY desc state, city;

   b. SELECT firstname, lastname FROM customers
      SORT BY desc state, city;

   c. SELECT firstname, lastname FROM customers
      ORDER BY state desc, city;

   d. SELECT firstname, lastname FROM customers
      ORDER BY state desc, city desc;

   e. SELECT firstname, lastname FROM customers
      ORDER BY 5 desc, 6 desc;

5. Which of the following will not return a customer with the last name of THOMPSON in its results?

   a. SELECT lastname FROM customers
      WHERE lastname = "THOMPSON";

   b. SELECT * FROM customers;

   c. SELECT lastname FROM customers
      WHERE lastname > 'R';

   d. SELECT * FROM customers
      WHERE lastname <'V';

6. Which of the following will display all books published by Publisher 1 with a retail price of at least $25.00?

   a. SELECT * FROM books
      WHERE pubid = 1 AND retail >= 25;

   b. SELECT * FROM books
      WHERE pubid = 1 AND retail > 25;

252

c.   SELECT * FROM books
        WHERE pubid = 1 AND WHERE retail > 25;

   d.   SELECT * FROM books
        WHERE pubid=1, retail >=25;

   e.   SELECT * FROM books
        WHERE pubid = 1, retail >= $25.00;

7. What is the default sort sequence for the ORDER BY clause?

   a.   ascending

   b.   descending

   c.   the order the records are stored in the table

   d.   There is no default sort sequence.

8. Which of the following will not display books published by Publisher 2 and having a retail price of at least $35.00?

   a.   SELECT * FROM books
        WHERE pubid = 2, retail >= $35.00;

   b.   SELECT * FROM books
        WHERE pubid = 2 AND NOT retail< 35;

   c.   SELECT * FROM books
        WHERE pubid IN (1, 2, 5) AND retail NOT BETWEEN 1
        AND 29.99;

   d.   All of the above will display the specified books.

9. Which of the following will include a customer with the first name of BONITA in the results?

   a.   SELECT * FROM customers WHERE firstname = 'B%';

   b.   SELECT * FROM customers WHERE firstname LIKE '%N%';

   c.   SELECT * FROM customers WHERE firstname = '%N%';

   d.   SELECT * FROM customers WHERE firstname LIKE '_B%';

10. Which of the following characters or symbols is used to represent exactly one character during a pattern search?

   a.   C

   b.   ?

   c.   _

   d.   %

   e.   none of the above

11. Which of the following will return the book *HANDCRANKED COMPUTERS* in the results?

   a.   SELECT * FROM books WHERE title = 'H_N_%';

   b.   SELECT * FROM books WHERE title LIKE "H_N_C%";

   c.   SELECT * FROM books WHERE title LIKE 'H_N_C%';

   d.   SELECT * FROM books WHERE title LIKE '_H%';

12. Which of the following clauses is used to present the results of a query in a sorted order?

    a. WHERE

    b. SELECT

    c. SORT

    d. ORDER

    e. none of the above

13. Which of the following SQL statements will return all books published after March 20, 2005?

    a. SELECT * FROM books WHERE pubdate> 03–20–2005;

    b. SELECT * FROM books WHERE pubdate> '03–20–2005';

    c. SELECT * FROM books WHERE pubdate> '20–MAR–05';

    d. SELECT * FROM books WHERE pubdate> 'MAR–20–05';

14. Which of the following will list all books published before June 2, 2004, *and* all books either published by Publisher 4 or in the Fitness category?

    a. SELECT * FROM books
       WHERE category = 'FITNESS' OR pubid = 4
       AND pubdate < '06–02–2004';

    b. SELECT * FROM books
       WHERE category = 'FITNESS' AND pubid = 4
       OR pubdate < '06–02–2004';

    c. SELECT * FROM books
       WHERE category = 'FITNESS' OR
       (pubid = 4 AND pubdate < '06–02–2004');

    d. SELECT * FROM books
       WHERE category = 'FITNESS'
       OR pubid = 4, pubdate < '06–02–04';

    e. none of the above

15. Which of the following will find all orders placed before April 5, 2005 but that have not yet shipped?

    a. SELECT * FROM orders WHERE orderdate < '04–05–05'
       AND shipdate = NULL;

    b. SELECT * FROM orders WHERE orderdate < '05–04–05'
       AND shipdate IS NULL;

    c. SELECT * FROM orders WHERE orderdate < 'Apr–05–05'
       AND shipdate IS NULL;

    d. SELECT * FROM orders WHERE orderdate < '05–Apr–05'
       AND shipdate IS NULL;

    e. none of the above

16. Which of the following symbols represents any number of characters in a pattern search?

   a.   *

   b.   ?

   c.   %

   d.   _

17. Which of the following will list books generating at least $12.00 in profit?

   a.   SELECT * FROM books WHERE retail-cost > 12;

   b.   SELECT * FROM books WHERE retail-cost < 12;

   c.   SELECT * FROM books WHERE profit => 12;

   d.   SELECT * FROM books WHERE retail-cost => 12.00;

   e.   none of the above

18. Which of the following will list each book having a profit of at least $10.00 in descending order by profit?

   a.   SELECT * FROM books
        WHERE profit => 10.00
        ORDER BY "Profit" desc;

   b.   SELECT title, retail-cost "Profit" FROM books
        WHERE profit => 10.00
        ORDER BY "Profit" desc;

   c.   SELECT title, retail-cost "Profit" FROM books
        WHERE "Profit" => 10.00
        ORDER BY "Profit" desc;

   d.   SELECT title, retail-cost profit FROM books
        WHERE retail-cost >= 10.00
        ORDER BY "PROFIT" desc;

   e.   SELECT title, retail-cost "Profit" FROM books
        WHERE profit => 10.00
        ORDER BY 3 desc;

19. Which of the following will include the book HOW TO GET FASTER PIZZA in its results?

   a.   SELECT * FROM books WHERE title LIKE '%AS_E%';

   b.   SELECT * FROM books WHERE title LIKE 'AS_E%';

   c.   SELECT * FROM books WHERE title = '%AS_E%'

   d.   SELECT * FROM books WHERE title = 'AS_E%';

20. Which of the following will return all books published after March 20, 2005?

   a.   SELECT * FROM books WHERE pubdate> 03–20–2005;

   b.   SELECT * FROM books WHERE pubdate> '03–20–2005';

   c.   SELECT * FROM books WHERE pubdate NOT< '20–MAR–05';

   d.   SELECT * FROM books WHERE pubdate NOT < 'MAR–20–05';

   e.   none of the above

Hands-On Assignments

To perform these assignments, refer to the tables in Appendix A.

Give the SQL statements and output for the following data requests:

1. Which customers live in New Jersey? List each customer's last name, first name, and state.

2. Which orders were shipped after April 1, 2005? List each order number and the date it shipped.

3. Which books are not in the Fitness Category? List each book title and category.

4. Which customers live in either Georgia or New Jersey? Put the results in ascending order by last name. List each customer's customer number, last name, and state.

5. Which orders were placed before April 2, 2005? List each order number and order date.

6. Retrieve all authors whose last name contains the letter pattern "IN." Put the results in order of last name, then first name. List each author's last name and first name.

7. Retrieve all customers who were referred to the bookstore by another customer. List each customer's last name and the number of the customer who made the referral.

8. Retrieve the book title and category for all books in the Children and Cooking Categories. Create three different queries to accomplish this task as follows: a) Use a search pattern operation, b) use a logical operator, and c) use another operator not used in a or b.

9. Use a search pattern to find any book where the title has an "A" for the second letter in the title, and an "N" for the fourth letter. List each book's ISBN and title. Sort the list by title in descending order.

10. List the title and publish date of any computer book that was published in 2005. Accomplish this task with three different queries as follows: a) Use a range operator, b) use a logical operator, and c) use a search pattern operation.

Advanced Challenge

To perform these activities, refer to the tables in Appendix A.

During the course of an afternoon at work, you receive various requests for data stored in the database. As you complete each request, you decide to document the SQL statements you used to obtain the data so you can tell each person how you obtained the results. The following are two of the requests that were made:

1. One of the managers at JustLee Books requests a list of the titles of all books that generate a profit of at least $10.00. The manager requested that the results be listed in descending order, based upon the profit returned by each book.

2. One of the customer service representatives is trying to identify all books that are either in the Computer or Family Life Category *and* were published by Publisher 1 or Publisher 3. However, the results should not include any book that sells for less than $45.00.

For each request, create a document that identifies the SQL statement used and the results.

Case Study: *City Jail*

Note: Run the CJ_08.sql file from the Chapter 8 folder to ensure that all necessary tables and constraints are available for this case study. This script will rebuild the City Jail database. Do not be concerned with errors from the DROP TABLE commands, which delete any existing tables of the same names.

The following list reflects common data requests received from city managers. Provide the SQL statement that would satisfy the request. If the query can be accomplished using different operators, provide the alternative solutions so the performance tuning group can test them and identify the more efficient statements. Test the statements and show execution results as well.

1. List all criminal aliases that begin with the letter B.

2. List all crimes that occurred (were charged) during the month of October 2005. List the crime id, criminal id, date charged, and classification.

3. List all crimes that have a status of CA (can appeal) or IA (in appeal). List the crime id, criminal id, date charged, and status.

4. List all crimes that are classified as a felony. List the crime id, criminal id, date charged, and classification.

5. List all crimes that have a hearing date more than 14 days after the date charged. List the crime id, criminal id, date charged, and hearing date.

6. List all criminals who have a zip code of 23510. List the criminal id, last name, and zip code. Sort the list by criminal id.

7. List all crimes that do not have a hearing date scheduled. List the crime id, criminal id, date charged, and hearing date.

8. List all sentences that have a probation officer assigned. List the sentence id, criminal id, and probation officer id. Sort the list by probation officer id and then the criminal id.

9. List all crimes that are classified as misdemeanors and are currently in appeal. List the crime id, criminal id, classification, and status.

10. List all crime charges that have a balance owed. List the charge id, crime id, fine amount, court fee, amount paid, and amount owed.

11. List all police officers who are assigned to either a precinct of OCVW or GHNT and have a status of active. List the officer id, last name, precinct, and status. Sort the list by precinct then officer last name.

CHAPTER 9

JOINING DATA FROM MULTIPLE TABLES

LEARNING OBJECTIVES

After completing this chapter, you should be able to do the following:

- Identify a Cartesian join
- Create an equality join using the WHERE clause
- Create an equality join using the JOIN keyword
- Create a non-equality join using the WHERE clause
- Create a non-equality join using the JOIN...ON approach
- Create a self-join using the WHERE clause
- Create a self-join using the JOIN keyword
- Distinguish an inner join from an outer join
- Create an outer join using the WHERE clause
- Create an outer join using the OUTER keyword
- Use set operators to combine the results of multiple queries

INTRODUCTION

The main advantage of using a relational database is that you can virtually eliminate data redundancy by structuring data in multiple tables. However, this structure requires that data rows from multiple tables be combined or joined before you can perform many kinds of queries. This chapter focuses on adding join conditions, instructions in queries that combine data from more than one table.

Traditionally, Oracle database users had to include join conditions in the WHERE clause to specify how data rows of different tables are related. Beginning with Oracle 9*i*, support for ANSI-compliant joins was introduced. An ANSI-compliant join uses the JOIN keyword in the FROM clause. The ANSI JOIN method has several advantages over using a traditional WHERE clause join. It has greater portability of SQL code between different DBMS platforms as most relational databases are ANSI compliant. Also, because the WHERE clause is reserved for conditions that restrict the rows from being returned, the ANSI JOIN method has improved statement clarity.

In this chapter, you'll examine several kinds of joins. For each join, you'll examine the syntax for creating the join, first using the traditional WHERE clause approach and then using the ANSI JOIN method. You will need to understand both approaches to creating joins to support existing code, to prepare new systems with greater portability goals, and to pass the Oracle 10g SQL exam. Because there are no performance advantages associated with either of these join methods, both methods are widely used. Figure 9-1 provides an overview of this chapter's topics.

ELEMENT	DESCRIPTION
Cartesian Join Also known as a Cartesian product or cross join	Replicates each row from the first table with every row from the second table. Creates a join between tables by displaying every possible record combination. Can be created by two methods: (1) not including a joining condition in a WHERE clause (2) using the JOIN method with the CROSS JOIN keywords
Equality Join Also known as equijoin, inner join, or simple join	Creates a join through a commonly named and defined column. Can be created by two methods: (1) using the WHERE clause (2) using the JOIN method with the NATURAL JOIN or JOIN...ON or JOIN...USING keywords
Non-Equality Join	Joins tables when there are no equivalent rows in the tables to be joined, for example, to match values in one column of a table with a range of values in another table. Can be created by two methods: (1) using the WHERE clause (2) using the JOIN method with the JOIN...ON keywords
Self-Join	Joins a table to itself. Can be created by two methods: (1) using the WHERE clause (2) using the JOIN method with the JOIN...ON keywords

FIGURE 9-1 Types of joins and set operators

ELEMENT	DESCRIPTION
Outer Join	Includes records of a table in output when there is no matching record in the other table. Can be created by two methods: (1) using the WHERE clause with a (+) operator (2) using the JOIN method with the OUTER JOIN keywords with the assigned type of LEFT, RIGHT, or FULL
Set Operators	Combines results of multiple SELECT statements.Includes the keywords UNION, UNION ALL, INTERSECT, and MINUS.

FIGURE 9-1 Types of joins and set operators (continued)

NOTE

Before attempting to work through the examples provided in this chapter, run the prech09.sql file to ensure that all necessary tables and constraints are available. This script will make additions that are needed to the JustLee Books database.

CARTESIAN JOINS

In a **Cartesian join**, also called a **Cartesian product** or **cross join**, each record in the first table is matched with each record in the second table. This type of join is useful when performing certain statistical procedures for data analysis. Thus, if you have three records in the first table and four in the second table, the first record from the first table would be matched with each of the four records in the second table. Then, the second record of the first table would be matched with each of the four records from the second table, and so on. You can always identify a Cartesian join because the results will display $m * n$ rows.

As shown in Figure 9-2, a Cartesian join of Table 1 (m) and Table 2 (n) would result in 12 records being displayed.

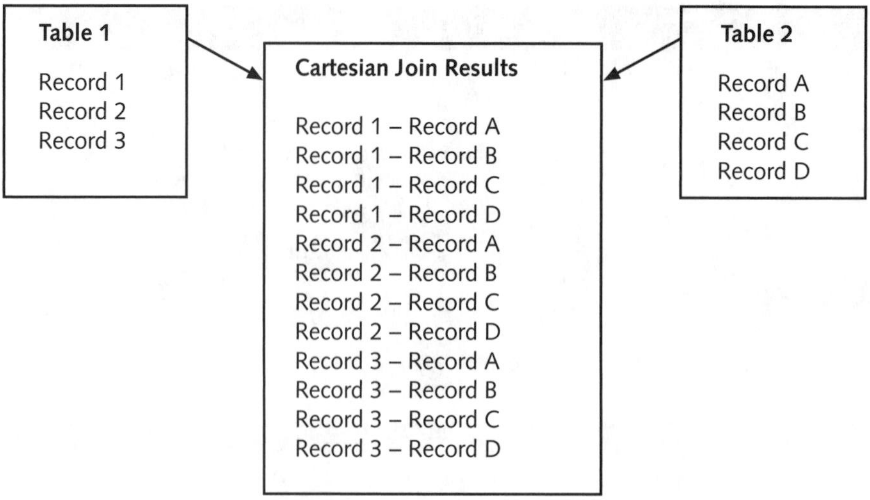

Table 1	Cartesian Join Results	Table 2
Record 1		Record A
Record 2	Record 1 – Record A	Record B
Record 3	Record 1 – Record B	Record C
	Record 1 – Record C	Record D
	Record 1 – Record D	
	Record 2 – Record A	
	Record 2 – Record B	
	Record 2 – Record C	
	Record 2 – Record D	
	Record 3 – Record A	
	Record 3 – Record B	
	Record 3 – Record C	
	Record 3 – Record D	

FIGURE 9-2 Results of a Cartesian join

Sometimes, a user intends to generate a Cartesian product, but this is usually the exception rather than the rule. Be aware that selecting data from multiple tables in a query and accidentally omitting a proper join also produces a Cartesian product, which, in this case, is an incorrect result. Most Cartesian products are produced in error.

Cartesian Join—Traditional Method

JustLee Books needs to perform a manual book inventory in all three of their book warehouses. The manager has requested an inventory sheet listing each book for each warehouse along with a column to record the physical count. In this case, the WAREHOUSES table can be joined with the BOOKS table to produce a Cartesian product. Figure 9-3 displays the query including the two tables along with the results. (Only the final screen of the results is shown to conserve space.)

```
SELECT isbn, title, location, '      ' Count
  FROM books, warehouses
  ORDER BY location, title;
```

(Execute) (Load Script) (Save Script) (Cancel)

ISBN	TITLE	LOCATION	COUNT
3957136468	HOLY GRAIL OF ORACLE	Norfolk	
0132149871	HOW TO GET FASTER PIZZA	Norfolk	
9247381001	HOW TO MANAGE THE MANAGER	Norfolk	
2491748320	PAINLESS CHILD-REARING	Norfolk	
0401140733	REVENGE OF MICKEY	Norfolk	
2147428890	SHORTEST POEMS	Norfolk	
0299282519	THE WOK WAY TO COOK	Norfolk	
8117949391	BIG BEAR AND LITTLE DOVE	San Diego	
1059831198	BODYBUILD IN 10 MINUTES A DAY	San Diego	
4981341710	BUILDING A CAR WITH TOOTHPICKS	San Diego	
3437212490	COOKING WITH MUSHROOMS	San Diego	
8843172113	DATABASE IMPLEMENTATION	San Diego	
9959789321	E-BUSINESS THE EASY WAY	San Diego	
1915762492	HANDCRANKED COMPUTERS	San Diego	
3957136468	HOLY GRAIL OF ORACLE	San Diego	
0132149871	HOW TO GET FASTER PIZZA	San Diego	
9247381001	HOW TO MANAGE THE MANAGER	San Diego	
2491748320	PAINLESS CHILD-REARING	San Diego	
0401140733	REVENGE OF MICKEY	San Diego	
2147428890	SHORTEST POEMS	San Diego	
0299282519	THE WOK WAY TO COOK	San Diego	

42 rows selected.

FIGURE 9-3 Producing a desired Cartesian product (partial output shown)

The results include 42 rows—3 rows (WAREHOUSES table) * 14 rows (BOOKS table) = 42 rows. In this case, a Cartesian product produces the desired results. However, suppose that you need to find the publisher's name for each book in inventory. The SELECT statement in Figure 9-4 instructs Oracle 10g to list the Title column, which is stored in the BOOKS table, and the Name column, which is stored in the PUBLISHER table. (Only the final screen of the results is shown to conserve space.)

```
SELECT title, name
  FROM books, publisher;
```

Execute   Load Script   Save Script   Cancel

HOW TO MANAGE THE MANAGER	READING MATERIALS INC.
SHORTEST POEMS	READING MATERIALS INC.
BODYBUILD IN 10 MINUTES A DAY	REED-N-RITE
REVENGE OF MICKEY	REED-N-RITE
BUILDING A CAR WITH TOOTHPICKS	REED-N-RITE
DATABASE IMPLEMENTATION	REED-N-RITE
COOKING WITH MUSHROOMS	REED-N-RITE
HOLY GRAIL OF ORACLE	REED-N-RITE
HANDCRANKED COMPUTERS	REED-N-RITE
E-BUSINESS THE EASY WAY	REED-N-RITE
PAINLESS CHILD-REARING	REED-N-RITE
THE WOK WAY TO COOK	REED-N-RITE
BIG BEAR AND LITTLE DOVE	REED-N-RITE
HOW TO GET FASTER PIZZA	REED-N-RITE
HOW TO MANAGE THE MANAGER	REED-N-RITE
SHORTEST POEMS	REED-N-RITE

70 rows selected.

FIGURE 9-4   Producing an undesired Cartesian product (partial output shown)

Although there are only 14 book titles in the database, 70 records were returned! This should lead you to be suspicious of the nature of the output. The problem with the SQL statement in Figure 9-4 is that you have specified the columns to be retrieved from the two tables, but not that the tables have a common field, which should be used to join the tables. Because the software does not "know" how the tables are related, it automatically replicates every possible combination of records, producing a Cartesian product. Later in this chapter, you will learn how to instruct the system to properly match or join rows of multiple tables based on column values.

Cartesian Join—JOIN Method

Beginning with Oracle 10*g*, the **CROSS** keyword, combined with the JOIN keyword, can be used in the FROM clause to explicitly instruct Oracle 10*g* to create a Cartesian, or cross, join. The CROSS keyword instructs Oracle 10*g* to create cross-products, using all the records of the tables listed. Figure 9-5 displays the book inventory listing from the earlier example produced using the JOIN method.

Notice the syntax of the SQL statement shown in Figure 9-5. In the FROM clause, the names of the tables to be used in the Cartesian join are separated by the **CROSS JOIN** keywords. Do not use commas to separate any parts of the FROM clause, as you would using the traditional method.

```
SELECT isbn, title, location, '      ' Count
  FROM books CROSS JOIN warehouses
  ORDER BY location, title;
```

| Execute | Load Script | Save Script | Cancel |

ISBN	TITLE	LOCATION	COUNT
2491748320	PAINLESS CHILD-REARING	Norfolk	
0401140733	REVENGE OF MICKEY	Norfolk	
2147428890	SHORTEST POEMS	Norfolk	
0299282519	THE WOK WAY TO COOK	Norfolk	
8117949391	BIG BEAR AND LITTLE DOVE	San Diego	
1059831198	BODYBUILD IN 10 MINUTES A DAY	San Diego	
4981341710	BUILDING A CAR WITH TOOTHPICKS	San Diego	
3437212490	COOKING WITH MUSHROOMS	San Diego	
8843172113	DATABASE IMPLEMENTATION	San Diego	
9959789321	E-BUSINESS THE EASY WAY	San Diego	
1915762492	HANDCRANKED COMPUTERS	San Diego	
3957136468	HOLY GRAIL OF ORACLE	San Diego	
0132149871	HOW TO GET FASTER PIZZA	San Diego	
9247381001	HOW TO MANAGE THE MANAGER	San Diego	
2491748320	PAINLESS CHILD-REARING	San Diego	
0401140733	REVENGE OF MICKEY	San Diego	
2147428890	SHORTEST POEMS	San Diego	
0299282519	THE WOK WAY TO COOK	San Diego	

42 rows selected.

FIGURE 9-5 Using a CROSS JOIN (partial output shown)

NOTE

If you receive an error message, make certain there is *not* a comma entered after the BOOKS table name in the FROM clause.

EQUALITY JOINS

The query in Figure 9-4 returned an undesired Cartesian join because the software did not know what data the two tables had in common. The most common type of join that you will use in the workplace is based upon two (or more) tables having equivalent data stored in a common column. Such joins are called **equality joins**. They are also referred to as **equijoins**, **inner joins**, or **simple joins**.

A **common column** is a column with equivalent data that exists in two or more tables. For example, the BOOKS and PUBLISHER tables both have a common column called Pubid that contains an identification code assigned to each publisher. Thus, when you wanted a list of publishers for each book in the BOOKS table, you wanted to match the publisher ID stored in the record of each book in the BOOKS table with the corresponding publisher ID in the PUBLISHER table. The results *should* have included only the name of the publisher whenever there was a match between the Pubid columns stored in each table.

To successfully master join operations, a solid understanding of the database structure is necessary. An E-R Model, as developed in Chapter 1, will identify table relationships. Figure 9-6 displays the E-R Model for JustLee Books. Each of the one-to-many relationship lines should be supported with a foreign key (FK) constraint to ensure the consistency of data for the common columns.

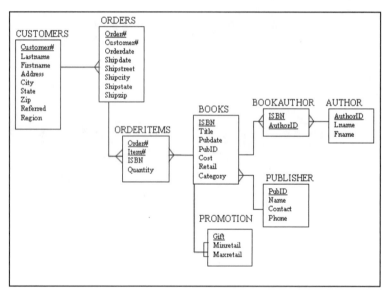

Note: Underlines denote primary key columns

FIGURE 9-6 JustLee Book's table structure

A review of referential integrity or foreign key (FK) constraints allows the identification of the common columns between tables. Keep in mind, even though common columns typically have the same name, this is not a requirement; you cannot depend on column names alone. Review the FK constraints in the JustLee Book's table creation script from Chapter 8 (prech08.sql) to identify the common column for each relationship line in the E-R Model. These common columns will be used in query join conditions to instruct the system how to logically relate the rows of multiple tables.

> **NOTE**
>
> Do not be concerned with the relationship between the BOOKS and PROMOTION tables at this point. This is addressed in the section on non-equality joins later in this chapter.

The following section demonstrates how to create joins based upon a common column containing equivalent data stored in multiple tables.

Equality Joins—Traditional Method

The traditional way to avoid the problem of an unintended Cartesian join is to use the WHERE clause. The WHERE clause is used to instruct Oracle 10g how to correctly join tables. We have previously used the WHERE clause to provide conditions that restrict the rows affected by the SQL statement. Accomplishing a traditional join adds the join conditions along with other needed conditions in the WHERE clause. In other words, the WHERE clause can perform two different activities: joining tables and providing conditions to limit the rows affected.

Using the same scenario described in the section on Cartesian joins (Figure 9-4), let's include the WHERE clause to retrieve the correct results, as shown in Figure 9-7.

The WHERE clause tells Oracle 10g that the BOOKS table and the PUBLISHER table are related by the Pubid column. The equal sign specifies that the contents of the Pubid column in each table must be exactly equal for the rows to be joined and returned in the results. Also notice that in the WHERE clause in Figure 9-7, the Pubid column names were prefixed with their corresponding table names. Any time Oracle 10g references multiple tables having the same column name, the column name *must* be prefixed with the table name. You will receive an error message if your query is ambiguous and does not specify exactly which column is the common column. By entering **publisher.pubid**, you are specifying the Pubid column within the PUBLISHER table. This is known as "qualifying" the column name. A **column qualifier** indicates the table containing the column being referenced.

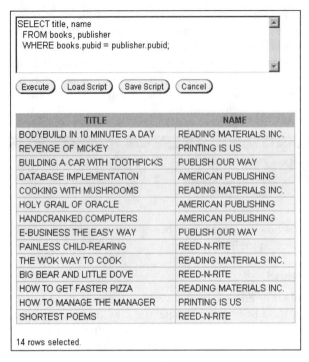

FIGURE 9-7   An equality join

Suppose that you wanted some additional information—the publisher ID—to be included in the output. If the publisher ID had also been listed in the SELECT clause to be included in the query output, the table name prefix would be needed. If the table prefix is omitted, you will receive a column ambiguity error message, as shown in Figure 9-8.

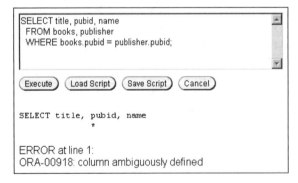

FIGURE 9-8   Column ambiguously defined error

In Figure 9-8, it does not matter whether the Pubid column is qualified with the BOOKS table or the PUBLISHER table. The Pubid column is the common column of these two tables and will be the same value within a proper join operation.

Search conditions can be added to the WHERE clause along with join conditions, as shown in Figure 9-9. Notice the AND logical operator in the WHERE clause. The inclusion of this operator limits query results to only those from Publisher 4. Any of the search conditions used in Chapter 8 can be issued in the WHERE clause when you are joining a table.

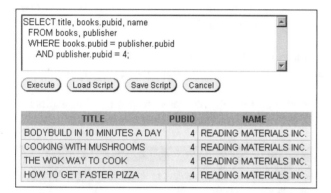

FIGURE 9-9   Including search and join conditions in a WHERE clause

Table aliases can be used to simplify the process of qualifying columns with the table name. In Figure 9-10, the SELECT statement requests the title, publisher's ID number, and publisher's name for any book costing less than $15.00 or any book from Publisher 1.

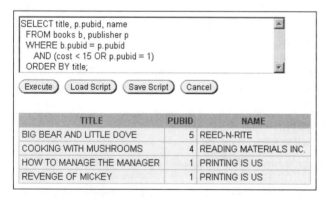

FIGURE 9-10   Equality join using table aliases

Let's take a look at some of the elements in Figure 9-10.

- The SELECT clause not only lists the columns to be displayed, it also includes a **table alias** for the PUBLISHER table (p). A period is used to separate a table alias from a column name, as it is when you qualify a column with the full table name. This alias was assigned in the FROM clause (and is discussed next).
- The table alias in the FROM clause works like a column alias by temporarily giving a different name to a table. Table aliases offer a couple of advantages. First, they improve processing efficiency, as the system no longer needs to identify which table a specified column is in. Second, coding is simplified when the whole table name does not have to be indicated (although a table alias can have as many as 30 characters). There is one important rule you must remember when using a table alias: *If a table alias is assigned in the FROM clause, it must be used any time the table is referenced in that SQL statement.*
- The WHERE clause includes not only the join condition for the BOOKS and PUBLISHER tables, but also other search conditions using the AND and OR logical operators.
- The statement concludes with an ORDER BY clause to display the results in a sorted order.

N O T E

Make certain to use the letter "p" (the alias for the PUBLISHER table) or the letter "b" (the alias for the BOOKS table) before the Pubid column name, or you will receive an error message.

Up to this point, our join examples have only included two tables. But suppose we need a list of all customer names along with all of the books each customer has purchased. The book titles are in the BOOKS table and the customer names are in the CUSTOMERS table. Would we join the BOOKS and CUSTOMERS tables to produce the needed list? NO! Joins need to follow logical relationships between tables. In this case, the BOOKS and CUS-TOMERS tables are not directly related. Database diagrams or E-R Models as presented in Chapter 1 are very helpful in planning join operations as table relationship lines are displayed.

In this scenario, the join operation needs to include four tables: CUSTOMERS, ORDERS, ORDERITEMS, and BOOKS. This requires three join operations as follows:

1. Join CUSTOMERS to ORDERS based on Customer#
2. Join ORDERS to ORDERITEMS based on Order#
3. Join ORDERITEMS to BOOKS based on ISBN

Multiple join operations are included in a WHERE clause by using the AND logical operator, as displayed in Figure 9-11.

Keep in mind that even though no columns are specifically selected for output from the ORDERS and ORDERITEMS tables, these tables are needed in the query to logically join tables together.

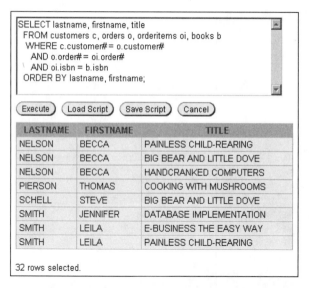

```
SELECT lastname, firstname, title
  FROM customers c, orders o, orderitems oi, books b
  WHERE c.customer# = o.customer#
    AND o.order# = oi.order#
    AND oi.isbn = b.isbn
  ORDER BY lastname, firstname;
```

(Execute) (Load Script) (Save Script) (Cancel)

LASTNAME	FIRSTNAME	TITLE
NELSON	BECCA	PAINLESS CHILD-REARING
NELSON	BECCA	BIG BEAR AND LITTLE DOVE
NELSON	BECCA	HANDCRANKED COMPUTERS
PIERSON	THOMAS	COOKING WITH MUSHROOMS
SCHELL	STEVE	BIG BEAR AND LITTLE DOVE
SMITH	JENNIFER	DATABASE IMPLEMENTATION
SMITH	LEILA	E-BUSINESS THE EASY WAY
SMITH	LEILA	PAINLESS CHILD-REARING

32 rows selected.

FIGURE 9-11   Joining four tables (partial output shown)

NOTE

The number of join operations needed is the number of tables in the query minus one.

Regardless of how many join operations are required, other conditions can still be added in the WHERE clause. For example, Figure 9-12 shows the previous query modified to select books only in the computer category. Notice that the results have reduced from 32 rows to 10 rows of output.

Equality Joins—JOIN Method

You can use three approaches to create an equality join that uses the JOIN keyword: NATURAL JOIN, JOIN...USING, and JOIN...ON.

1. The **NATURAL JOIN** keywords automatically create a join between two tables based on common named fields.
2. The **USING** clause allows you to create joins based on a column that has the same name and definition in both tables.
3. When the tables to be joined in a USING clause do not have a commonly named and defined field, you must add the **ON** clause to the JOIN keyword to specify how the tables are related.

The query in Figure 9-13 uses the NATURAL JOIN keywords to instruct Oracle 10g to list the book title of each book in the BOOKS table—and the corresponding publisher ID number and publisher name.

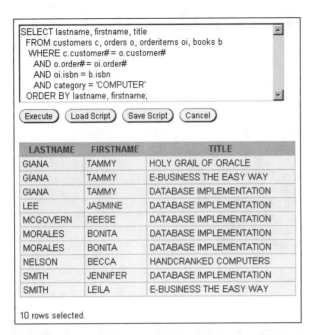

FIGURE 9-12   Multiple joins combined with a search condition

FIGURE 9-13   Using a NATURAL JOIN

Because both the BOOKS and the PUBLISHER tables contain the Pubid column, this column is a common column and should be used to relate the two tables. When using the NATURAL JOIN keywords, you are not required to specify which column(s) the two tables have in common. The NATURAL keyword implies that the two specified tables have at least one column in common with the same name and contain the same data type. Oracle 10g will compare the two tables and use the common column(s) to join the table.

Unlike the traditional method, you are not allowed to use a column qualifier for the column used to create the join. In essence, because the data value in a column is equivalent in both tables when the records match, it does not make sense to identify the column from only one of the tables. Thus, Oracle 10g returns an error message if a qualifier is used anywhere in a SELECT statement that includes the NATURAL JOIN keywords, as shown in Figure 9-14.

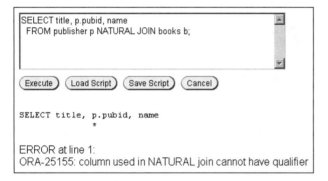

FIGURE 9-14   Column qualifier error with a NATURAL JOIN

Most developers avoid using the NATURAL JOIN because it can cause unexpected results. What if a column named "Description" is added to both the BOOKS and PUBLISHER tables but the columns are not related to each other? A NATURAL JOIN attempts to use these columns in a join operation even though they have no relationship. Developers prefer a join that explicitly specifies what column(s) are to be used to join the rows. This is accomplished by including a USING clause immediately after the FROM clause. Figure 9-15 reissues the previous query using the JOIN...USING keywords. Notice that the query returns the same results as the NATURAL JOIN keywords.

As with the NATURAL JOIN keywords, a column referenced by a USING clause cannot contain a column qualifier anywhere in the SELECT statement. In addition, the column referenced in the USING clause must be enclosed in parentheses.

There might be instances in which you have created tables with common columns, but without a common name. When there are no commonly named columns to use in a join, you need to use the JOIN keyword in the FROM clause and add an ON clause immediately after the FROM clause to specify which fields are related. To demonstrate this join, we created a PUBLISHER2 table, which contains everything in the PUBLISHER table except that the Pubid column is named Id. Figure 9-16 uses the same scenario from the previous examples, listing the Title, the Publisher ID, and the publisher Name columns from the BOOKS and PUBLISHER2 tables. The ON clause in this case instructs the system to join

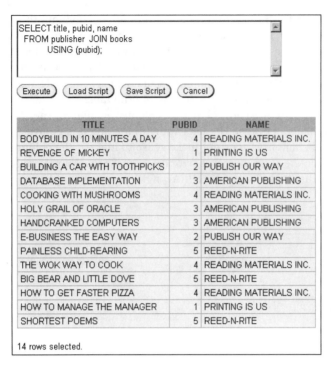

```
SELECT title, pubid, name
  FROM publisher JOIN books
       USING (pubid);
```

Execute   Load Script   Save Script   Cancel

TITLE	PUBID	NAME
BODYBUILD IN 10 MINUTES A DAY	4	READING MATERIALS INC.
REVENGE OF MICKEY	1	PRINTING IS US
BUILDING A CAR WITH TOOTHPICKS	2	PUBLISH OUR WAY
DATABASE IMPLEMENTATION	3	AMERICAN PUBLISHING
COOKING WITH MUSHROOMS	4	READING MATERIALS INC.
HOLY GRAIL OF ORACLE	3	AMERICAN PUBLISHING
HANDCRANKED COMPUTERS	3	AMERICAN PUBLISHING
E-BUSINESS THE EASY WAY	2	PUBLISH OUR WAY
PAINLESS CHILD-REARING	5	REED-N-RITE
THE WOK WAY TO COOK	4	READING MATERIALS INC.
BIG BEAR AND LITTLE DOVE	5	REED-N-RITE
HOW TO GET FASTER PIZZA	4	READING MATERIALS INC.
HOW TO MANAGE THE MANAGER	1	PRINTING IS US
SHORTEST POEMS	5	REED-N-RITE

14 rows selected.

FIGURE 9-15 Performing a join with the JOIN...USING keywords

rows using the Pubid column of the BOOKS table (using the alias "b") and the Id column of the PUBLISHER2 table (using the alias "p"). Because it is possible that ambiguity can exist when referencing columns with the JOIN...ON keywords, Oracle 10g requires the use of column qualifiers to avoid ambiguity.

NOTE

The ON clause is very similar to the traditional WHERE clause join, and, like it, requires a table alias before the column names in the ON clause, or you receive an ambiguity error.

Using the ON clause or USING clause in a SELECT statement provides you with the freedom of using the WHERE clause exclusively for restricting the rows to be included in the results. This can improve the readability of complex SELECT statements. Figure 9-17 shows how to use the JOIN...ON approach to return the title, publisher ID (Pubid), and publisher name for all books published by Publisher 4.

There are two main differences between using the USING and ON clauses with the JOIN keyword.

1. The USING clause can be used *only* if the tables being joined have a common column with the same name. This is not a requirement for the ON clause.
2. A condition is specified in the ON clause; this is not allowed in the USING clause. The USING clause can contain only the name of the common column.

```
SELECT title, pubid, name
  FROM publisher2 p JOIN books b
     ON p.id = b.pubid;
```

(Execute) (Load Script) (Save Script) (Cancel)

TITLE	PUBID	NAME
BODYBUILD IN 10 MINUTES A DAY	4	READING MATERIALS INC.
REVENGE OF MICKEY	1	PRINTING IS US
BUILDING A CAR WITH TOOTHPICKS	2	PUBLISH OUR WAY
DATABASE IMPLEMENTATION	3	AMERICAN PUBLISHING
COOKING WITH MUSHROOMS	4	READING MATERIALS INC.
HOLY GRAIL OF ORACLE	3	AMERICAN PUBLISHING
HANDCRANKED COMPUTERS	3	AMERICAN PUBLISHING
E-BUSINESS THE EASY WAY	2	PUBLISH OUR WAY
PAINLESS CHILD-REARING	5	REED-N-RITE
THE WOK WAY TO COOK	4	READING MATERIALS INC.
BIG BEAR AND LITTLE DOVE	5	REED-N-RITE
HOW TO GET FASTER PIZZA	4	READING MATERIALS INC.
HOW TO MANAGE THE MANAGER	1	PRINTING IS US
SHORTEST POEMS	5	REED-N-RITE

14 rows selected.

FIGURE 9-16   Performing a join with the JOIN...ON keywords

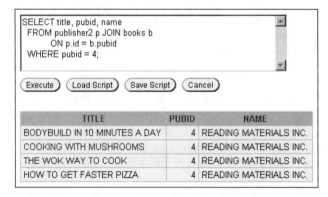

```
SELECT title, pubid, name
  FROM publisher2 p JOIN books b
     ON p.id = b.pubid
  WHERE pubid = 4;
```

(Execute) (Load Script) (Save Script) (Cancel)

TITLE	PUBID	NAME
BODYBUILD IN 10 MINUTES A DAY	4	READING MATERIALS INC.
COOKING WITH MUSHROOMS	4	READING MATERIALS INC.
THE WOK WAY TO COOK	4	READING MATERIALS INC.
HOW TO GET FASTER PIZZA	4	READING MATERIALS INC.

FIGURE 9-17   The JOIN method reserves the WHERE clause for search conditions

To demonstrate how to accomplish ANSI joins involving more than two tables, let's return to the request in which we need a list of all customer names with all the book titles each customer has purchased from the computer category. To peform this join operation, we need to include four tables: CUSTOMERS, ORDERS, ORDERITEMS, and BOOKS. The query in Figure 9-18 shows the required join operation with the USING clause.

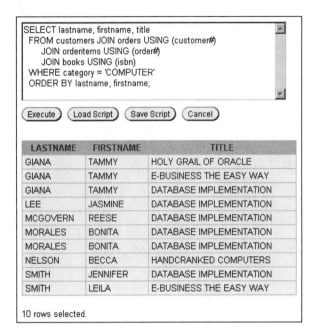

```
SELECT lastname, firstname, title
  FROM customers JOIN orders USING (customer#)
      JOIN orderitems USING (order#)
      JOIN books USING (isbn)
  WHERE category = 'COMPUTER'
  ORDER BY lastname, firstname;
```

(Execute) (Load Script) (Save Script) (Cancel)

LASTNAME	FIRSTNAME	TITLE
GIANA	TAMMY	HOLY GRAIL OF ORACLE
GIANA	TAMMY	E-BUSINESS THE EASY WAY
GIANA	TAMMY	DATABASE IMPLEMENTATION
LEE	JASMINE	DATABASE IMPLEMENTATION
MCGOVERN	REESE	DATABASE IMPLEMENTATION
MORALES	BONITA	DATABASE IMPLEMENTATION
MORALES	BONITA	DATABASE IMPLEMENTATION
NELSON	BECCA	HANDCRANKED COMPUTERS
SMITH	JENNIFER	DATABASE IMPLEMENTATION
SMITH	LEILA	E-BUSINESS THE EASY WAY

10 rows selected.

FIGURE 9-18 Multiple joins combined with a search condition

NON-EQUALITY JOINS

With an equality join, the data value of a record stored in the common column for the first table must match the data value in the second table. However, there are many cases in which there will be no exact match. A **non-equality join** is used when the related columns cannot be joined through the use of an equal sign—when there are no equivalent rows in the tables to be joined. For example, the shipping fee charged by many freight companies is based on the weight of the item being shipped. To use a database table to determine shipping fees, you could store every possible weight and its corresponding fee in a table. Then, whenever an item is shipped, you could use an equality join to match the weight of that particular item to the equivalent weight stored in the table and find the correct fee. However, most shipping fees are based on a scale, or range, of weights. For example, an item weighing between three and five pounds might have one fee, whereas an item weighing between five and eight pounds might have another fee.

A non-equality join enables you to store the minimum value for a range in one column of a record, and the maximum value for the range in another column. Thus, instead of finding a column-to-column match, you can use a non-equality join to determine whether the item being shipped falls between minimum and maximum ranges in the columns. If the join does find a matching range for the item, the corresponding shipping fee can be returned in the results.

As with the traditional method of equality joins, a non-equality join can be performed in a WHERE clause. In addition, the JOIN keyword can be used with the ON clause to specify the relevant columns for the join. Let's first look at creating a non-equality join with the WHERE clause, and then with the JOIN...ON approach.

Non-Equality Joins—Traditional Method

Once a year, JustLee Books offers a weeklong promotion in which customers receive a gift based on the value of each book purchased. If a customer purchases a book with a retail price of $12 or less, the customer receives a bookmark. If the retail price is more than $12 but less than or equal to $25, the customer receives a box of book-owner labels. For books retailing for more than $25 and less than or equal to $56, the customer is entitled to a free book cover. For books retailing for more than $56, the customer reveives free shipping. Figure 9-19 shows the PROMOTION table.

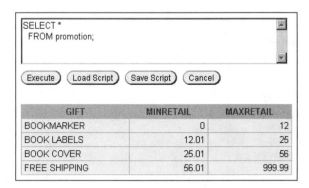

FIGURE 9-19   Contents of the PROMOTION table

Because the rows in the BOOKS and PROMOTION tables do not contain equivalent values, you are required to use a non-equality join to determine which gift a customer will receive during the promotion, as shown in Figure 9-20.

As shown in Figure 9-20, the BETWEEN operator is used in the WHERE clause to determine the range in which the retail price of the book falls. Then, based on the range set between the Minretail and Maxretail columns, the query determines which gift is appropriate for each purchase.

Note that when you use a non-equality join to determine where a value falls within a range, *you must make certain none of the values overlap*. If you select all the records from the PROMOTION table to see the values stored in each field, you will notice that the Minretail value in one row of the PROMOTION table does not equal the Maxretail value in another row. If any of the values did overlap, a customer could be returned twice in the results (and receive two gifts rather than one). You should check output from a non-equality join to make certain that rows that appear more than once aren't the result of overlapping values in your ranges.

Non-Equality Joins—JOIN Method

A non-equality join using the JOIN keyword has the same syntax as an equality join with the JOIN keyword. The only difference is that an equal sign is not used to establish the relationship in the ON clause. Figure 9-21 displays the gift for each book using the JOIN keyword. Notice that the joining condition used in the ON clause is the same as the one used in the WHERE clause in Figure 9-20.

FIGURE 9-20   A traditional non-equality join

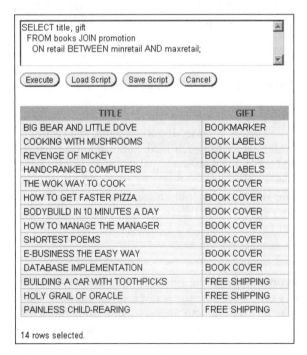

FIGURE 9-21   A JOIN method non-equality join

SELF-JOINS

Sometimes, data in one column of a table has a relationship with another column within the same table. For example, customers who refer a new customer to JustLee Books receive a discount certificate for a future purchase. The Referred column of the CUSTOMERS table stores the customer number of the individual who referred the new customer. If you need to determine the name of the customer who referred another customer, you face a problem: The CUSTOMERS table serves as the master table for all customer information. Thus, the Referred column in the CUSTOMERS table relates to other rows within the same table. To retrieve all the information you need, you must join a table to itself. In this case, you would need to join the Referred and Customer# columns of the CUSTOMERS table, as depicted in Figure 9-22. This is known as a **self-join**. You can create a self-join using either a WHERE clause or the JOIN keyword with the ON clause.

Customer 1003 (Leila Smith) has referred two customers (Tammy Giana and Jorge Perez)

CUSTOMER#	LASTNAME	FIRSTNAME	ADDRESS	CITY	STATE	ZIP	REFERRED
1001	MORALES	BONITA	P.O. BOX 651	EASTPOINT	FL	32328	
1002	THOMPSON	RYAN	P.O. BOX 9835	SANTA MONICA	CA	90404	
1003	SMITH	LEILA	P.O. BOX 66	TALLAHASSEE	FL	32306	
1004	PIERSON	THOMAS	69821 SOUTH AVENUE	BOISE	ID	83707	
1005	GIRARD	CINDY	P.O. BOX 851	SEATTLE	WA	98115	
1006	CRUZ	MESHIA	82 DIRT ROAD	ALBANY	NY	12211	
1007	GIANA	TAMMY	9153 MAIN STREET	AUSTIN	TX	78710	1003
1008	JONES	KENNETH	P.O. BOX 137	CHEYENNE	WY	82003	
1009	PEREZ	JORGE	P.O. BOX 8564	BURBANK	CA	91510	1003
1010	LUCAS	JAKE	114 EAST SAVANNAH	ATLANTA	GA	30314	
1011	MCGOVERN	REESE	P.O. BOX 18	CHICAGO	IL	60606	
1012	MCKENZIE	WILLIAM	P.O. BOX 971	BOSTON	MA	02110	
1013	NGUYEN	NICHOLAS	357 WHITE EAGLE AVE.	CLERMONT	FL	34711	1006

Customer 1006 (Meshia Cruz) has referred one customer (Nicholas Nguyen)

FIGURE 9-22   Two columns of the same table are related (Partial table shown)

Self-Joins—Traditional Method

To perform a self-join, you list the CUSTOMERS table twice in the FROM statement. However, you must make it appear as if the query is referencing two different tables. To do this, you must assign each listing of the CUSTOMERS table a different table alias. In Figure 9-23, the query uses the table alias "c" to identify the table containing the information for the new customer, and the table alias "r" to identify the table storing the individual who referred the new customer. Because the table aliases are different, Oracle 10g operates as if two copies of the CUSTOMERS table exist and is able to examine and match up different records within the same table while executing the query.

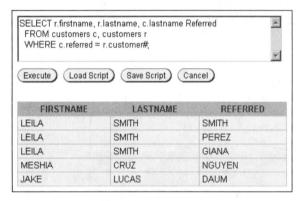

FIGURE 9-23   A self-join constructed in the WHERE clause

NOTE

If an error message is returned, make certain the CUSTOMERS table is listed twice in the FROM clause, each with a different table alias. Also, remember to precede each column with the table alias so there are no ambiguity errors.

Self-Joins—JOIN Method

Regardless of the method used, the concept behind a self-join is the same—to use table aliases to make it appear that you are joining two different tables. To demonstrate a self-join using the JOIN keyword, let's use the same circumstances as discussed in the previous section, in which we need to determine which customers referred new customers to JustLee Books. The self-join query using the JOIN keyword is shown in Figure 9-24.

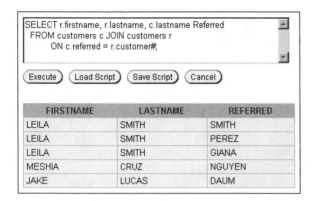

```
SELECT r.firstname, r.lastname, c.lastname Referred
    FROM customers c JOIN customers r
        ON c.referred = r.customer#;
```

(Execute) (Load Script) (Save Script) (Cancel)

FIRSTNAME	LASTNAME	REFERRED
LEILA	SMITH	SMITH
LEILA	SMITH	PEREZ
LEILA	SMITH	GIANA
MESHIA	CRUZ	NGUYEN
JAKE	LUCAS	DAUM

FIGURE 9-24 A self-join using the JOIN...ON keywords

As previously discussed, we listed the table twice in the FROM clause, but gave each listing a different table alias to mimic using two different tables. The columns used to relate the two occurrences of the table are identified in the ON clause. When you use the JOIN...ON approach to create a self-join, you can still place row restrictions in a WHERE clause. This makes it easy for anyone viewing the statement to know which portion of the statement is being used to join the tables, and which portion is limiting the number of rows being returned.

OUTER JOINS

When performing the preceding equality, non-equality, and self-joins, a row was only returned if there was a corresponding record in each table queried. These types of joins can be categorized as **inner joins** because records are listed in the results only if a match is found in each table. In fact, the default **INNER** keyword can be included with the JOIN keyword to specify that only records having a matching row in the corresponding table should be returned in the results.

However, suppose that you wanted a list of *all* customers (not just the ones who've placed an order) and the order number(s) of orders the customers have recently placed? (Recall that the CUSTOMERS table lists all customers who have ever placed an order, but the ORDERS table lists just the current month's orders and unfilled orders from previous months.) An inner join might not give you the exact results you desire, because some customers might not have placed a recent order.

The query in Figure 9-25 shows an equality join that will return all order numbers stored in the ORDERS table and the name of the customer placing the order.

```
SELECT lastname, firstname, order#
  FROM customers c, orders o
  WHERE c.customer# = o.customer#
  ORDER BY c.customer#;
```

(Execute) (Load Script) (Save Script) (Cancel)

LASTNAME	FIRSTNAME	ORDER#
MORALES	BONITA	1003
MORALES	BONITA	1018
SMITH	LEILA	1006
SMITH	LEILA	1016
PIERSON	THOMAS	1008
GIRARD	CINDY	1000
GIRARD	CINDY	1009
GIANA	TAMMY	1007
GIANA	TAMMY	1014
JONES	KENNETH	1020
LUCAS	JAKE	1001
LUCAS	JAKE	1011
MCGOVERN	REESE	1002
LEE	JASMINE	1013
SCHELL	STEVE	1017
NELSON	BECCA	1012
MONTIASA	GREG	1005
MONTIASA	GREG	1019
SMITH	JENNIFER	1010
FALAH	KENNETH	1004
FALAH	KENNETH	1015

FIGURE 9-25 An inner join drops nonmatching rows

Although this query identifies any customer who has placed an order that is stored in the ORDERS table, it does not list customers who have not recently placed an order. Your instructions specified a list of *all* customers, so you will need to make a change to the query you just issued. When you need to include records in the results of a joining query that exist in one table but do not have a corresponding row in the other table, you need to use an outer join. The keywords **OUTER JOIN** instruct Oracle 10g to include records of a table in the output even if there is no matching record in the other table. In essence, Oracle 10g will join the "dangling" record to a NULL record in the other table. An outer join can be created by using either the WHERE clause with an **outer join operator** (+) or the OUTER JOIN keywords.

Outer Joins—Traditional Method

To tell Oracle 10g to create NULL rows for records that do not have a matching row, use an outer join operator, which is a plus sign within parentheses (+). It is placed in the joining condition of the WHERE clause immediately after the column name of the table that is missing the corresponding row (this tells the software to create a NULL row in that table to join with the row in the other table).

Figure 9-26 corrects the problem in the customer query from Figure 9-25. It shows a list of all customers, and for those who have placed orders, it shows the corresponding order number(s).

```
SELECT lastname, firstname, order#
  FROM customers c, orders o
  WHERE c.customer# = o.customer#(+)
  ORDER BY c.customer#;
```

(Execute) (Load Script) (Save Script) (Cancel)

LASTNAME	FIRSTNAME	ORDER#
MORALES	BONITA	1003
MORALES	BONITA	1018
THOMPSON	RYAN	
SMITH	LEILA	1006
SMITH	LEILA	1016
PIERSON	THOMAS	1008
GIRARD	CINDY	1000
GIRARD	CINDY	1009
CRUZ	MESHIA	
GIANA	TAMMY	1007
GIANA	TAMMY	1014
JONES	KENNETH	1020
PEREZ	JORGE	
LUCAS	JAKE	1001
LUCAS	JAKE	1011
MCGOVERN	REESE	1002
MCKENZIE	WILLIAM	
NGUYEN	NICHOLAS	
LEE	JASMINE	1013
SCHELL	STEVE	1017
DAUM	MICHELL	
NELSON	BECCA	1012
MONTIASA	GREG	1005
MONTIASA	GREG	1019
SMITH	JENNIFER	1010
FALAH	KENNETH	1004
FALAH	KENNETH	1015

FIGURE 9-26   A traditional outer join using the (+) operator

If a customer in the CUSTOMERS table has not placed a recent order, the customer doesn't appear in the ORDERS table—we refer to the ORDERS table as the deficient table (that is, the table with the missing data). Therefore, the outer join operator (+) is placed immediately after the reference to the deficient ORDERS table in the WHERE clause.

You need to remember two rules when working with the traditional approach to outer joins:

1. The outer join operator can be used for only *one* table in the joining condition. In other words, using the traditional approach, you cannot create NULL rows in both tables at the same time.
2. A condition that includes the outer join operator cannot use the IN or the OR operator because this would imply that a row should be shown in the results if it matches a row in the other table, or if it matches some other given condition.

283

Joining Data from Multiple Tables

Outer Joins—JOIN Method

When creating a traditional outer join with the outer join operator, the outer join can be applied to only one table—not both. However, with the JOIN keyword, you can create a full outer join, as well as a left or right outer join. A full outer join will keep all rows from both tables in the results no matter which table is deficient when matching rows (that is, it performs a combination of left and right outer joins). *By default, use of the JOIN keyword creates an inner join.* To use the JOIN keyword to create an outer join, you include the keyword **LEFT**, **RIGHT**, or **FULL** with the JOIN keyword to identify the join type. You can also include the OUTER keyword to request an outer join; however, it is optional.

Figure 9-27 re-creates the query used in Figure 9-26 by using the LEFT OUTER JOIN keywords.

FIGURE 9-27   Using the LEFT OUTER JOIN keywords

The use of the LEFT JOIN keywords means that if the table listed on the left side of the joining condition given in the ON clause has an unmatched record, it should be matched with a NULL record and displayed in the results. As shown in Figure 9-28, had you used the RIGHT JOIN keywords, Oracle 10g would have interpreted the query to mean the results should include any order that did not have a corresponding match in the CUSTOMERS table (this would happen only if the customer record had been deleted from the

CUSTOMERS table, but the order placed still existed in the ORDERS table). Because a customer exists for every order that has been placed recently, no NULL rows were created. However, notice that the customers who have not placed an order currently in the ORDERS table are no longer displayed (Thompson, Cruz, McKenzie, and so on). These customers are no longer displayed because the CUSTOMERS table is referenced in the left side of the joining condition, and this statement is executing a RIGHT OUTER JOIN query.

```
SELECT lastname, firstname, order#
  FROM customers c RIGHT OUTER JOIN orders o
    ON c.customer# = o.customer#
  ORDER BY c.customer#;
```

(Execute) (Load Script) (Save Script) (Cancel)

LASTNAME	FIRSTNAME	ORDER#
MORALES	BONITA	1018
MORALES	BONITA	1003
SMITH	LEILA	1016
SMITH	LEILA	1006
PIERSON	THOMAS	1008
GIRARD	CINDY	1009
GIRARD	CINDY	1000
GIANA	TAMMY	1014
GIANA	TAMMY	1007
JONES	KENNETH	1020
LUCAS	JAKE	1011
LUCAS	JAKE	1001
MCGOVERN	REESE	1002
LEE	JASMINE	1013
SCHELL	STEVE	1017
NELSON	BECCA	1012
MONTIASA	GREG	1019
MONTIASA	GREG	1005
SMITH	JENNIFER	1010
FALAH	KENNETH	1015
FALAH	KENNETH	1004

21 rows selected.

FIGURE 9-28   Using the RIGHT OUTER JOIN keywords

Substituting the FULL JOIN keywords instructs Oracle 10g to return records from either table that do not have a matching record in the other table. A FULL JOIN is not anavailable option when creating an outer join with the operator in the WHERE clause for the traditional approach; it is only available with the JOIN keyword.

N O T E

When joining three or more tables, keep in mind that each join operation will, by default, drop nonmatching rows. To retain nonmatching rows, you might need an outer join operation for each join in the query.

SET OPERATORS

Set operators are used to combine the results of two (or more) SELECT statements. Valid set operators in Oracle 10g are UNION, UNION ALL, INTERSECT, and MINUS. When used with two SELECT statements, the **UNION** set operator returns the results of both queries. However, if there are any duplicates, they are removed, and the duplicated record is listed only once. To include duplicates in the results, use the **UNION ALL** set operator. **INTERSECT** lists only records that are returned by both queries; the **MINUS** set operator removes the results of the second query from the output. Figure 9-29 provides a summary of the set operators.

SET OPERATOR	DESCRIPTION
UNION	Returns the results of the combined SELECT statements and suppresses duplicates
UNION ALL	Returns the results of the combined SELECT statements but does not suppress duplicates
INTERSECT	Returns only the rows included in the results of both SELECT statements
MINUS	Removes the results of the second query that are also found in the first query and displays only the rows that are uniquely returned by only the first query

FIGURE 9-29   List of set operators

Suppose that you want a list of all author IDs with books that are in either the children or family life categories. As previously mentioned, the UNION set operator displays all the rows returned by either query. In Figure 9-30, the authors with books in the family life category are retrieved in the first SELECT and the authors with books in the children category are retrieved in the second SELECT. Because the UNION set operator combines the two SELECT statements, each book is listed only once, even if a number appears several times in the ORDERS table. For example, author R100 has a book in each category but is only listed once in the results.

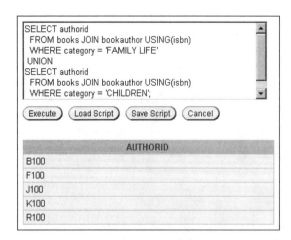

FIGURE 9-30   Producing an unduplicated combined list using the UNION set operator

Unlike the UNION set operator, the UNION ALL set operator displays every row returned by the combined SELECT statements. In Figure 9-31, the first SELECT retrieves a list of all books and the second SELECT retrieves all the books that have been ordered. The results show five rows for ISBN 0401140733. This indicates that four orders include this book (the fifth row is from the BOOKS table). The UNION ALL set operator does not suppress duplicate rows, so ISBNs are displayed more than once for books that have been ordered.

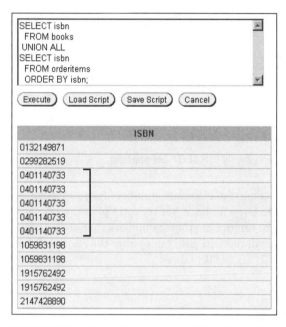

FIGURE 9-31   Producing a combined list with duplication using the UNION ALL set operator (partial output displayed)

The query in Figure 9-32 contains two SELECT statements. The first SELECT statement asks for all the customer numbers in the CUSTOMERS table—basically a list of all customers. The second SELECT statement lists all the customer numbers for customers who have recently placed an order. By using the INTERSECT set operator to combine the two SELECT statements, you instruct Oracle 10g to provide a list of all customers who have recently placed an order and who exist in the CUSTOMERS table. In other words, only customers who are retrieved in both of the SELECTS will be in the results.

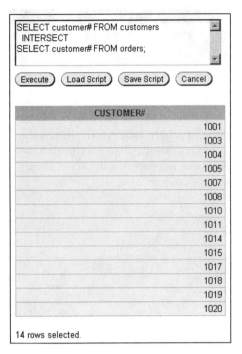

FIGURE 9-32   Identifying overlapping values using the INTERSECT set operator

The query shown in Figure 9-33 requests a list of customer numbers for those customers who are stored in the CUSTOMERS table but who have not recently placed an order. To do this, you use the MINUS set operator to remove customer numbers that are returned by the second SELECT statement (customers in the ORDERS table) from the results of the first SELECT statement (customers in the CUSTOMERS table).

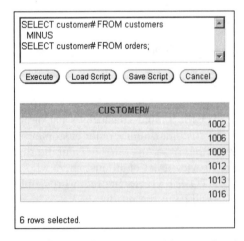

FIGURE 9-33   Subtracting result sets using the MINUS set operator

Twenty customer numbers were returned in response to the first SELECT statement; however, only 14 of those customers had placed orders. Because you used the MINUS set operator, the 14 customers who had placed orders were deleted from the results. As you can see in the output, there are six remaining customers. These are the ones who have not recently placed an order—they are listed in the CUSTOMERS table but not in the ORDERS table.

NOTE

Another set operator, EXISTS, is also available and is discussed in Chapter 12.

Chapter Summary

- Data stored in multiple tables regarding a single entity can be linked together through the use of joins.

- A Cartesian join between two tables returns every possible combination of rows from the tables. The resulting number of rows is always $m * n$.

- An equality join is created when the data joining the records from two different tables are an exact match (that is, an equality condition creates the relationship). The traditional approach uses an equal sign as the comparison operator in the WHERE clause. The JOIN approach can use the NATURAL JOIN, JOIN...USING, or JOIN...ON keywords.

- The NATURAL JOIN keywords do not require a condition to establish the relationship between two tables. However, a common column must exist. Column qualifiers cannot be used with the NATURAL JOIN keywords.

- The JOIN...USING approach is similar to the NATURAL JOIN approach, except that the common column is specified in the USING clause. A condition cannot be given in the USING clause to indicate how the tables are related. Column qualifiers cannot be used for the common column specified in the USING clause.

- The JOIN...ON approach joins tables based upon a specified condition. The JOIN keyword in the FROM clause indicates the tables to be joined, and the ON clause indicates how the two tables are related. This approach must be used if the tables being joined do not have a common column with the same column name in each table.

- A non-equality join establishes a relationship based upon anything other than an equal condition.

- Self-joins are used when a table must be joined to itself to retrieve needed data.

- Broadly speaking, a join can be either an inner join, in which the only records returned in the results have a matching record in all tables, or an outer join, in which records can be returned regardless of whether there is a matching record in the join.

- Inner joins are categorized as being equality, non-equality, or self-joins.

- An outer join is created when records need to be included in the results without having corresponding records in the join tables. The record is matched with a NULL record so it will be included in the output.

- When using the WHERE clause to create an outer join, records from only one table can be matched to a NULL record. The outer join operator (+) is placed next to the table that does not contain rows that match existing rows in the other table.

- With the OUTER JOIN keyword, you have the FULL option available that will include records from either table that do not have a corresponding record in the other table.

- Set operators such as UNION, UNION ALL, INTERSECT, and MINUS can be used to combine the results of multiple queries.

Chapter 9 Syntax Summary

The following tables present a summary of the syntax and information that you have learned in this chapter. You can use the tables as a study guide and reference.

SYNTAX GUIDE		
Element	Description	Example
Kinds of Joins		

WHERE clause	In the traditional approach, the WHERE clause can be used to indicate which column(s) should be used to join tables.	```SELECT columnname [,...]```
		 ```FROM tablename1, tablename2```
		 ```WHERE tablename1.columnname```
		 ```<comparison operator>```
		 ```tablename2.columnname;```

NATURAL JOIN keywords	These keywords are used in the FROM clause to join tables containing a common column with the same name and definition.	```SELECT columnname [,...]```
		 ```FROM tablename1 NATURAL```
		 ```JOIN tablename2;```

JOIN...USING	The JOIN keyword is used in the FROM clause and, combined with the USING clause, it identifies the common column to be used to join the tables. It is normally used if the tables have more than one commonly named column, and only one is being used for the join.	```SELECT columnname [,...]```
		 ```FROM tablename1```
		 ```JOIN tablename2```
		 ```USING (columnname);```

JOIN...ON	The JOIN keyword is used in the FROM clause. The ON clause identifies the column to be used to join the tables.	```SELECT columnname [,...]```
		 ```FROM tablename1```
		 ```JOIN tablename2```
		 ```ON tablename1.columnname```
		 ```<comparison operator>```
		 ```tablename2.columnname;```

OUTER JOIN Can be a RIGHT, LEFT, or FULL OUTER JOIN	This indicates that at least one of the tables does not have a matching row in the other table.	```SELECT columnname [,...]```
		 ```FROM tablename1```
		 ```[RIGHT|LEFT|FULL] OUTER```
		 ```JOIN tablename2```
		 ```ON tablename1.columnname =```
		 ```tablename2.columnname;```

Element	Description	Example
Kinds of Joins		

Cartesian Join
• Also known as a Cartesian product or cross join
• Matches each record in one table with each record in another table
	Example
	`SELECT title, name`
	`FROM books, publisher;`
		Uses keywords **CROSS JOIN**
		Example
		`SELECT title, name`
		`FROM books CROSS JOIN`
		`publisher;`

Equality Join
• Also known as an equijoin, inner join, or simple join
• Joins data in tables having equivalent data in a common column
• Might need to create table aliases
	Uses the keyword **WHERE**
	Example
	`SELECT title, books.pubid,`
	`name`
	`FROM publisher, books`
	`WHERE publisher.pubid =`
	`books.pubid`
	`AND publisher.pubid = 4;`
		Uses keywords NATURAL JOIN or JOIN...USING to create a join between tables having a commonly defined field;
		Uses keywords JOIN...ON when tables don't have a commonly defined field. The column qualifier ON tells Oracle 10*g* how tables are related
		Examples:
		`SELECT title, pubid, name`
		`FROM publisher NATURAL`
		`JOIN books;`
		
		`SELECT title, pubid, name`
		`FROM publisher JOIN books`
		`USING (pubid);`
		
		`SELECT title, name`
		`FROM books b JOIN`
		`publisher p ON b.pubid= p.pubid;`

Non-Equality Join
• Joins tables when there are no equivalent rows in the tables to be joined (that is, to match values in one column with a range of values in another column)
• Can use any comparison operator *except* the equal sign (=)
	Uses the keyword **WHERE**
	Example:
	`SELECT title, gift`
	`FROM books, promotion`
	`WHERE retail`
	`BETWEEN minretail AND`
	`maxretail;`
		Uses keywords JOIN...ON—and the same syntax as JOIN method equality join
		Example
		`SELECT title, gift`
		`FROM books JOIN promotion`
		`ON retail BETWEEN`
		`minretail AND maxretail;`

SYNTAX GUIDE

Element	Description	Example
Kinds of Joins		

Self-Join
• Joins a table to itself so columns within the table can be joined
• Must create table alias
	Uses the keyword **WHERE**
	Example:
	`SELECT r.firstname,`
	`r.lastname, c.lastname`
	`referred`
	`FROM customers c,`
	`customers r`
	`WHERE c.referred =`
	`r.customer#;`
		Uses the keyword **JOIN...ON**
		Example:
		`SELECT r.firstname, r.lastname,`
		`c.lastname`
		`referred`
		`FROM customers c`
		`JOIN customers r ON`
		`c.referred = r.customer#;`

Outer Join
• Includes records of a table in output when there is no matching record in the other table
	Uses the keyword **WHERE**;
	Uses the outer join operator (+) to create NULL rows in the deficient table for records that do not have a matching row
	Example:
	`SELECT lastname,`
	`firstname, order#`
	`FROM customers c,`
	`orders o`
	`WHERE`
	`c.customer#=o.customer#(+)`
	`ORDER BY c.customer#;`
		Includes the keyword **LEFT, RIGHT,** or **FULL** with the **OUTER JOIN** keyword
		Example:
		`SELECT lastname, firstname,`
		`order#`
		`FROM customers c LEFT`
		`OUTER JOIN orders o`
		`ON c.customer# =`
		`o.customer#`
		`ORDER BY c.customer#;`

SYNTAX GUIDE

Operator	Description	Example

Set Operators
Includes UNION, UNION ALL, INTERSECT, MINUS
	Combines results of multiple SELECT statements	`SELECT customer# FROM`
		`customers`
		`UNION`
		`SELECT customer# FROM`
		`orders;`

Review Questions

To answer these questions, refer to the tables in Appendix A.

1. Explain the difference between an inner join and an outer join.

2. How many rows will be returned in a Cartesian join between one table having five records and a second table having ten records?

3. Describe the problems you might encounter when using the NATURAL JOIN keywords to perform join operations.

4. Why are the NATURAL JOIN keywords not an option for producing a self-join? (*Hint:* Think about what happens if you use a table alias with the NATURAL JOIN keywords.)

5. What is the purpose of a column qualifier? When are you required to use a column qualifier?

6. In an OUTER JOIN query, the outer join operator (+) is placed next to which table?

7. What is the difference between the UNION and UNION ALL set operators?

8. How many join conditions are needed for a query that joins five tables?

9. What is the difference between an equality and a non-equality join?

10. What are the differences between the JOIN...USING and JOIN...ON approaches for joining tables?

Multiple Choice

To answer the following questions, refer to the tables in Appendix A.

1. Which of the following queries creates a Cartesian join?

   a. SELECT title, authorid FROM books, bookauthor;

   b. SELECT title, name FROM books CROSS JOIN publisher;

   c. SELECT title, gift FROM books NATURAL JOIN promotion;

   d. all of the above

2. Which of the following operators is not allowed in an outer join?

   a. AND

   b. =

   c. OR

   d. >

3. Which of the following is an example of an equality join?

   a. SELECT title, authorid FROM books, bookauthor
      WHERE books.isbn = bookauthor.isbn AND retail > 20;

   b. SELECT title, name FROM books CROSS JOIN publisher;

   c. SELECT title, gift FROM books, promotion
      WHERE retail>=minretail AND retail<=maxretail;

   d. none of the above

4. Which of the following is an example of a non-equality join?

   a. SELECT title, authorid FROM books, bookauthor
      WHERE books.isbn = bookauthor.isbn AND retail > 20;

   b. SELECT title, name FROM books JOIN publisher USING
      (pubid);

   c. SELECT title, gift FROM books, promotion
      WHERE retail>=minretail AND retail<=maxretail;

   d. none of the above

5. The following SQL statement represents which type of query?

    SELECT title, order#, quantity
    FROM books FULL JOIN orderitems
    ON books.isbn = orderitems.isbn;

    a. equality

    b. self-join

    c. non-equality

    d. outer join

6. Which of the following queries is valid?

    a. SELECT b.title, b.retail, o.quantity
       FROM books b NATURAL JOIN orders od
       NATURAL JOIN orderitems o
       WHERE od.order#=1005;

    b. SELECT b.title, b.retail, o.quantity
       FROM books b, orders od, orderitems o
       WHERE orders.order#=orderitems.order#
       AND orderitems.isbn=book.isbn
       AND od.order#=1005;

    c. SELECT b.title, b.retail, o.quantity
       FROM books b, orderitems o
       WHERE o.isbn = b.isbn AND o.order#=1005;

    d. none of the above

7. Given the following query:

    SELECT zip, order# FROM customers NATURAL JOIN orders;

    Which of the following queries is equivalent?

    a. SELECT zip, order#
       FROM customers JOIN orders
       WHERE customers.customer#=orders.customer#;

    b. SELECT zip, order#
       FROM customers, orders
       WHERE customers.customer#=orders.customer#;

    c. SELECT zip, order#
       FROM customers, orders
       WHERE customers.customer#=orders.customer#(+);

    d. none of the above

8. Which line in the following SQL statement contains an error?

    1 SELECT name, title
    2 FROM books NATURAL JOIN publisher
    3 WHERE category = 'FITNESS'

4 OR
5 books.pubid=4
6 /

a. line 1

b. line 2

c. line 3

d. line 4

e. line 5

9. Given the following query:

SELECT lastname, firstname, order#
FROM customers c LEFT OUTER JOIN orders o
ON c.customer# = o.customers#
ORDER BY c.customers#;

Which of the following queries will return the same results?

a. SELECT lastname, firstname, order#
FROM customers c OUTER JOIN orders o
ON c.customer# = o.customers#
ORDER BY c.customers#;

b. SELECT lastname, firstname, order#
FROM orders o RIGHT OUTER JOIN customers c
ON c.customer# = o.customers#
ORDER BY c.customers#;

c. SELECT lastname, firstname, order#
FROM customers c, orders o
ON c.customer# = o.customers#(+)
ORDER BY c.customers#;

d. none of the above

10. Given the following query:

SELECT DISTINCT zip, category
FROM customers NATURAL JOIN orders NATURAL JOIN orderitems
NATURAL JOIN books;

Which of the following queries is equivalent?

a. SELECT zip FROM customers
UNION
SELECT category FROM books;

b. SELECT DISTINCT zip, category
FROM customers c, orders o, orderitems oi, books b
WHERE c.customer#=o.customer# AND o.order#=oi.order#
AND oi.isbn=b.isbn;

c.   SELECT DISTINCT zip, category
     FROM customers c JOIN orders o JOIN orderitems oi
     JOIN books b
     ON c.customer#=o.customer# AND o.order#=oi.order#
     AND oi.isbn=b.isbn;

d.   all of the above

e.   none of the above

11. Which line in the following SQL statement contains an error?

     1 SELECT name, title
     2 FROM books JOIN publisher
     3 WHERE books.pubid = publisher.pubid
     4 AND
     5 cost <45.95
     6 /

     a.   line 1

     b.   line 2

     c.   line 3

     d.   line 4

     e.   line 5

12. Given the following query:

     SELECT title, gift FROM books CROSS JOIN promotion;

     Which of the following queries is equivalent?

     a.   SELECT title, gift FROM books NATURAL JOIN promotion;

     b.   SELECT title FROM books INTERSECT SELECT gift FROM
          promotion;

     c.   SELECT title FROM books UNION ALL SELECT gift FROM
          promotion;

     d.   all of the above

13. If the PRODUCTS table contains seven records and the INVENTORY table has eight
     records, how many records would the following query produce?

     SELECT * FROM products CROSS JOIN inventory;

     a.   0

     b.   8

     c.   7

     d.   15

     e.   56

14. Which of the following SQL statements is not valid?

    a.   SELECT b.isbn, p.name FROM books b NATURAL JOIN
           publisher p;

    b.   SELECT isbn, name
           FROM books b, publisher p
           WHERE b.pubid = p.pubid;

    c.   SELECT isbn, name
           FROM books b JOIN publisher p
           ON b.pubid=p.pubid;

    d.   SELECT isbn, name
           FROM books JOIN publisher
           USING (pubid);

    e.   None—all of the above are valid SQL statements.

15. Which of the following will produce a list of all books published by Printing Is Us?

    a.   SELECT title FROM books NATURAL JOIN publisher
           WHERE name LIKE 'PRIN%';

    b.   SELECT title FROM books, publisher
           WHERE pubname = 1;

    c.   SELECT * FROM books b, publisher p
           JOIN tables ON b.pubid = p.pubid;

    d.   none of the above

16. Which of the following SQL statements is not valid?

    a.   SELECT isbn FROM books
           MINUS
           SELECT isbn FROM orderitems;

    b.   SELECT isbn, name FROM books, publisher
           WHERE books.pubid (+) = publisher.pubid (+);

    c.   SELECT title, name FROM books NATURAL JOIN publisher

    d.   None—all of the above SQL statements are valid.

17. Which of the following statements regarding an outer join between two tables is true?

    a.   If the relationship between the tables is established through a WHERE clause, both tables can include the outer join operator.

    b.   To include unmatched records in the results, the record is paired with a NULL record in the deficient table.

    c.   The RIGHT, LEFT, and FULL keywords are equivalent keywords.

    d.   all of the above

    e.   none of the above

18. Which line in the following SQL statement contains an error?

    1 SELECT name, title
    2 FROM books b, publisher p
    3 WHERE books.pubid = publisher.pubid
    4 AND
    5 retail > 25 OR retail-cost > 18.95;

    a. line 1

    b. line 3

    c. line 4

    d. line 5

19. What is the maximum number of characters allowed in a table alias?

    a. 10

    b. 30

    c. 255

    d. 256

20. Which of the following SQL statements is valid?

    a. SELECT books.title, orderitems.quantity
       FROM books b, orderitems o
       WHERE b.isbn= o.ibsn;

    b. SELECT title, quantity
       FROM books b JOIN orderitems o;

    c. SELECT books.title, orderitems.quantity
       FROM books JOIN orderitems
       ON books.isbn = orderitems.isbn;

    d. none of the above

Hands-On Assignments

To perform these assignments, refer to the tables in Appendix A.

Generate and test two SQL queries for each of the following tasks: (a) the SQL statement needed to perform the stated task using the traditional approach, and (b) the SQL statement needed to perform the stated task using the JOIN keyword.

1. Create a list that displays the title of each book and the name and phone number of the person at the publisher's office whom you would need to contact to reorder each book.

2. Determine which orders have not yet shipped and the name of the customer who placed each order. Sort the results by the date on which the order was placed.

3. List the customer number and names of all individuals who have purchased books in the fitness category.

4. Determine which books Jake Lucas has purchased. Perform the search using the customer name, not the customer number.

5. Determine the profit of each book sold to Jake Lucas. Sort the results by the date of the order. If more than one book was ordered, sort the results by the profit amount in descending order. Perform the search using the customer name, not the customer number.

6. Which book was written by an author with the last name of Adams? Perform the search using the author name.

7. What gift will a customer who orders the book *Shortest Poems* receive?

8. Identify the author(s) of the books ordered by Becca Nelson. Perform the search using the customer name.

9. Display a list of all books in the BOOKS table. If a book has been ordered by a customer, also list the corresponding order number(s) and the state in which the customer resides.

10. Produce a list of all customers who live in the state of Florida and have ordered books about computers.

Advanced Challenge

To perform this activity, refer to the tables in Appendix A.

The Marketing Department of JustLee Books is preparing for its annual sales promotion. Each customer who places an order during the promotion will receive a free gift with each book purchased. The gift received will be based upon the retail price of the book. JustLee Books also participates in co-op advertising programs with certain publishers. If the publisher's name is included in advertisements, JustLee Books is reimbursed a certain percentage of the advertisement costs. To determine the projected costs of this year's sales promotion, the Marketing Department needs the publisher's name, the profit amount, and the free gift description for each book held in inventory by JustLee Books.

Also, the Marketing Department is analyzing books that do not sell. A list of ISBNs for all books with no sales recorded is needed. Use a set operation to complete this task.

Create a document that includes a synopsis of the request and the necessary SQL statement and the output requested by the Marketing Department.

Case Study: *City Jail*

Note: It is assumed that the City Jail database creation script from Chapter 8 has been executed. That script makes all database objects available to complete this case study.

The following list reflects the current data requests received from city managers. Provide the SQL statements that satisfy the requests. For each request, provide a solution using the traditional method and one using an ANSI join statement. Test the statements and show execution results as well.

1. Provide a list of all criminals along with each of the crime charges filed. The report needs to include the criminal id, name, crime code, and fine amount.

2. Provide a list of all criminals along with each crime status and appeal status (if applicable). The reports need to include the criminal id, name, crime classification, date charged, appeal filing date, and appeal status. Show all criminals regardless of whether they have filed an appeal.

3. Provide a list of all criminals along with crime information. The report needs to include the criminal id, name, crime classification, date charged, crime code, and fine amount. Include only crimes that are classified as "Other." Sort the list by criminal id and date charged.

4. Provide an alphabetical list of all criminals including criminal id, name, violent offender status, parole status, and any known aliases.

5. A table named PROB_CONTACT contains the required frequency of contact with a probation officer based on the length of the probation period (that is, the number of days assigned to probation). Review the data in this table, which indicates ranges for the number of days and applicable contact frequencies. Provide a list that contains each criminal who has been assigned a probation period (sentence applied is probation and the probation start and end date is assigned) and the required contact frequency. The list should contain the criminal name, crime id, probation start date, probation end date, and required frequency of contact. Sort the list by criminal name and probation start date.

6. A column named Mgr_id has been added to the PROB_OFFICERS table and contains the probation officer ID for the probation supervisor for each officer. Produce a list showing each probation officer name and the name of her supervisor. Sort the list alphabetically by the probation officer name.

CHAPTER 10

SELECTED SINGLE-ROW FUNCTIONS

LEARNING OBJECTIVES

After completing this chapter, you should be able to do the following:

- Use the UPPER, LOWER, and INITCAP functions to change the case of field values and character strings
- Extract a substring using the SUBSTR function
- Locate a substring within a character string with the INSTR function
- Nest functions inside other functions
- Determine the length of a character string using the LENGTH function
- Use the LPAD and RPAD functions to pad a string to a certain width
- Use the LTRIM and RTRIM functions to remove specific character strings
- Substitute character string values with the REPLACE and TRANSLATE functions
- Round and truncate numeric data using the ROUND and TRUNC functions
- Return the remainder only of a division operation using the MOD function
- Calculate the number of months between two dates using the MONTHS_ BETWEEN function
- Manipulate date data using the ADD_MONTHS, NEXT_DAY, TO_DATE, and ROUND functions
- Extend pattern matching capabilities with regular expressions
- Identify and correct problems associated with calculations involving NULL values using the NVL function
- Display dates and numbers in a specific format using the TO_CHAR function
- Perform condition processing similar to an IF statement with the DECODE function
- Use the SOUNDEX function to identify character phonetics
- Use the DUAL table to test functions

INTRODUCTION

In this chapter, you will learn about single-row SQL functions. A **function** is a predefined block of code that accepts one or more **arguments**—values listed within parentheses—and then returns a single value as output. The nature of an argument depends on the **syntax**, or structure, of the function being executed. **Single-row functions** return one row of results for each record processed. By contrast, **multiple-row functions** return only one result per group or category of rows processed, such as counting the number of books published by each publisher. (Multiple-row functions are presented in Chapter 11.)

Single-row functions range from performing character case conversions to calculating the number of months between two dates. The functions presented in this chapter have been grouped into character functions (case conversion functions and character manipulation functions), number functions, date functions, regular expressions, and other functions. Figure 10-1 presents an overview of the functions discussed in this chapter.

TYPE OF FUNCTIONS	FUNCTIONS
Case conversion functions	UPPER, LOWER, INITCAP
Character manipulation functions	SUBSTR, INSTR, LENGTH, LPAD/RPAD, RTRIM/LTRIM, REPLACE, TRANSLATE, CONCAT
Numeric functions	ROUND, TRUNC, MOD, ABS
Date functions	MONTHS_BETWEEN, ADD_MONTHS, NEXT_DAY, TO_DATE, ROUND, CURRENT_DATE
Regular expressions	REGEXP_LIKE, REGEXP_SUBSTR
Other functions	NVL, NVL2, TO_CHAR, DECODE, SOUNDEX, TO_NUMBER

FIGURE 10-1　Functions covered in this chapter

Oracle 10*g* supports a wide variety of single-row functions. This chapter covers those that are most commonly used. However, you should review the Functions chapter in the SQL Reference provided by Oracle to become familiar with the variety of functions available. All reference books can be found in the documentation area of the Oracle Technology Network Web site (*otn.oracle.com*).

> **NOTE**
>
> Before attempting to work through the examples provided in this chapter, run the prech10.sql file to make additions needed to the JustLee Books database. This assumes you have already executed the prech08. sql script as instructed in Chapter 8.

CASE CONVERSION FUNCTIONS

You can use **character functions** to change the case of characters (for example, to convert uppercase letters to lowercase letters) or to manipulate characters (such as substituting one character for another). Although most database administrators rarely need to use character functions, application developers frequently include them to create user-friendly database interfaces. Let's first examine case conversion functions.

Case conversion functions alter the case of data stored in a field or character string. Used in a query, the case conversion is only temporary—it changes how data is viewed by Oracle 10*g* during the execution of a specific query, but it does not affect how data is stored. A case conversion function could be used in an INSERT statement to set the case of a stored value. The character conversion functions supported by Oracle 10*g* are LOWER, UPPER, and INITCAP.

LOWER Function

The data in JustLee Books' database is in uppercase letters. However, when users perform data searches, they might enter character strings for search conditions in lowercase letters if they don't know that everything is in uppercase. If this happens, Oracle does not return any rows because the data is in uppercase letters. You can solve this problem in several ways. Perhaps the easiest way is to use the LOWER function. The **LOWER** function converts character strings to lowercase letters. You can use the LOWER function to temporarily convert a table's data to lowercase characters during a query's execution, as shown in Figure 10-2.

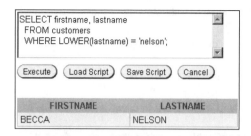

FIGURE 10-2 LOWER function in a WHERE clause

In Figure 10-2, you are searching for a customer with the last name Nelson. The syntax for the LOWER function is LOWER(c), where c is the field or character string to be converted. In this case, you need the data in the Lastname field converted to lowercase characters during the search. Therefore, the Lastname field is inserted between the parentheses.

Notice that the results of the query in Figure 10-2 are still displayed in uppercase letters. Because the LOWER function was not used in the SELECT clause of the SELECT statement, the first and last name are displayed in the same case in which they are stored. If you want data to be displayed in lowercase characters, simply include the LOWER function for each field to be converted in the SELECT clause. Figure 10-3 shows the LOWER function in a SELECT and a WHERE clause.

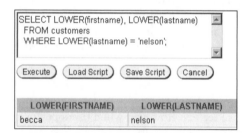

FIGURE 10-3 LOWER function in SELECT and WHERE clauses

In Figure 10-3, the LOWER function in the SELECT clause makes Oracle display the query results in lowercase characters. However, you still must include the LOWER function in the WHERE clause because the character string 'nelson' is, for comparison purposes, still entered in its lowercase form. In other words, *when a function is used in a SELECT clause, it affects only how the data is displayed in the results*. By contrast, when a function is used in a WHERE clause, it is used only during the specified comparison operation.

Because the columns contain the results of the LOWER function, the column headings are displayed as the actual function used in the SELECT clause. Oracle 10g includes the function in the column heading to indicate that the data was manipulated or altered before it was listed in the output. If you do not consider such a column heading desirable, simply include a column alias after the function in the SELECT clause, and the alias will be displayed as the column header in the results.

UPPER Function

The **UPPER** function converts the characters indicated into uppercase characters. This function can be used in the same way as the LOWER function to affect the display of characters (used in a SELECT clause) and to modify the case of characters for a search condition (used in a WHERE clause). The syntax for the UPPER function is UPPER(c), where c is the character string or field to be converted into uppercase characters. As discussed in Chapter 5, substitution variables allow user input to be used to complete a SQL statement. If the user input provides a value to be used in a search condition, the UPPER function could be used to ensure that the user value is converted to uppercase characters prior to the completion of the comparison in the search condition. For example, the query shown in Figure 10-4 requests that a last name value be provided.

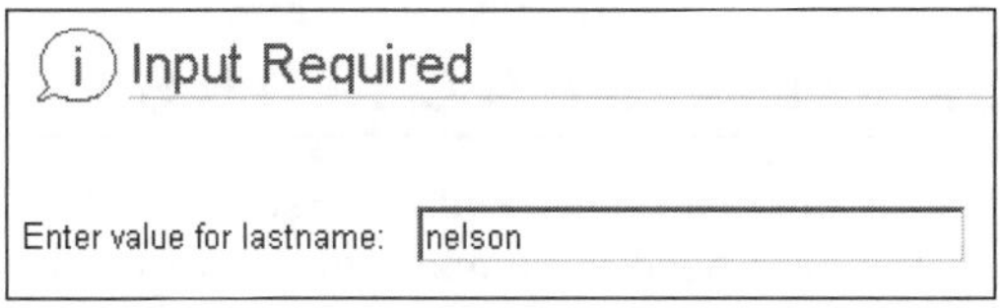

FIGURE 10-4   Using the UPPER function to manage user input

In Figure 10-4, the UPPER function is included in the condition portion of the WHERE clause to instruct Oracle 10g to convert the user-entered character string of *nelson* to uppercase characters while executing the SELECT query. It is much more efficient to convert the single-search condition to the same case as the data stored in the table for the following reasons:

1. You don't need to be concerned with whether the user knows the exact case to use when entering search strings.
2. Oracle 10g does not need to convert the case of each row value contained within the field during the execution of the query, thus reducing the processing burden placed on the Oracle server.

INITCAP Function

Although having the table data and the search criteria in the same case enables you to find the record(s) you seek, the output might not be presented in an appealing manner. It is generally easier for most people to read data displayed in mixed case letters, rather than in all uppercase or lowercase letters. Oracle 10g includes the **INITCAP** function to convert character strings to mixed case, with each word beginning with a capital letter—for example, *Great Mushroom Recipes*. The syntax of the INITCAP function is INITCAP(c), where c represents the field or character string to be converted. Thus, the INITCAP function converts the first (INITial) letter of each word in the character string to uppercase (CAPital letter) and the remaining characters into lowercase, as shown in Figure 10-5.

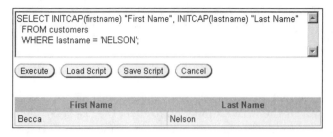

```
SELECT INITCAP(firstname) "First Name", INITCAP(lastname) "Last Name"
  FROM customers
  WHERE lastname = 'NELSON';
```

(Execute) (Load Script) (Save Script) (Cancel)

First Name	Last Name
Becca	Nelson

FIGURE 10-5   INITCAP function in a SELECT clause modifies display

In Figure 10-5, the INITCAP function in the SELECT clause makes Oracle convert data in the Firstname and Lastname columns to mixed case, with the first letter of the word in each field being in uppercase and the remaining letters in lowercase. Although the INIT-CAP function can also be used in a WHERE clause, this is rare because all users might not be consistent in the case they use for entering data. Also notice column aliases are used for the headings of the Firstname and Lastname columns in the output.

CHARACTER MANIPULATION FUNCTIONS

Although most data needed by the management of JustLee Books is already stored in the appropriate form in the database, sometimes data might need to be manipulated to yield the desired query output. For example, there might be times when you will need to determine the length of a string, extract portions of a string, or reposition a string, using **manipulation functions**.

The following sections explain some of the more commonly used manipulation functions; however, you should realize that Oracle 10g supports almost 200 different functions. You can read documentation of the functions supported by Oracle 10g in the SQL reference available on the Oracle Technology Web site.

SUBSTR Function

You can use the **SUBSTR** function to return a **substring**, or portion of a string. Many organizations code data values such as their inventory, benefits information, and general ledger accounts according to some type of coding scheme. For example, a benefit code might contain three characters and the second character might indicate the health plan selected by the employee. The area code of a customer's telephone number indicates a region within a state.

One way to determine where a customer resides is to look at the first three numbers of a customer's zip code. The United States Postal Service assigns the same first three digits of the zip code for a geographical distribution area within each state. The Marketing Department can use this data to determine where to concentrate certain promotional campaigns. Thus, they can use the SUBSTR function to extract the first three digits of the zip code stored for each customer.

The syntax for this function is SUBSTR(c, p, l), where c represents the character string, p represents the beginning character position for the extraction, and l represents the length of the string to be returned in the results of the query, as shown in Figure 10-6.

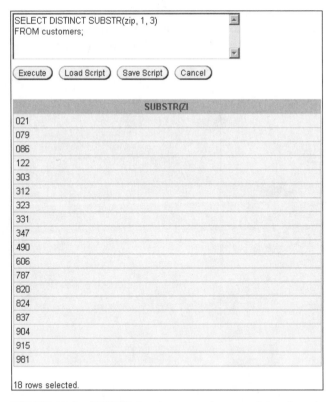

```
SELECT DISTINCT SUBSTR(zip, 1, 3)
FROM customers;
```

Execute   Load Script   Save Script   Cancel

SUBSTR(ZI
021
079
086
122
303
312
323
331
347
490
606
787
820
824
837
904
915
981

18 rows selected.

FIGURE 10-6   SUBSTR function extracting part of the zip code

309

NOTE

The column header in Figure 10-6 is truncated to show only a portion of the SUBSTR command. The header is truncated to the default width setting to display the data. Column width settings can be modified to display the entire header. Formatting column widths is covered in Chapter 14.

In Figure 10-6, the SELECT clause contains the DISTINCT keyword to eliminate duplication in the results. The arguments *(zip, 1, 3)* instruct the software to begin at the first character position of the Zip column, extract three characters as the substring, and then return them in the results.

The function can also extract substrings from the end of the data stored in the field. For example, if a negative 3 (–3) had been entered to indicate the beginning position, the software would establish the beginning position by counting backward three positions from the end of the field. Similarly, if *(zip, -3, 2)* had been entered as the parameters for the SUBSTR function, the third and fourth digits of the zip code would have been returned. Notice that in Figure 10-7, the entire zip code is displayed for each unique zip code stored in the CUSTOMERS table. The second column of the output contains just the first three digits of the zip code, and the third column contains just the third and fourth digits of the zip code, as requested by the SUBSTR(zip, -3, 2) portion of the SELECT clause.

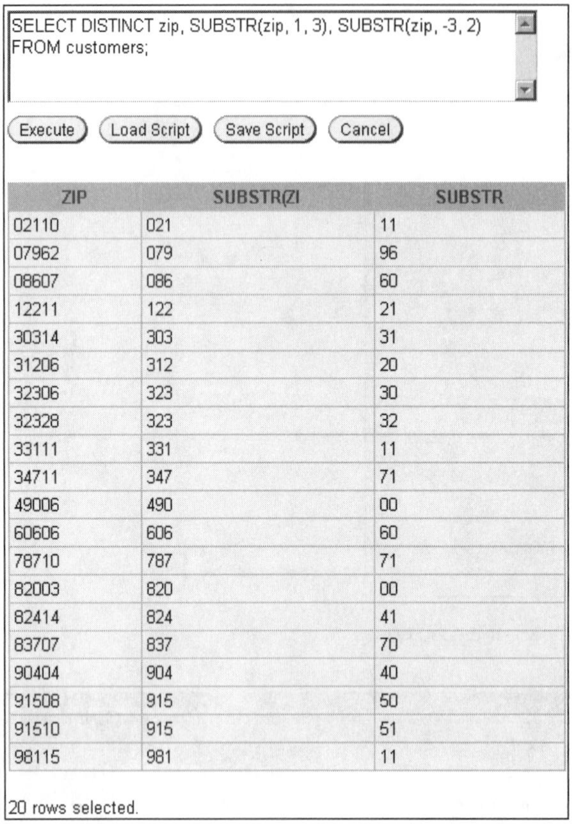

```
SELECT DISTINCT zip, SUBSTR(zip, 1, 3), SUBSTR(zip, -3, 2)
FROM customers;
```

(Execute) (Load Script) (Save Script) (Cancel)

ZIP	SUBSTR(ZI	SUBSTR
02110	021	11
07962	079	96
08607	086	60
12211	122	21
30314	303	31
31206	312	20
32306	323	30
32328	323	32
33111	331	11
34711	347	71
49006	490	00
60606	606	60
78710	787	71
82003	820	00
82414	824	41
83707	837	70
90404	904	40
91508	915	50
91510	915	51
98115	981	11

20 rows selected.

FIGURE 10-7 Comparison of SUBSTR arguments

INSTR Function

The **INSTR** (Instring) function searches a string for a specified set of characters or substrings. The function returns a numeric value representing the first character position in which the substring is found. If the substring does not exist within the string value, a zero is returned. Two arguments must be provided to the INSTR function: the string value to be searched for and the set of characters or substring to be located. Two optional arguments are also available: start position, which indicates the character of the string value on which to begin the search, and occurrence, which is the occurrence of the search value to be located (that is, the first occurrence, second occurrence, and so on). By default, the search will begin at the beginning of the string value and the position of the first occurrence will be located. Figure 10-8 demonstrates the use of different arguments in the INSTR function.

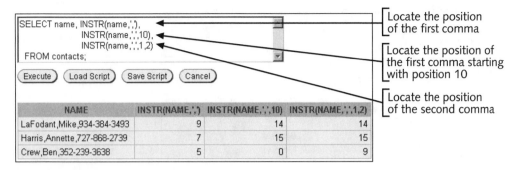

FIGURE 10-8 Comparison of INSTR arguments

The INSTR function is often used in conjunction with the SUBSTR function. For example, in Figure 10-8 the INSTR function assists in identifying the location of commas in the Name field. What if, however, you want to extract the first and last name from the Name field? In this scenario, you can nest the INSTR and SUBSTR to accomplish the task. **Nesting** simply means that one function is used as an argument inside another function. Figure 10-9 shows code to nest these functions in a SELECT clause to extract the first name from the Name field.

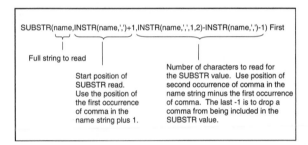

FIGURE 10-9 Nested functions—INSTR nested inside SUBSTR

Any of the single-row functions can be nested inside other single-row functions. When nesting functions, you should remember the following important rules:

1. All arguments required for each function must be provided.
2. For every opening parenthesis, there must be a corresponding closing parenthesis.
3. The nested, or inner, function is evaluated first. The result of the inner function is passed to the outer function, and then the outer function is executed.

Figure 10-10 displays a SELECT statement that uses nested functions to extract the first and last name from the Name field.

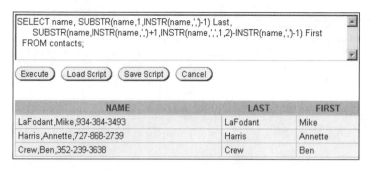

```
SELECT name, SUBSTR(name,1,INSTR(name,',')-1) Last,
    SUBSTR(name,INSTR(name,',')+1,INSTR(name,',',1,2)-INSTR(name,',')-1) First
  FROM contacts;
```

(Execute) (Load Script) (Save Script) (Cancel)

NAME	LAST	FIRST
LaFodant,Mike,934-384-3493	LaFodant	Mike
Harris,Annette,727-868-2739	Harris	Annette
Crew,Ben,352-239-3638	Crew	Ben

FIGURE 10-10 Using nested functions in a query

LENGTH Function

When you plan the width of table columns, design text areas for forms, or determine the size of mailing labels, you might ask, "What is the greatest number of characters that will be entered on this line?" For example, suppose that you are creating mailing labels. You'll need to have labels that are wide enough to accommodate the longest mailing address. To determine the number of characters in a string, you can use the **LENGTH** function. The syntax of the LENGTH function is LENGTH(c), where c represents the character string to be analyzed. Figure 10-11 shows the LENGTH function.

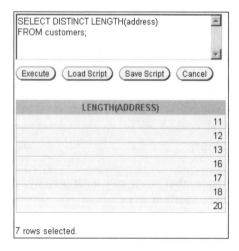

```
SELECT DISTINCT LENGTH(address)
FROM customers;
```

(Execute) (Load Script) (Save Script) (Cancel)

LENGTH(ADDRESS)
11
12
13
16
17
18
20

7 rows selected.

FIGURE 10-11 Checking data width using the LENGTH function

The (address) argument of the LENGTH function in Figure 10-11 determines the number of characters, or the length of the data, contained in the Address field for each customer. The DISTINCT keyword instructs the software not to include duplicate values in the results. As shown by the output in Figure 10-11, a mailing label that accommodates at least 20 characters is required to send mail to current customers.

Using the LENGTH function on a column with a CHAR datatype always returns the total width or size of the column.

LPAD and RPAD Functions

Have you ever received a check in which the amount of the check is preceded by a series of asterisks? Many companies fill in the blank spaces on checks and forms with symbols to make it difficult for someone to alter the numbers listed. The **LPAD** function can be used to pad, or fill in, the area to the left of a character string with a specific character—or even a blank space.

Before attempting to use this function in the iSQL*Plus interface, you need to modify the default output setting to produce pure text output. By default, iSQL*Plus presents output in an HTML table that does not show blank padded characters. To change the default output setting, perform the following steps. The client SQL*Plus interface displays output in text format by default, so no change is required for this interface.

1. Click the **Preferences** link.
2. Click the **System Configuration – Script Formatting** link.
3. Move down to the Preformatted Output setting, and set it to **On**.
4. Click the **Apply** button and move back to the Workspace area.

Now that we have text formatted output, let's try using the LPAD function. The syntax of the LPAD function is LPAD(c, l, s), where c represents the character string to be padded, l represents the length of the character string *after* being padded, and s represents the symbol or character to be used as padding, as shown in Figure 10-12.

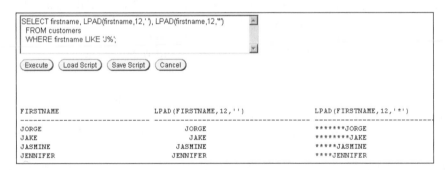

FIGURE 10-12 Using the LPAD function

In the example in Figure 10-12, the LPAD function is used twice on the Firstname column using different padding characters. The first LPAD function contains the arguments (firstname, 12, ' ').

- The first argument, firstname, informs the software that the Firstname column field will be padded.
- The second argument, 12, means that the data contained in the Firstname column should be padded to a total length of 12 spaces—the total length includes both the data and the padding symbol.
- The third argument, ' ', is an instruction to use a blank space as the padding symbol. Because the LPAD function is used, blank spaces are placed to the left of the customer's first name, until the total width of the data is 12 spaces. Notice that in Figure 10-12, because blank spaces are used to "left pad" the data, the customers' first names are right-aligned in the second column of output.

The only difference in the second LPAD function is that the padding character is set to an asterisk (*) rather than a blank.

Oracle 10g also provides an **RPAD** function that uses a symbol to pad the right side of a character string to a specific width. The syntax of the RPAD function is RPAD(c, l, s), where c represents the character string to be padded, l represents the total length of the character string *after* being padded, and s represents the symbol or character to be used as padding.

LTRIM and RTRIM Functions

You can use the **LTRIM** function to remove a specific string of characters from the left side of data values. The syntax for the LTRIM function is LTRIM(c, s), where c represents the data to be affected and s represents the string to be removed from the left of the data.

For example, suppose that the preprinted forms used by JustLee Books contain the string "P.O. Box". However, some of your customers have "P.O. Box" as part of their address in the CUSTOMERS table. The LTRIM function in Figure 10-13 removes the character string "P.O. Box" from each customer's address before it is displayed in the output. This prevents "P.O. Box" from being displayed twice on the form.

Oracle 10g also supports the **RTRIM** function to remove specific characters from the right side of a set of data. The syntax for the RTRIM function is RTRIM(c, s), where c represents the data to be affected and s represents the string to be removed from the right side of the data.

REPLACE Function

The **REPLACE** function is similar to the "search and replace" function used in some programs. The REPLACE function looks for the occurrence of a specified string of characters and, if found, substitutes it with another set of characters. The syntax for the REPLACE function is REPLACE(c, s, r), where c represents the data or column to be searched, s represents the

```
SELECT firstname, lastname, address, LTRIM(address, 'P.O. BOX')
FROM customers
WHERE address LIKE 'P.O. BOX%';
```

Execute   Load Script   Save Script   Cancel

FIRSTNAME	LASTNAME	ADDRESS	LTRIM(ADDRESS,'P.O.BOX')
BONITA	MORALES	P.O. BOX 651	651
RYAN	THOMPSON	P.O. BOX 9835	9835
LEILA	SMITH	P.O. BOX 66	66
CINDY	GIRARD	P.O. BOX 851	851
KENNETH	JONES	P.O. BOX 137	137
JORGE	PEREZ	P.O. BOX 8564	8564
REESE	MCGOVERN	P.O. BOX 18	18
WILLIAM	MCKENZIE	P.O. BOX 971	971
JASMINE	LEE	P.O. BOX 2947	2947
STEVE	SCHELL	P.O. BOX 677	677
BECCA	NELSON	P.O. BOX 563	563
JENNIFER	SMITH	P.O. BOX 1151	1151
KENNETH	FALAH	P.O. BOX 335	335

13 rows selected.

FIGURE 10-13   Using the LTRIM function

string of characters to be found, and r represents the string of characters to be substituted for s. In Figure 10-14, every occurrence of "P.O." in a customer's address has been replaced with the words "POST OFFICE," using REPLACE(address, 'P.O.', 'POST OFFICE') to indicate that the character string POST OFFICE should be substituted in the display every time Oracle 10g encounters the string P.O. in the address column of a customer.

TRANSLATE Function

The **TRANSLATE** function is used to replace characters within a string with a new value. It is different from the REPLACE function in that it modifies single characters rather than a character string. In addition, with TRANSLATE you can make more than one substitution operation within a single use of the function. The TRANSLATE function has three arguments: the string value to be used, the character to be searched for, and the substitution character. Figure 10-15 displays two examples of using the TRANSLATE function.

In the first query in Figure 10-15, only one character substitution is indicated in the TRANSLATE function. This operation will search the Name field for every comma and change it to a dash. If the string contained four commas, each one would be changed to a dash. The second query includes two substitution operations. Notice that the character substitution arguments are listed in order. The first search character is a comma, which is to be replaced with a dash, as indicated by the first substitution character. The second search character, an uppercase "A," will be replaced with a lowercase "a," which is the second substitution character listed.

```
SELECT address, REPLACE(address, 'P.O. ', 'POST OFFICE ')
FROM customers;
```

(Execute) (Load Script) (Save Script) (Cancel)

ADDRESS	REPLACE(ADDRESS,'P.O.','POSTOFFICE')
P.O. BOX 651	POST OFFICE BOX 651
P.O. BOX 9835	POST OFFICE BOX 9835
P.O. BOX 66	POST OFFICE BOX 66
69821 SOUTH AVENUE	69821 SOUTH AVENUE
P.O. BOX 851	POST OFFICE BOX 851
82 DIRT ROAD	82 DIRT ROAD
9153 MAIN STREET	9153 MAIN STREET
P.O. BOX 137	POST OFFICE BOX 137
P.O. BOX 8564	POST OFFICE BOX 8564
114 EAST SAVANNAH	114 EAST SAVANNAH
P.O. BOX 18	POST OFFICE BOX 18
P.O. BOX 971	POST OFFICE BOX 971
357 WHITE EAGLE AVE.	357 WHITE EAGLE AVE.
P.O. BOX 2947	POST OFFICE BOX 2947
P.O. BOX 677	POST OFFICE BOX 677
9851231 LONG ROAD	9851231 LONG ROAD
P.O. BOX 563	POST OFFICE BOX 563
1008 GRAND AVENUE	1008 GRAND AVENUE
P.O. BOX 1151	POST OFFICE BOX 1151
P.O. BOX 335	POST OFFICE BOX 335

20 rows selected.

FIGURE 10-14   Using the REPLACE function

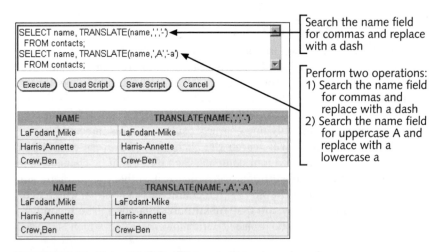

```
SELECT name, TRANSLATE(name,',','-')
  FROM contacts;
SELECT name, TRANSLATE(name,'A','-a')
  FROM contacts;
```

(Execute) (Load Script) (Save Script) (Cancel)

Search the name field for commas and replace with a dash

NAME	TRANSLATE(NAME,',','-')
LaFodant,Mike	LaFodant-Mike
Harris,Annette	Harris-Annette
Crew,Ben	Crew-Ben

Perform two operations:
1) Search the name field for commas and replace with a dash
2) Search the name field for uppercase A and replace with a lowercase a

NAME	TRANSLATE(NAME,',A','-A')
LaFodant,Mike	LaFodant-Mike
Harris,Annette	Harris-annette
Crew,Ben	Crew-Ben

FIGURE 10-15   Using TRANSLATE to substitute character values

CONCAT Function

You have learned how to use vertical bars (||) to concatenate, or combine, the data from various columns with string literals. The **CONCAT** function can also be used to concatenate the data from two columns. The main difference between the concatenation operator and the CONCAT function is that you can use the concatenation operator (||) to combine a long list of columns and string literals; by contrast, you can use the CONCAT function only to combine *two items* (columns or string literals). The concatenation operator (||) is typically preferred for concatenation because it is not limited to two items. If you need to combine more than two items using the CONCAT function, you must nest a CONCAT function inside another CONCAT function.

The syntax for the CONCAT function is CONCAT(*c1, c2*), where *c1* represents the first item to be included in the concatenation and *c2* represents the second item to be included in the operation. Both *c1* and *c2* can be either a column name or a string literal.

In Figure 10-16, a label has been added to each customer's customer number, so someone reading the output will know that the column is a customer number without having to look at the column heading. A column alias has also been added to the column to identify its contents.

FIGURE 10-16 Using the CONCAT function

NUMBER FUNCTIONS

Oracle 10g provides a set of functions that are specifically designed to address the manipulation of numeric data. In many organizations' daily operations, some of the most needed number functions are the ROUND, TRUNC, MOD, and ABS functions.

ROUND Function

The **ROUND** function is used to round numeric fields to the stated precision. The syntax of the ROUND function is ROUND(n, p), where *n* represents the numeric data, or field, to be rounded and *p* represents the position of the digit to which the data should be rounded. If the value of *p* is a positive number, the function refers to the right side of the decimal. However, if a negative value is entered, Oracle 10g rounds to the left side of the decimal position.

In the third column of the results in Figure 10-17, the retail price of each book has been rounded to the nearest tenth, or dime. As shown in the second column of output, the actual retail price of most books ends with .95. In Oracle 10g (as in most programs), values of five or more are rounded up, and values of less than five are rounded down. In Figure 10-17, the retail price for most of the books appears to be rounded to the nearest dollar. However, notice the book titled *The Wok Way to Cook*. The retail price of this book is $28.75, and when rounded to the nearest tenth, the price is $28.80.

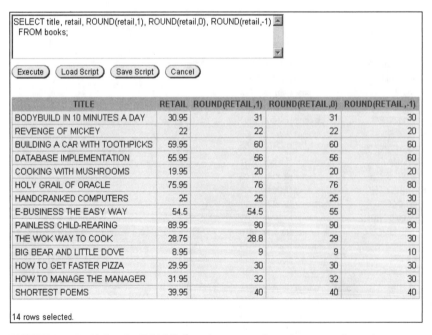

```
SELECT title, retail, ROUND(retail,1), ROUND(retail,0), ROUND(retail,-1)
  FROM books;
```

Execute | Load Script | Save Script | Cancel

TITLE	RETAIL	ROUND(RETAIL,1)	ROUND(RETAIL,0)	ROUND(RETAIL,-1)
BODYBUILD IN 10 MINUTES A DAY	30.95	31	31	30
REVENGE OF MICKEY	22	22	22	20
BUILDING A CAR WITH TOOTHPICKS	59.95	60	60	60
DATABASE IMPLEMENTATION	55.95	56	56	60
COOKING WITH MUSHROOMS	19.95	20	20	20
HOLY GRAIL OF ORACLE	75.95	76	76	80
HANDCRANKED COMPUTERS	25	25	25	30
E-BUSINESS THE EASY WAY	54.5	54.5	55	50
PAINLESS CHILD-REARING	89.95	90	90	90
THE WOK WAY TO COOK	28.75	28.8	29	30
BIG BEAR AND LITTLE DOVE	8.95	9	9	10
HOW TO GET FASTER PIZZA	29.95	30	30	30
HOW TO MANAGE THE MANAGER	31.95	32	32	30
SHORTEST POEMS	39.95	40	40	40

14 rows selected.

FIGURE 10-17   Using the ROUND function to round numbers to various places

The fourth column in Figure 10-17 shows the retail price of the books rounded to the nearest dollar using ROUND(retail, 0). The "0" indicates that the retail price should be rounded to no decimal places. The last column in Figure 10-17 shows the results of rounding the retail price to the nearest tens of dollars, using -1 to indicate the amount to the left of the decimal position that should be rounded.

TRUNC Function

There might be times when you need to truncate, rather than round, numeric data. You can use the **TRUNC** (Truncate) function to truncate a numeric value to a specific position. Any number(s) after that position is simply removed, or truncated. The syntax for the TRUNC function is TRUNC(n, p), where n represents the numeric data or field to be truncated, and p represents the position of the digit from which the data should be truncated. As with the ROUND function, entering a positive value for p indicates a position to the right of the decimal, whereas a negative number indicates a position to the left of the decimal.

Figure 10-18 demonstrates the TRUNC function using the same arguments shown previously for the ROUND function. The third column of the output in Figure 10-18 displays the results of TRUNC(retail, 1). Compare the results of the TRUNC function with the results of the ROUND function. Unlike the ROUND function, the results of the TRUNC function did not depend on what value came after the tenths' position; the TRUNC function simply dropped any value beyond the first number after the decimal, without changing the value of the first number.

```
SELECT title, retail, TRUNC(retail,1), TRUNC(retail,0), TRUNC(retail,-1)
  FROM books;
```

Execute   Load Script   Save Script   Cancel

TITLE	RETAIL	TRUNC(RETAIL,1)	TRUNC(RETAIL,0)	TRUNC(RETAIL,-1)
BODYBUILD IN 10 MINUTES A DAY	30.95	30.9	30	30
REVENGE OF MICKEY	22	22	22	20
BUILDING A CAR WITH TOOTHPICKS	59.95	59.9	59	50
DATABASE IMPLEMENTATION	55.95	55.9	55	50
COOKING WITH MUSHROOMS	19.95	19.9	19	10
HOLY GRAIL OF ORACLE	75.95	75.9	75	70
HANDCRANKED COMPUTERS	25	25	25	20
E-BUSINESS THE EASY WAY	54.5	54.5	54	50
PAINLESS CHILD-REARING	89.95	89.9	89	80
THE WOK WAY TO COOK	28.75	28.7	28	20
BIG BEAR AND LITTLE DOVE	8.95	8.9	8	0
HOW TO GET FASTER PIZZA	29.95	29.9	29	20
HOW TO MANAGE THE MANAGER	31.95	31.9	31	30
SHORTEST POEMS	39.95	39.9	39	30

14 rows selected.

FIGURE 10-18 Using the TRUNC function to truncate numbers at various places

Again refer to the book *The Wok Way to Cook*. The retail price of the book is $28.75. After the retail price is truncated, notice the result is $28.7 rather than the $28.8 value received after the price was rounded. The value after the tenths' position had no effect on the result returned—it was simply removed from the retail price.

When a zero is included as the second argument of the TRUNC function, no decimal positions are displayed for the retail price in the output. However, remember that the dollar amount displayed is not rounded; the decimals are simply dropped.

MOD Function

The **MOD** function returns the remainder only of a division operation. Two arguments are needed for the function: the numerator and denominator of the division operation. Let's say we have a figure representing the total ounces of a liquid product, and we need to convert the amount to pounds and ounces. The first query in Figure 10-19 shows the result of a division operation, which produces a result of 14.6875 pounds (16 is used as the denominator because 16 ounces equal 1 pound). The result shows the remaining ounces over 14 pounds in decimal form (.6875 of a pound). This decimal amount might not be easily used by someone needing to know the weight in pounds and ounces to package the liquid for shipping. The second query uses the MOD function to capture the remainder amount of 11, which represents the ounces over 14 pounds.

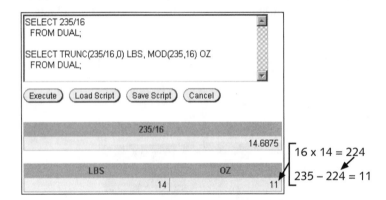

FIGURE 10-19 Using the MOD function to return the remainder

ABS Function

The **ABS** (Absolute) function returns the absolute or positive value of the numeric value provided as the argument. For example, when subtracting two date values to determine the difference in terms of days, you might have noticed that the result is negative if the more recent date is subtracted from an earlier date. Suppose we perform the operation of 18-OCT-05 minus 20-OCT-05: The result would be negative 2. If you are concerned only with the difference of two dates in terms of days and do not want to be concerned with the order of the dates in the calculation, use the ABS function to always return the absolute value of the result. The query in Figure 10-20 demonstrates the effect of using the ABS function on a date calculation.

```
SELECT pubdate, SYSDATE, ROUND(pubdate-SYSDATE) Days,
    ABS(ROUND(pubdate-SYSDATE)) "ABS Days"
 FROM books
 WHERE category = 'CHILDREN';
```

(Execute) (Load Script) (Save Script) (Cancel)

PUBDATE	SYSDATE	DAYS	ABS Days
18-MAR-05	07-AUG-05	-143	143
08-NOV-04	07-AUG-05	-273	273

FIGURE 10-20 The effect of using the ABS function

DATE FUNCTIONS

By default, Oracle 10g displays date values in a DD–MON–YY format that represents a two-digit number for the day of the month, a three-letter month abbreviation, and a two-digit number for the year. For example, the date of February 2, 2007 would be displayed as 02–FEB–07. Although users reference a date as a nonnumeric field (that is, a character string that must be enclosed in single quotation marks), it is actually stored in a numeric format that includes century, year, month, day, hours, minutes, and seconds. The valid range of dates that Oracle 10g can reference is January 1, 4712 B.C. to December 31, 9999 A.D.

NOTE

A default date format is set in the Oracle database configuration. The DBA can set a different date format in the configuration files and, therefore, some installations might not display dates by default in the DD–MON–YY format.

Although dates appear as nonnumeric fields, users can perform calculations with dates because they are stored as numeric data. The numeric version of a date used by Oracle 10g is a **Julian date**, which represents the number of days that have passed between a specified date and January 1, 4712 B.C. For example, if you need to calculate the number of days between two dates, Oracle 10g would first convert the dates to the Julian date numeric format and then determine the difference between the two dates. If Oracle 10g did not have a numeric equivalent for a date, there would be a problem trying to derive the solution for the arithmetic expression '02–JAN–06' - '08–SEP–05'. Look at the calculation with date columns in Figure 10-21.

```
SELECT order#, shipdate, orderdate, (shipdate-orderdate) Delay
 FROM orders
 WHERE order# = 1004;
```

(Execute) (Load Script) (Save Script) (Cancel)

ORDER#	SHIPDATE	ORDERDATE	DELAY
1004	05-APR-05	01-APR-05	4

FIGURE 10-21 Calculations with date columns

In Figure 10-21, order 1004 did not ship the same day as it was ordered. To determine how many days shipment was delayed, the Orderdate column is subtracted from the Shipdate column. If calculations are performed between two date fields, the results are returned in terms of days because the dates are internally stored as numeric values.

By contrast, if you need the results returned in terms of weeks rather than days, simply divide the results by seven. For example, to have the results of the query in Figure 10-21 reported in terms of weeks (or portion of a week in this case), the equation in the SELECT clause would have been changed to (shipdate-orderdate)/7. However, there are times when it is difficult to convert date calculation results to a unit that you need. For example, suppose that you needed to know the number of months between two dates: Do you divide by 30 or by 31? What number would you use for February? As shown in the following sections, Oracle 10g provides various functions to assist you with these kinds of calculations.

> **NOTE**
>
> The calculation shown in Figure 10-21 has a more recent date (Shipdate) subtracted from an earlier date (Orderdate). If the order of the calculation was reversed (Orderdate-Shipdate), the result would be the same number of days except the resulting number would be a negative number.

MONTHS_BETWEEN Function

Suppose that management wants to know whether customers are ordering more recently released books or more books that were published several months (or even years) ago. To find an answer to this question, you might simply subtract the publication date (Pubdate) from the order date (Orderdate) for a book to determine how many days a book had been available to the public before it was ordered. However, as mentioned in the previous section, what number should be used to convert days to months? Oracle 10g provides the **MONTHS_BETWEEN** function to determine the number of months between two dates. The syntax for the function is MONTHS_BETWEEN(d1, d2), where d1 and d2 are the two dates in question and d2 is subtracted from d1. Figure 10-22 shows this function.

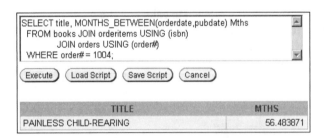

FIGURE 10-22   Using the MONTHS_BETWEEN function

In Figure 10-22, the user wants to determine how many months the book from order 1004 was available before the customer placed this order. Notice that the results of the query do not just provide a whole number to indicate the number of months that have elapsed. Instead, the digits after the decimal indicate portions of a month. To remove the portions of a month, include the MONTHS_BETWEEN function nested inside a TRUNC or ROUND function.

ADD_MONTHS Function

The management of JustLee Books renegotiates all contract pricing for books five years after they are published and stocks books for only seven years after they are published. Management believes that after seven years, sales for most books will decline to such a level that it is no longer profitable to keep them in inventory. Thus, management periodically requests a list of the current books in inventory, along with the date on which each book contract should be renegotiated and the date the book should be dropped from inventory. One method of calculating the "renegotiate date" is to add 1825 (365*5) days to the publication date of each book. However, an even better approach is to use the **ADD_MONTHS** function, as shown in Figure 10-23.

FIGURE 10-23 Using the ADD_MONTHS function

In Figure 10-23, the ADD_MONTHS function is applied to the Pubdate column for each book to determine its "renegotiate date" and "drop date." The syntax for the ADD_MONTHS function is ADD_MONTHS(d, m), where d represents the beginning date for the calculation and m represents the number of months to add to the date. As shown in the output of the query, the result of the ADD_MONTHS function is a new date with the correct number of months added to the old date.

NEXT_DAY Function

JustLee Books' policy is that books must be shipped by the first Monday after they receive a customer's order. Whenever an order is received, the customer is informed of the latest date that order is expected to ship. The **NEXT_DAY** function can determine the next occurrence of a specific day of the week after a given date. The syntax for the NEXT_DAY function is NEXT_DAY(d, DAY), where d represents the starting date and DAY represents the day of the week to be identified. Figure 10-24 shows the NEXT_DAY function.

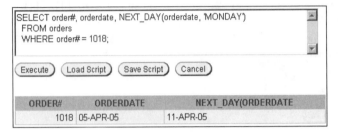

FIGURE 10-24   Using the NEXT_DAY function

In Figure 10-24, order 1018 was ordered on April 5, 2005. Because JustLee's policy is to inform a customer of the latest possible ship date, the query requests the date of the first Monday following the date of the order. As shown in the results, the customer can expect the order to be shipped by April 11—the first Monday after April 5.

> **NOTE**
>
> If the order date in this example happened to be on a Monday, the NEXT_DAY function would return the date for the following Monday or one week away.

TO_DATE Function

The **TO_DATE** function is of particular interest to application developers. Many database users might be uncomfortable entering a date in the format of DD–MON–YY and prefer to enter a date as MM/DD/YY or Month DD, YYYY. The TO_DATE function allows users to enter a date in any format, and then it converts the entry into the default format used by Oracle 10g (two-digit day, three-letter month abbreviation, and two-digit year). The syntax for the TO_DATE function is TO_DATE(d, f), where d represents the date being entered by the user and f is the format for the date that was entered. Figure 10-25 shows valid formats for entering a date in the TO_DATE function.

> **NOTE**
>
> For a list of additional format model elements, see the discussion of the TO_CHAR function later in this chapter.

DATE FORMATS		
Element	Description	Example
MONTH	Name of the month spelled out—padded with blank spaces to a total width of nine spaces	APRIL
MON	Three-letter abbreviation for the name of the month	APR
MM	Two-digit numeric value of the month	04
RM	Roman numeral month	IV
D	Numeric value for the day of the week	Wednesday = 4
DD	Numeric value for the day of the month	28
DDD	Numeric value for the day of the year	December 31 = 365
DAY	Name of the day of the week—padded with blank spaces to a length of nine characters	Wednesday
DY	Three-letter abbreviation for the day of the week	WED
YYYY	Four-digit numeric value of the year	2004
YYY or YY or Y	The last three, two, or single digit(s) of the year	2004 = 004; 2004 = 04; 2004 = 4
YEAR	Spelled-out version of the year	TWO THOUSAND FOUR
B.C. or A.D.	Value indicating B.C. or A.D.	2004 A.D.

FIGURE 10-25 Data format elements

When working with the TO_DATE function, the user enters the date specified as the first argument of the function. The second argument is a format model that allows Oracle 10g to distinguish the different parts of the date. Because both the date entered and the format model are character strings, each argument must be enclosed in single quotation marks.

Suppose that a user needs a list of the orders placed on March 31, 2005, and shipped in April. For illustrative purposes, suppose that the user enters the order date in the format of Month DD, YYYY. As shown in Figure 10-26, using the TO_DATE function in a WHERE clause enables the user to enter the preferred format for the desired order date, and then includes the format model necessary for Oracle 10g to interpret the order date. Notice that literals such as commas are included in the format string.

NOTE

The TO_DATE function is used routinely in INSERT statements to convert user date input into values Oracle 10g identifies as dates for inserting into DATE columns.

ROUND Function

Even though the ROUND function is typically associated with numeric data, you can use it on date values as well. The syntax of the ROUND function is ROUND(d, fmt), where d represents the date data, or field, to be rounded and fmt represents the format model to be used.

```
SELECT order#, orderdate, shipdate
 FROM orders
 WHERE orderdate = TO_DATE('March 31, 2005', 'Month DD, YYYY');
```

Execute   Load Script   Save Script   Cancel

ORDER#	ORDERDATE	SHIPDATE
1000	31-MAR-05	02-APR-05
1001	31-MAR-05	01-APR-05
1002	31-MAR-05	01-APR-05

FIGURE 10-26   Using the TO_DATE function

A date can be rounded by month or year. Figure 10-27 shows each book's publish date rounded by month and year. If the day of the month is 16 or later, the date is rounded up to the first of the next month. If the day of the month is less than 16, the date is rounded down to the first of that month. The year rounding works similarly, with July 16th being the cut-off to determine rounding.

```
SELECT pubdate, ROUND(pubdate,'MONTH'), ROUND(pubdate,'YEAR')
 FROM books;
```

Execute   Load Script   Save Script   Cancel

PUBDATE	ROUND(PUBDATE,'MON	ROUND(PUBDATE,'YEA
21-JAN-01	01-FEB-01	01-JAN-01
14-DEC-01	01-DEC-01	01-JAN-02
18-MAR-05	01-APR-05	01-JAN-05
04-JUN-05	01-JUN-05	01-JAN-05
28-FEB-06	01-MAR-06	01-JAN-06
31-DEC-05	01-JAN-06	01-JAN-06
21-JAN-05	01-FEB-05	01-JAN-05
01-MAR-04	01-MAR-04	01-JAN-04
17-JUL-00	01-AUG-00	01-JAN-01
11-SEP-00	01-SEP-00	01-JAN-01
08-NOV-04	01-NOV-04	01-JAN-05
11-NOV-02	01-NOV-02	01-JAN-03
09-MAY-05	01-MAY-05	01-JAN-05
01-MAY-05	01-MAY-05	01-JAN-05

14 rows selected.

FIGURE 10-27   Rounding dates by MONTH and YEAR

TRUNC Function

The TRUNC function, used for truncating numeric results, can also be quite useful in dealing with date calculations. The query previously presented in Figure 10-22 returned the number of months between the date of an order and the publication date of the book ordered.

However, the output included portions of a month. To eliminate the decimal portion of the output and return only the number of whole months between the two dates, you can nest the MONTHS_BETWEEN function inside the TRUNC function, as shown in Figure 10-28.

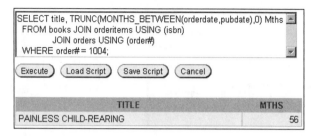

FIGURE 10-28   Using the TRUNC function on data calculation results

Remember that the TRUNC function has two arguments—the value to be truncated and the position at which the truncation should occur. The value that you need truncated is the result of the MONTHS_BETWEEN function. Therefore, the MONTHS_BETWEEN function is entered as the first argument of the TRUNC function in Figure 10-28. The closing parenthesis after the Pubdate column name completes the MONTHS_BETWEEN function.

The nested MONTHS_BETWEEN function is followed by a comma, a zero, and a parenthesis. These serve to complete the TRUNC function. The zero indicates that there should be no decimal positions after the truncation occurs, and a parenthesis closes the TRUNC function. The value calculated by the inner function (MONTHS_BETWEEN) is used to complete the outer function (TRUNC), and the result is the whole number of months between the publication date of the book and the date the book was ordered.

CURRENT_DATE Versus SYSDATE

As many companies are conducting transactions in many parts of the world, time zone recognition is essential. You should know the differences between the CURRENT_DATE and SYSDATE functions, which both identify the current date and time. Whereas the SYSDATE function returns the current date and time that is set on the operating system on which the database resides, the CURRENT_DATE function returns the current date and time from the user session. The client SQL*Plus software will be used in this example because this will allow the session time zone setting to be different from the database time zone setting. Because iSQL*Plus is run from the database server, it is not actually a client session and it does not allow us to identify a different client session time zone.

To demonstrate, the local computer is set to a different time zone than the database server (this was done by resetting the time zone in Windows). Figure 10-29 displays a SQL*Plus connection from the local computer to the Oracle10g server. The first query verifies the time zone settings of the client (SESSIONTIMEZONE) and the database (SYSTIMESTAMP). The time zone is indicated in terms of Greenwich mean time (also referred to as coordinated universal time). The second query displays the SYSDATE and CURRENT_DATE setting to highlight the difference in time recorded due to the difference in time zones of the client session and the database server.

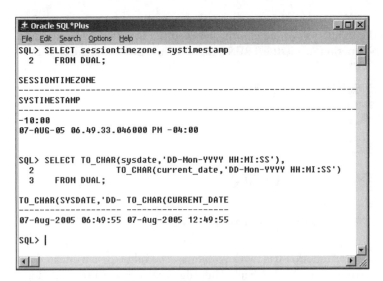

```
Oracle SQL*Plus                                          _ |□| X|
File  Edit  Search  Options  Help
SQL> SELECT sessiontimezone, systimestamp
  2      FROM DUAL;

SESSIONTIMEZONE
--------------------------------------------------------------
SYSTIMESTAMP
--------------------------------------------------------------
-10:00
07-AUG-05 06.49.33.046000 PM -04:00


SQL> SELECT TO_CHAR(sysdate,'DD-Mon-YYYY HH:MI:SS'),
  2              TO_CHAR(current_date,'DD-Mon-YYYY HH:MI:SS')
  3      FROM DUAL;

TO_CHAR(SYSDATE,'DD-  TO_CHAR(CURRENT_DATE
-------------------- --------------------
07-Aug-2005 06:49:55 07-Aug-2005 12:49:55

SQL> |
```

FIGURE 10-29   Identifying time zone information

REGULAR EXPRESSIONS

Oracle 10g recognizes **regular expressions**, which allow the description of complex patterns in textual data. Although these tools have roots in the UNIX environment, they are used in many languages. Regular expressions adhere to POSIX (Portable Operating System Interface for UNIX) standards—a set of *IEEE* and *ISO* standards that define an *interface* between *programs* and *operating systems*.

We already know that simple pattern matching can be accomplished by using the LIKE operator. However, the LIKE operator is limited to the _ and % wildcard characters and must reference a whole string value rather than just a portion of it. Let's assume that JustLee Books is researching the possibility of expanding operations into used book sales. Data providing a list of used book dealers has been purchased and loaded into a table named SUPPLIERS. Let's review the data in this table, as displayed in Figure 10-30.

Notice that the Description column might or might not contain a phone number. In addition, the phone number might be recorded in various formats, such as with dashes or periods (and one phone number is missing an area code). What if JustLee Books wants to produce a list of only the vendors that have a complete phone number recorded? The REGEXP_LIKE function will address this task, as it can reference an extended set of pattern operators available to regular expression operations. Let's take a look at the regular expression in a query to accomplish this task and then review the pattern operators being used. Figure 10-31 shows the REGEXP_LIKE function.

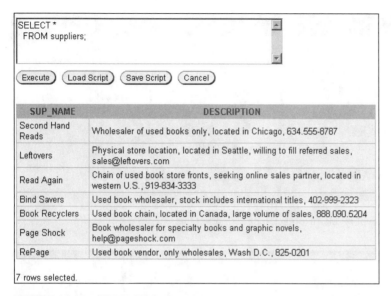

FIGURE 10-30 SUPPLIERS table data

FIGURE 10-31 Using REGEXP_LIKE to identify complete phone numbers

Notice that the REGEXP_LIKE function has two arguments. The first argument indicates which value should be searched for (the Description column in this case). The second argument identifies the pattern that it is attempting to locate in the search value (a 12-character string in this case). Keep in mind that the search pattern needs to match only a part of the entire value being searched. It does not need to reflect the contents of the entire search string as is needed with the LIKE operator. Figure 10-32 breaks down the pattern operators used to identify the goal of each as it searches for a 12-character string.

PATTERN OPERATOR	DESCRIPTION
[0-9]{3}	The [0-9] operator indicates that it is looking for a single character that must be a digit. The {3} indicates the previous operator is to be repeated three times. In this case, these two operators together search for a three-digit number.
[-.]	This operator indicates that the next character must be either a dash (-) or a period (.).
[0-9]{3}	The next three places must contain three digits.
[-.]	The next place can be either a dash (-) or period (.).
[0-9]{4}	The last four places in the string must contain four digits.

FIGURE 10-32   Description of regular expression pattern operators

NOTE

If you wanted a list of the suppliers that DO NOT have a complete phone number recorded, the NOT operator could be used in the WHERE clause to identify these records.

What if JustLee Books wanted a list of just the supplier names and phone numbers rather than the entire description? The REGEXP_SUBSTR function can be used to extend the capabilities of the SUBSTR function using the same pattern matching operators. Figure 10-33 shows a SELECT clause modified to extract the phone number from the description column using REGEXP_SUBSTR.

FIGURE 10-33   Using REGEXP_SUBSTR to extract phone numbers

This discussion serves only to introduce the concept of regular expressions, as this is an extensive topic. There are numerous other pattern matching operators available, which provide a tremendously flexible tool in addressing complex pattern searches. Additional information regarding regular expressions can be found in the SQL reference available on the Oracle Web site.

OTHER FUNCTIONS

Some functions provided by Oracle 10g do not fall neatly into a character, numeric, or date category. However, these functions are very important and are widely used in business. We'll discuss the following six functions in this section: NVL, NVL2, TO_CHAR, DECODE, TO_NUMBER, and SOUNDEX.

NVL Function

You can use the **NVL** function to address problems that can be caused when performing arithmetic operations with fields that may contain NULL values. (Recall that a NULL value is the absence of data, not a blank space or a zero.) When a NULL value is used in a calculation, the result is always a NULL value. The NVL function is used to substitute a value for the existing NULL so the calculation can be completed. The syntax for the NVL function is NVL(x, y), where y represents the value to be substituted if x is NULL. In many cases, the substitute for a NULL value in a calculation is zero (0).

In practice, the NVL function is commonly used to calculate an individual's gross pay as "salary + commission." But what happens when an individual's sales are not high enough to earn a commission? If you add the individual's salary to a NULL commission, the resulting gross pay is NULL—no paycheck! Unfortunately, you will probably realize that the error has occurred about the same time the individual storms into your office, angry over the absence of a paycheck. To avoid this problem, rather than calculating gross pay as salary plus commission, use salary + NVL(commission, 0). The NVL function simply substitutes a zero whenever the commission is NULL, and the person still gets paid, because "salary + 0" still equals "salary."

There are times, however, when substituting a simple zero for a NULL value does not return the desired results. In such cases, you might need to use a little more imagination. For example, suppose that management requests a report that reflects shipment delays for orders. Some of those orders might not have shipped yet, but they are expected to ship on the following Monday. If the Shipdate column for an order is left blank for the pending shipments, management might assume that there was no delay in the shipment of the order. However, if a zero is substituted for the missing ship date, strange results might be displayed in the query's output, as you can see in Figure 10-34.

In Figure 10-34, the Delay between the Orderdate and Shipdate columns for each order is calculated. Because some orders have not yet been shipped, the actual ship date is a NULL value, and the Delay column is left blank. If management simply skims over the report, it would appear that the longest delay is four days (order 1004). However, this is not accurate. Because management's policy is that orders must be shipped by the first Monday after an order is received, you can substitute the expected Shipdate into the Delay calculation. See Figure 10-35.

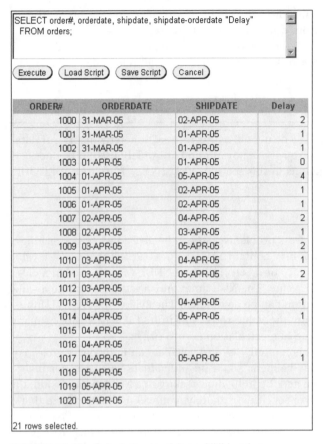

FIGURE 10-34 Calculation involving a NULL value

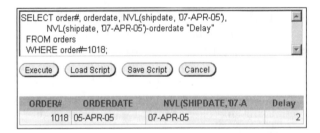

FIGURE 10-35 Using NVL to substitute a value for a NULL value

Order 1018 in Figure 10-35 has not yet shipped. To calculate the anticipated shipping delay for that order, Figure 10-35 shows using the NVL function to substitute the date for the first Monday (07-APR-05) after the date of the order to estimate the shipping date and the resulting delay. The first NVL function given in the SELECT clause is `NVL(shipdate, '07-APR-05')`.

The purpose of this function is to substitute the anticipated shipping date for the NULL value in the results. However, simply because the substitution was made in the third column of the output does not mean that it will affect the calculation of the delay in the shipment of the order. The portion of the SELECT clause that is listed on line 2 of the query is what is actually calculating the shipment delay: `NVL(shipdate, '07-APR-05')-orderdate "Delay"`. This portion of the SELECT clause instructs Oracle 10g that if the Shipdate column is NULL, substitute April 7, 2005 as the shipping date for the order, and then subtract the order date from the assigned date of shipment. After the NVL function has made the substitution, the subtraction is performed and the delay is calculated. The "Delay" heading is the alias for that column.

NVL2 Function

The **NVL2** function is a variation of the NVL function that allows different options based on whether a NULL value exists. The syntax for the NVL2 function is `NVL2(x, y, z)`, where y represents what should be substituted if x is not NULL, and z represents what should be substituted if x is NULL. This allows the user a little more flexibility when working with NULL values.

In reference to the gross pay calculation described in the preceding section, rather than using the equation `salary+NVL(commission, 0)` to substitute a zero whenever the commission is NULL, the user could have used `NVL2(commission, salary, salary+commission)`. The NVL2 function would be read as, "If the commission IS NOT NULL, then the gross pay is just salary. If the commission IS NULL, then calculate gross pay as salary plus commission."

Suppose that management needs a report describing the shipment status of orders. The report should list an order as being shipped or not shipped. You could use the NVL2 function to display the status of each order, based on whether the Shipdate column contains a NULL value. Because a date indicates that an order has been shipped, and a NULL value indicates that an order has not yet shipped, this is the ideal situation for using the NVL2 function. See Figure 10-36.

In Figure 10-36, the NVL2 function has one of two character strings displayed, depending on whether the Shipdate column contains a NULL value. Because character strings are being used in the function, the strings must be enclosed in single quotation marks. If the strings are not enclosed in single quotation marks, Oracle 10g assumes an existing column is being referenced and it returns an error because such columns do not exist in the table.

TO_CHAR Function

The **TO_CHAR** function is widely used to convert dates and numbers to a formatted character string. It is the opposite of the TO_DATE function discussed previously in regard to handling date data. The TO_DATE function allows a user to *enter* a date in any type of format, whereas the TO_CHAR function is used to have Oracle 10g *display* dates in a

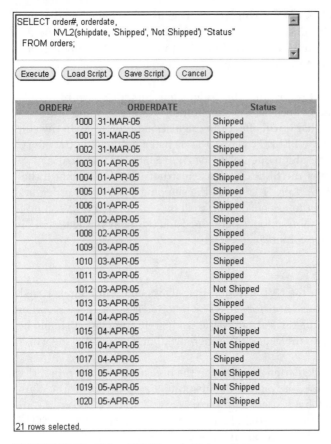

FIGURE 10-36   Using NVL2 to substitute values

particular format. The syntax of the TO_CHAR function is `TO_CHAR(n, 'f')`, where *n* is the date or number to be formatted and *f* is the format model to be used. A format model consists of a series of elements that represents exactly how the data should appear and must be entered within single quotation marks, as shown in Figure 10-37.

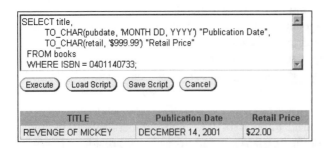

FIGURE 10-37   Formatting values for display with TO_CHAR

The query in Figure 10-37 contains two TO_CHAR functions.

The first TO_CHAR function is used to convert the publication date (pubdate) to a specific date format (Month DD, YYYY) that spells out the month of the year, followed by the day of the month, a comma, and then the four-digit year. Notice that there is an unusual amount of space after the word DECEMBER. If you want to eliminate insignificant spaces or zeros in the display, simply enter an *fm* at the beginning of the model to be used to display the data. The *fm* instructs Oracle 10g to turn off the "fill mode," which adds blank spaces to create a fixed width for the name of the month. To eliminate the space from DECEMBER in the Pubdate column, the correct function and argument are `TO_CHAR(pubdate, 'fmMONTH DD, YYYY')`. If you prefer to have the name of the month displayed in mixed case (for example, December), simply use that case in the format model—`('fmMonth DD, YYYY')`.

The second TO_CHAR function in Figure 10-37 is used to format the retail price (retail) of the book to display a dollar sign and two decimal positions. Without the format model, Retail Price would have been displayed as 22—without the dollar sign or any decimals.

Oracle 10g provides a wide variety of elements you can use to create format models for dates and numbers. The table in Figure 10-38 describes some commonly used format elements.

NOTE

An "RR" format was previously used to address potential problems raised by Y2K and storing only two-digit years. However, most individuals in the industry use the YYYY format model element to specify the exact century for a date.

DECODE Function

The **DECODE** function takes a specified value and compares it to values in a list. If a match is found, the specified result is returned. If no match is found, a default result is returned. If no default result is defined, a NULL is returned as the result. The DECODE function allows the user to specify different actions to be taken, depending on the circumstances (for example, the exact value is or is not contained within a column). It saves the user from having to enter multiple statements for each possible scenario. The syntax for the DECODE function is `DECODE(V, L1, R1, L2, R2,..., D)`, where V is the value you are searching for, $L1$ represents the first value in the list, $R1$ represents the result to be returned if $L1$ and V are equivalent, and so on, and D is the default result to return if no match is found.

NOTE

The DECODE function is similar to the CASE or IF...THEN...ELSE structures found in many programming languages.

FORMATS		
Element	Description	Example
Date Elements		
MONTH	Name of the month spelled out—padded with blank spaces to a total width of nine spaces	APRIL
MON	Three-letter abbreviation for the name of the month	APR
MM	Two-digit numeric value of the month	04
RM	Roman numeral month	IV
D	Numeric value for the day of the week	Wednesday = 4
DD	Numeric value for the day of the month	28
DDD	Numeric value for the day of the year	December 31 = 365
DAY	Name of the day of the week—padded with blank spaces to a length of nine characters	Wednesday
DY	Three-letter abbreviation for the day of the week	WED
YYYY	Numeric value for the four-digit year	2004
YYY or YY or Y	Numeric value for the last three, two, or single digit(s) of year	2004 = 004; 2004=04; the 2004 = 4
YEAR	Spelled-out version of the year	TWO THOUSAND FOUR
BC or AD	Value indicating B.C. or A.D.	2004 A.D.
Time Elements		
SS	Seconds	Value between 0–59
SSSS	Seconds past midnight	Value between 0–86399
MI	Minutes	Value between 0–59
HH or HH12	Hours	Value between 1–12
HH24	Hours	Value between 0–23
A.M. or P.M.	Value indicating morning or evening hours	A.M. (before noon) or P.M. (after noon)
Number Elements		
9	Indicates width of display with a series of 9s, but insignificant leading zeros are not displayed	99999
0	Displays insignificant leading zeros	0009999
$	Displays a floating dollar sign	$99999
.	Indicates number of decimals to display	999.99
,	Displays a comma in the position indicated	9,999

FIGURE 10-38 Format model elements

FORMATS		
Element	Description	Example
Other Elements		
, . (punctuation symbols)	Displays indicated punctuation	DD, YYYY = 24, 2001
"string"	Displays the exact character string inside the double quotation marks	"of the year" YYYY = of the year 2001
TH	Displays the ordinal number	DDTH = 8th
SP	Spells out the number	DDSP = EIGHT
SPTH	Spell out the ordinal number	DDSPTH = EIGHTH

FIGURE 10-38   Format model elements (continued)

JustLee Books is required to collect sales tax from customers who live in Florida and California, but not from customers residing in other states. Suppose Florida's sales tax is 7%, while California's is 8%. This means that if a customer resides in California, the customer must pay 8% of the total order price as sales tax; if the customer lives in Florida, the customer must pay 7% sales tax; if the customer lives in any other state, the customer does not pay any sales tax.

To determine the sales tax rate that applies to each customer, the DECODE function can be used to compare the state in which each customer lives to a list of states. If a match occurs, the sales tax rate that applies to that state is returned. However, if the customer lives in a state that is not listed, a default sales rate of 0 is applied, as shown in Figure 10-39.

In Figure 10-39, the DECODE function begins on line 2. The State column is identified as the value to be compared against the list. The state of California (CA) is the first item against which the value of the State column is compared. If the State column contains the value of CA, a sales tax rate of .08 is returned, and the DECODE function is processed again for the next customer. If the value in the State column is not CA, the value is compared against the next item in the list. (Note that in Figure 10-39, the second listed item has been placed on line 3 to improve the readability of the function.) If the value for the State column is equal to FL, a sales tax rate of .07 is returned. If the value in the State column is not equal to the two items listed (CA or FL), the default value is assigned, which, in this case, is zero.

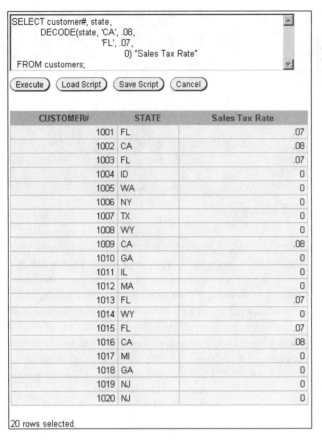

```
SELECT customer#, state,
       DECODE(state, 'CA', .08,
              'FL', .07,
                    0) "Sales Tax Rate"
FROM customers;
```

Execute   Load Script   Save Script   Cancel

CUSTOMER#	STATE	Sales Tax Rate
1001	FL	.07
1002	CA	.08
1003	FL	.07
1004	ID	0
1005	WA	0
1006	NY	0
1007	TX	0
1008	WY	0
1009	CA	.08
1010	GA	0
1011	IL	0
1012	MA	0
1013	FL	.07
1014	WY	0
1015	FL	.07
1016	CA	.08
1017	MI	0
1018	GA	0
1019	NJ	0
1020	NJ	0

20 rows selected.

FIGURE 10-39   Using DECODE to determine tax rate by state

SOUNDEX Function

Many government agencies and organizations perform searches for information based on the phonetic pronunciation of words rather than their actual spelling. For example, in many states, the first set of four characters and numbers of a driver's license number represents the phonetic sound of the individual's last name at the time the license was issued (it does not change if the person changes her last name). Oracle 10g can reference the phonetic sound or representation of words using the **SOUNDEX** function. The syntax of the function is SOUNDEX(c), where c is the character string being referenced.

As an example, assume a customer has called to make an inquiry regarding an order. The customer does not have the order number, so you enter a search based on his last name, which is Smyth. No records are returned, as no customers are recorded in the database with this last name. What do you do next? A secondary search using the SOUNDEX function can search for any names recorded with a pronunciation similar to Smyth, as shown in Figure 10-40.

Notice that two possible matches are found, both with a last name of Smith. In this case, an employee mistakenly entered the last name with an "i" rather than a "y."

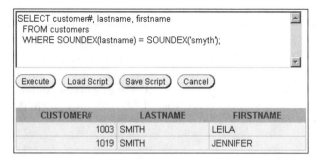

FIGURE 10-40   Using the SOUNDEX function

TO_NUMBER Function

The **TO_NUMBER** function converts a value to a numeric datatype if possible. For example, the string value of '2006' could be converted to a numeric datatype to be used in calculations. If the string that is being converted contains nonnumeric characters, the function returns an error. Let's say we need to calculate how old each book in the computer category is in terms of years. We can subtract the current date and book publication date, divide by 365, and then use rounding. Another method is to identify the year of each date and use subtraction. Figure 10-41 shows the TO_NUMBER function being used to convert the year value of each date to a numeric datatype to be used in the subtraction operation.

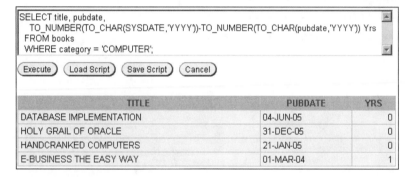

FIGURE 10-41   Using the TO_NUMBER function to convert a string to a numeric datatype

DUAL TABLE

Any of the single-row functions presented in this chapter can be used with the DUAL table. Although the DUAL table is rarely used in the industry, it can be valuable for someone learning how to work with functions or testing new functions. For example, if you want to practice rounding numbers or determining the length of character strings, you can enter a specific value in the appropriate function and reference the DUAL table in the FROM clause, as shown in Figure 10-42.

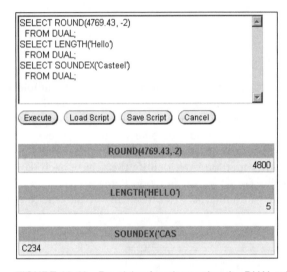

FIGURE 10-42   Practicing functions using the DUAL table

Chapter Summary

- Single-row functions return a result for each row or record processed.
- Character case conversion functions such as UPPER, LOWER, and INITCAP can be used to alter the case of character strings.
- Character manipulation functions can be used to extract substrings or portions of a string, identify the position of a substring within a string, replace occurrences of a string with another string, determine the length of a character string, and trim spaces or characters from strings.
- Nesting one function within another allows multiple operations to be performed on data.
- Simple number functions such as ROUND and TRUNC can round or truncate a number on both the left and right side of a decimal.
- The MOD function is used to return the remainder of a division operation.
- Date functions can be used to perform calculations with dates or to change the format of dates entered by a user.
- Regular expressions enable complex pattern matching operations.
- The NVL and NVL2 functions are used to address problems encountered with NULL values.
- The TO_CHAR function lets a user present numeric data and dates in a specific format.
- The DECODE function allows an action to be taken to be determined by a specific value.
- The SOUNDEX function looks for records based on the phonetic representation of characters.
- The DUAL table can be helpful when testing functions.

Chapter 10 Syntax Summary

The following table presents a summary of the syntax that you have learned in this chapter. You can use the table as a study guide and reference.

SYNTAX GUIDE		
Element	Description	Example
Case Conversion Functions		
LOWER	Converts characters to lowercase letters.	LOWER(c) c = character string or field to be converted to lowercase
UPPER	Converts characters to uppercase letters.	UPPER(c) c = character string or field to be converted to uppercase
INITCAP	Converts words to mixed case, initial capital letters.	INITCAP(c) c = character string or field to be converted to mixed case

SYNTAX GUIDE

Element	Description	Example
Character Manipulation Functions		
SUBSTR	Returns a substring, or portion of a string, in output.	SUBSTR *(c, p, l)* *c* = character string *p* = position (beginning) for the extraction *l* = length of output string
INSTR	Identifies position of search string	INSTR *(c, s, p, o)* *c* = character string to be searched *s* = search string *p* = search starting position *o* = occurrence of search string to identify
LENGTH	Returns the numbers of characters in a string.	LENGTH *(c)* *c* = character string to be analyzed
LPAD / RPAD	Pads, or fills in, the area to the Left (or **R**ight) of a character string, using a specific character—or even a blank space.	LPAD *(c, l, s)* *c* = character string to be padded *l* = length of character string after being padded *s* = symbol or character to be used as padding
RTRIM / LTRIM	Trims, or removes, a specific string of characters from the Right (or Left) of a set of data.	LTRIM *(c, s)* *c* = characters to be affected *s* = string to be removed from the left of the data
REPLACE	Used to perform a search and replace of displayed results.	REPLACE *(c, s, r)* *c* = data or column to be searched *s* = string of characters to be found *r* = string of characters to be substituted for *s*
TRANSLATE	Convert single characters to a substitution value	TRANSLATE *(c, s, r)* *c* = character string to be searched *r* = search character *s* = substitution character
CONCAT	Used to concatenate two data items.	CONCAT *(c1, c2)* *c1* = first data item to be concatenated *c2* = second data item to be included in the concatenation
Number Functions		
ROUND	Rounds numeric fields.	ROUND *(n, p)* *n* = numeric data or field to be rounded *p* = position of the digit to which the data should be rounded

SYNTAX GUIDE

Element	Description	Example
Number Functions		
TRUNC	Truncates, or cuts, numbers to a specific position.	TRUNC(*n*, *p*) *n* = numeric data or field to be truncated *p* = position of the digit to which the data should be truncated
MOD	Returns remainder of division operation	MOD(*n*, *d*) *n* = numerator *d* = denominator
ABS	Returns absolute value of a numeric value	ABS(*n*) *n* = numeric value
Date Functions		
MONTHS_BETWEEN	Determines the number of months between two dates.	MONTHS_BETWEEN(*d1*, *d2*) *d1* and *d2* = dates in question *d2* is subtracted from *d1*
ADD_MONTHS	Adds months to a date to signal a target date in the future.	ADD_MONTHS(*d*, *m*) *d* = date (beginning) for the calculation *m* = months—the number of months to add to the date
ROUND	Rounds date fields by month or year	ROUND(*d*, *fmt*) *d* = Date value *fmt* = Date format (YEAR or MONTH)
NEXT_DAY	Determines the next day—a specific day of the week after a given date.	NEXT_DAY(*d*, *DAY*) *d* = date (starting) *DAY* = the day of the week to be identified
TO_DATE	Converts a date in a specified format to the default date format.	TO_DATE(*d*, *f*) *d* = date entered by the user *f* = format of the entered date
Regular Expressions		
REGEXP_LIKE	Searches for pattern values within a character string	REGEXP_LIKE(*c*,*p*) *c* = character string *p* = pattern operators
REGEXP_SUBSTR	Extracts a part of a string which matches a pattern of values	REGEXP_SUBSTR(*c*,*p*) *c* = character string *p* = pattern operators

SYNTAX GUIDE		
Element	Description	Example
Other Functions		
NVL	Solves problems arising from arithmetic operations having fields that may contain NULL values. When a NULL value is used in a calculation, the result is a NULL value. The NVL function is used to substitute a value for the existing NULL.	NVL *(x, y)* *y* = value to be substituted if *x* is NULL
NVL2	Provides options based on whether a NULL value exists.	NVL2 *(x, y, z)* *y* = what should be substituted if *x* is not NULL *z* = what should be substituted if x is NULL
TO_CHAR	Converts dates and numbers to a formatted character string.	TO_CHAR *(n, 'f')* *n* = number or date to be formatted *f* = format model to be used
DECODE	Takes a given value and compares it to values in a list. If a match is found, the specified result is returned. If no match is found, a default result is returned. If no default result is defined, a NULL is returned as the result.	DECODE *(V, L1, R1, L2, R2, ..., D)* *V* = value sought *L1* = first value in the list *R1* = result to be returned if *L1* and *V* match *D* = default result to return if no match is found
SOUNDEX	Converts alphabetic characters to their phonetic representation, using an alphanumeric algorithm.	SOUNDEX *(c)* *c* = characters to be phonetically represented

Review Questions

1. Why are the functions in this chapter referred to as "single-row" functions?
2. What is the difference between the NVL and NVL2 functions?
3. What is the difference between the TO_CHAR and TO_DATE functions when working with date values?
4. How is the TRUNC function different from the ROUND function?
5. What functions can be used to search character strings for specific patterns of data?
6. What is the difference between using the CONCAT function and the concatenation operator (||) in a SELECT clause?
7. Which functions can be used to convert the case of character values?
8. Describe a scenario which would call for using the DECODE function.
9. What format model would you use to display the date 25-DEC-07 as Dec. 25?
10. Why would the function NVL(shipdate, 'Not Shipped') return an error message?

Multiple Choice

To answer the following questions, refer to the tables in Appendix A.

1. Which of the following is a valid SQL statement?

    a. `SELECT SYSDATE;`

    b. `SELECT UPPER(Hello) FROM dual;`

    c. `SELECT TO_CHAR(SYSDATE, 'fmMONTH DD, YYYY')`
       `FROM dual;`

    d. all of the above

    e. none of the above

2. Which of the following functions can be used to extract a portion of a character string?

    a. EXTRACT

    b. TRUNC

    c. SUBSTR

    d. INITCAP

3. Which of the following will determine how long ago orders were received that have not been shipped?

    a. `SELECT order#, shipdate-orderdate delay`
       `FROM orders;`

    b. `SELECT order#, SYSDATE - orderdate`
       `FROM orders`
       `WHERE shipdate IS NULL;`

    c. `SELECT order#, NVL(shipdate, 0)`
       `FROM orders`
       `WHERE orderdate is NULL;`

    d. `SELECT order#, NULL(shipdate)`
       `FROM orders;`

4. Which of the following SQL statements will produce Hello World as the output?

    a. `SELECT "Hello World"`
       `FROM dual;`

    b. `SELECT INITCAP('HELLO WORLD')`
       `FROM dual;`

    c. `SELECT LOWER('HELLO WORLD')`
       `FROM dual;`

    d. both A and B

    e. none of the above

5. Which of the following functions can be used to substitute a value for a NULL value?

   a. NVL

   b. TRUNC

   c. NVL2

   d. SUBSTR

   e. both A and D

   f. both A and C

6. Which of the following is not a valid format model for displaying the current time?

   a. 'HH:MM:SS'

   b. 'HH24:SS'

   c. 'HH12:MI:SS'

   d. All of the above are valid.

7. Which of the following will list only the last four digits of the contact person's phone number at American Publishing?

   a. ```
      SELECT EXTRACT(phone, -4, 1)
      FROM publisher
      WHERE name = 'AMERICAN PUBLISHING';
      ```

   b. ```
      SELECT SUBSTR(phone, -4, 1)
      FROM publisher
      WHERE name = 'AMERICAN PUBLISHING';
      ```

   c. ```
      SELECT EXTRACT(phone, -1, 4)
      FROM publisher
      WHERE name = 'AMERICAN PUBLISHING';
      ```

   d. ```
      SELECT SUBSTR(phone, -4, 4)
      FROM publisher
      WHERE name = 'AMERICAN PUBLISHING';
      ```

8. Which of the following functions can be used to determine how many months a book has been available?

   a. MONTH

   b. MON

   c. MONTH_BETWEEN

   d. none of the above

9. Which of the following will display the order date for order 1000 as 03/31?

   a. ```
      SELECT TO_CHAR(orderdate, 'MM/DD')
      FROM orders
      WHERE order# = 1000;
      ```

   b. ```
      SELECT TO_CHAR(orderdate, 'fmMM/DD')
      FROM orders
      WHERE order# = 1000;
      ```

c.
```
SELECT TO_CHAR(orderdate, 'fmMONTH/YY')
FROM orders
WHERE order# = 1000;
```

d. both A and B

e. none of the above

10. Which of the following functions includes different options depending on the value of a specified column?

a. NVL

b. DECODE

c. UPPER

d. SUBSTR

11. Which of the following SQL statements is not valid?

a.
```
SELECT TO_CHAR(orderdate, '99/9999')
FROM orders;
```

b.
```
SELECT INITCAP(firstname), UPPER(lastname)
FROM customers;
```

c.
```
SELECT cost, retail, TO_CHAR(retail-cost, '$999.99') profit
FROM books;
```

d. all of the above

12. Which function can be used to add spaces to a column until it is a specific width?

a. TRIML

b. PADL

c. LWIDTH

d. none of the above

13. Which of the following SELECT statements will return 30 as the result?

a. `SELECT ROUND(24.37, 2) FROM dual;`

b. `SELECT TRUNC(29.99, 2) FROM dual;`

c. `SELECT ROUND(29.01, -1) FROM dual;`

d. `SELECT TRUNC(29.99, -1) FROM dual;`

14. Which of the following is a valid SQL statement?

a. `SELECT TRUNC(ROUND(125.38, 1), 0) from dual;`

b. `SELECT ROUND(TRUNC(125.38, 0) from dual;`

c. `SELECT LTRIM(LPAD(state, 5, ' '), 4, -3, "*") from dual;`

d. `SELECT SUBSTR(ROUND(14.87, 2, 1), -4, 1) from dual;`

15. Which of the following functions cannot be used to convert the case of a character string?

   a. UPPER

   b. LOWER

   c. INITIALCAP

   d. All of the above can be used for case conversion.

16. Which of the following format elements will cause months to be displayed in a three-letter abbreviated format?

   a. MMM

   b. fmMONTH

   c. MON

   d. none of the above

17. Which of the following SQL statements will display a customer's name in all uppercase characters?

   a. `SELECT UPPER('firstname', 'lastname') FROM customers;`

   b. `SELECT UPPER(firstname, lastname) FROM customers;`

   c. `SELECT UPPER(lastname, ',' firstname) FROM customers;`

   d. none of the above

18. Which of the following functions can be used to display the character string FLORIDA in the results of a query whenever FL is encountered in the State field?

   a. SUBSTR

   b. NVL2

   c. REPLACE

   d. TRUNC

   e. none of the above

19. What is the name of the dummy table provided by Oracle 10*g*:

   a. DUMDUM

   b. DUAL

   c. ORAC

   d. SYS

20. If an integer is multiplied by a NULL value, the result is:

   a. an integer

   b. a whole number

   c. a NULL value

   d. None of the above—a syntax error message is returned.

Hands-On Assignments

To perform the following assignments, refer to the tables in Appendix A.

1. Obtain a list of all customer names in which the first letter of the first and last names are in uppercase and the rest are in lowercase.

2. Create a list of all customer numbers along with text that indicates whether the customer has been referred by another customer. Display the characters 'NOT REFERRED' if the customer was not referred to JustLee Books by another customer or 'REFERRED' if the customer has been referred.

3. Determine the amount of profit generated by the book purchased on order 1002. Display the book title and profit. The profit should be formatted to display a dollar sign and two decimal places.

4. Display a list of all book titles and the percentage of markup for each book. The percentage of markup should be displayed as a whole number (that is, multiplied by 100) with no decimal position, followed by a percent sign (for example, .2793 = 28%). (The percentage of markup should reflect the difference of the retail and cost amounts as a percent of the cost.)

5. Display the current day of the week, hour, minutes, and seconds of the current date setting on the computer you are using.

6. Create a list of all book titles and cost. Precede the cost of each book with asterisks so that the width of the displayed cost field is 12.

7. Determine the length of the data stored in the ISBN field of the BOOKS table. Make certain to have the length only display once (not once for each book).

8. Using today's date, determine the age (in months) of each book that JustLee sells. Make certain that only whole months are displayed, rather than portions of months. Display the book title, publication date, current date, and age.

9. Determine the calendar date of the next occurrence of Wednesday based on today's date.

10. Produce a list of each customer number and the third and fourth digits of their zip code. The query should also display the position of the first occurrence of a 3 in the customer number if it exists.

349

Advanced Challenge

To perform this activity, refer to the tables in Appendix A.

Management is proposing to increase the price of each book. The amount of the increase will be based on each book's category, according to the following scale: computer books, 10%; fitness books, 15%; self-help books, 25%; all other categories, 3%. Create a list that displays each book's title, category, current retail price, and revised retail price. The prices should be displayed with two decimal places. The column headings for the output should be as follows: Title, Category, Current Price, Revised Price. Sort the results by category. If there is more than one book in a category, a secondary sort should be performed on the book's title.

Create a document that provides management with the SELECT statement used to generate the results as well as the result of the statement.

Case Study: *City Jail*

Note: It is assumed that the City Jail database creation script from Chapter 8 has been executed. That script makes all database objects available to complete this case study.

The following list reflects the current data requests received from city managers. Provide the SQL statement that would satisfy the request. Test the statements and show execution results as well.

1. List the following information for all crimes that have a period between the date charged and the hearing date that is greater than 14 days: crime id, classification, date charged, hearing date, and the number of days between the date charged and the hearing date.

2. Produce a list showing each active police officer and his community assignment. The second letter of the precinct code indicates the officer's community assignment. Display the community description listed in the table that follows based on the second letter of the precinct code.

2ND LETTER OF PRECINCT CODE	DESCRIPTION
A	Shady Grove
B	Center City
C	Bay Landing

3. Produce a list of criminal sentencing information to include criminal id, name (display as all uppercase), sentence id, sentence start date, and the length in months of the sentence. The number of months should be shown as a whole number. The start date should be displayed in the format of "December 17, 2006."

4. A list of all amounts owed is needed. Produce a list showing each criminal name, charge id, total amount owed (fine amount plus court fee), amount paid, amount owed, and payment due date. If nothing has been paid to date, the amount paid is NULL. Include only criminals who owe some amount of money. Display the dollar amounts with a dollar sign and two decimals.

5. Display the criminal name and probation start date for all criminals who have a probation period greater than two months. Also display the date that is two months from the beginning of the probation period that will serve as a review date.

6. An INSERT statement is needed to support users adding a new appeal. Create an INSERT using substitution variables. Note that users will be entering dates in the format of two-digit month, two-digit day, and four-digit year such as "12 17 2006." In addition, a sequence named APPEALS_ID_SEQ exists to provide values for the Appeal_id column and the default setting for the Status column should take effect (that is, the DEFAULT option on the column should be used). Test the statement by adding the following appeal: crime_id = 25344031, filing date = 02 13 2006, hearing date = 02 27 2006.

CHAPTER **11**

GROUP FUNCTIONS

LEARNING OBJECTIVES

After completing this chapter, you should be able to do the following:

- Differentiate between single-row and multiple-row functions
- Use the SUM and AVG functions for numeric calculations
- Use the COUNT function to return the number of records containing non-NULL values
- Use COUNT(*) to include records containing NULL values
- Use the MIN and MAX functions with nonnumeric fields
- Determine when to use the GROUP BY clause to group data
- Identify when the HAVING clause should be used
- List the order of precedence for evaluating WHERE, GROUP BY, and HAVING clauses
- State the maximum depth for nesting group functions
- Nest a group function inside a single-row function
- Calculate the standard deviation and variance of a set of data, using the STDDEV and VARIANCE functions
- Understand the concept of multidimensional analysis
- Perform enhanced aggregation grouping with GROUPING SETS, CUBE, and ROLLUP
- Use composite columns and concatenated groupings in grouping operations

INTRODUCTION

Group functions, also called **multiple-row functions**, return one result per group of rows processed.

Multiple-row functions presented in this chapter include SUM, AVG, COUNT, MIN, MAX, STDDEV, and

VARIANCE. This chapter presents the GROUP BY clause to identify the group(s) of records to bepro-

cessed and the HAVING clause to restrict the groups returned in the query results. You will also learn about statistical functions. The last section of the chapter introduces enhanced aggregation capabilities provided by GROUPING SETS, CUBE, and ROLLUP operations.

Figure 11-1 provides an overview of this chapter's contents.

GROUP (MULTIPLE-ROW) FUNCTIONS		
Function (and Syntax)	Description	Example
GROUP BY Extensions		
SUM([DISTINCT|ALL] n)	Returns the sum or total value of the selected numeric field. Ignores NULL values.	SELECT SUM (retail-cost) FROM books;
AVG([DISTINCT|ALL] n)	Returns the average value of the selected numeric field. Ignores NULL values.	SELECT AVG(cost) FROM books;
COUNT(*|[|DISTINCT| ALL] c)	Returns the number of rows that contain a value in the identified field. Rows containing NULL values in the field will not be included in the results. To count rows containing NULL values, use an * rather than a field name.	SELECT COUNT(*) FROM books; or SELECT COUNT (shipdate) FROM orders;
MAX([DISTINCT|ALL] c)	Returns the highest (maximum) value from the selected field. Ignores NULL values.	SELECT MAX (customer#) FROM customers;
MIN([DISTINCT|ALL] c)	Returns the lowest (minimum) value from the selected field. Ignores NULL values.	SELECT MIN (retail-cost) FROM books;
STDDEV([DISTINCT|ALL] n)	Returns the standard deviation of the numeric field selected. Ignores NULL values.	SELECT STDDEV (retail) FROM books;
VARIANCE([DISTINCT|ALL] n)	Returns the variance of the numeric field selected. Ignores NULL values.	SELECT VARIANCE (retail) FROM books;

FIGURE 11-1   Selected group functions and GROUP BY extensions

GROUP (MULTIPLE-ROW) FUNCTIONS		
Function (and Syntax)	**Description**	**Example**
GROUP BY Extensions		
GROUPING SETS	Enables multiple GROUP BY clauses to be performed with a single query.	```SELECT name, category, AVG(retail) FROM publisher JOIN books USING (pubid) GROUP BY GROUPING SETS (name, category, (name,category),());```
CUBE	Performs aggregations for all possible combinations of columns included.	```SELECT name, category, AVG(retail) FROM publisher JOIN books USING (pubid) GROUP BY CUBE(name, category) ORDER BY name, category;```
ROLLUP	Performs increasing levels of cumulative subtotals based on the column list provided.	```SELECT name, category, AVG(retail) FROM publisher JOIN books USING (pubid) GROUP BY ROLLUP(name, category) ORDER BY name, category;```

FIGURE 11-1   Selected group functions and GROUP BY extensions (continued)

353

NOTE

If you have executed the prech08.sql file as instructed in Chapter 8, you will be able to work through the queries shown in this chapter.

GROUP FUNCTIONS

Multiple-row functions are commonly referred to as group functions because they process groups of rows. Because these functions return only one result per group of data, they are also known as **aggregate functions**. We will use two new clauses in the SELECT statement to add grouping capabilities: GROUP BY and HAVING. Figure 11-2 shows the position of these clauses in the SELECT statement.

```
SELECT *|columnname, columnname...
FROM tablename
[WHERE condition]
[GROUP BY columnname, columnname...]
[HAVING group condition];
```

FIGURE 11-2   Location of multiple-row functions in the SELECT statement

Keep the following two rules in mind as you work with group functions:

1. Use the DISTINCT keyword to include only unique values. *The ALL keyword is the default*, and it instructs Oracle 10g to include all values (except nulls).
2. All group functions except the COUNT(*) function ignore NULL values. To include NULL values, nest the NVL function within the group function.

SUM Function

The **SUM** function is used to calculate the total amount stored in a numeric field for a group of records. The syntax of the SUM function is SUM([DISTINCT|ALL] n), where n is a column containing numeric data. The optional DISTINCT keyword instructs Oracle 10g to include only *unique* numeric values in its calculation. The ALL keyword instructs Oracle 10g to include *multiple* occurrences of numeric values when totaling a field. If the DISTINCT or ALL keywords are not included when the SUM function is used in a query, Oracle 10g assumes the ALL keyword by default and uses all the numeric values in the field when the query is executed, as shown in Figure 11-3.

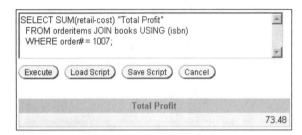

```
SELECT SUM(retail-cost) "Total Profit"
  FROM orderitems JOIN books USING (isbn)
  WHERE order# = 1007;
```

Execute   Load Script   Save Script   Cancel

Total Profit

73.48

FIGURE 11-3   Using the SUM function to calculate order profit

In Figure 11-3, the query calculates the total profit from books sold in order 1007. The SUM function in the SELECT clause uses the argument of **retail - cost** to instruct Oracle 10g to calculate the profit generated by each book before totaling the profit. Notice that the SELECT clause also includes a column alias to describe the output. As with single-row functions, if a group function is used in a SELECT clause, the actual function is displayed as the column heading unless a column alias is assigned.

The WHERE clause in Figure 11-3 restricts the rows used in the calculation to only those books that appear in order 1007. Because the books ordered are identified in the ORDER-ITEMS table, and the cost and retail price of the books are stored in the BOOKS table, the two tables are joined in the FROM clause. The difference between the cost and retail price for each book is calculated, then individually calculated profits are totaled, and, finally, the total profit for the order is presented as a single output for the query, as shown in Figure 11-3.

Suppose that management wants to determine total sales for one day—April 2, 2005. You might assume that you could simply query the ORDERS table; however, this table does not include the total "amount due" for an order. For example, one customer's order might have been for two copies of *Revenge of Mickey* and one copy of *Handcranked Computers*. The only way to calculate how much that customer owes for his order is to multiply the quantity of books purchased by the retail price of each book (quantity * retail).

This is commonly referred to as an "extended price." The extended prices are then totaled to yield a total amount due for the customer. (You will learn how to determine a customer's amount due later in the chapter.) However, in this case, management wants to know just the total sales for April 2, 2005. To calculate that total, simply add the extended prices for all orders placed on that day. The query in Figure 11-4 determines the day's total sales.

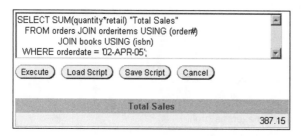

```
SELECT SUM(quantity*retail) "Total Sales"
  FROM orders JOIN orderitems USING (order#)
       JOIN books USING (isbn)
 WHERE orderdate = '02-APR-05';
```

Execute   Load Script   Save Script   Cancel

Total Sales
387.15

FIGURE 11-4   Using the SUM function to calculate total sales for a specified date

Let's look at each of the clauses in Figure 11-4. The SELECT clause uses the SUM function to calculate total sales by adding the extended price for each item ordered. The argument of the SUM function calculates the extended price for each book ordered. Notice that in the FROM clause, three tables must be joined. WHY? The date of the order is stored in the ORDERS table, the actual books and quantity ordered are stored in the ORDERITEMS table, and the retail price of each book is stored in the BOOKS table. To calculate the day's sales, each of these tables must be included in the FROM clause. The WHERE clause is included to restrict the calculation to only the orders that were placed on April 2, 2005. As shown in Figure 11-4, total sales for that date amounted to $387.15.

AVG Function

The **AVG** function calculates the average of the numeric values in a specified column. The syntax of the AVG function is AVG([DISTINCT|ALL] n), where n is a column containing numeric data.

For example, if the management of JustLee Books wants to know the average profit generated by all books in the Computer category, the WHERE clause can be used to restrict the rows processed to those containing the value "COMPUTER" in the Category column. As with the SUM function, first the profit for each book is calculated and then totaled. That total is then divided by the number of records that contain non-NULL values in the specified field, as shown in Figure 11-5.

As shown in the first query in Figure 11-5, the average profit returned by books in the Computer Category is $18.2625. When a query includes division, the resulting display might include more than two decimal positions. Because management prefers the results to be displayed with only two decimal positions, the TO_CHAR function is included in the second query in Figure 11-5 to specify the correct format. When the TO_CHAR function is used on numeric data, any excess decimals are rounded (not truncated) to the specified number of digits. As shown in this example, group functions can be nested inside single-row functions.

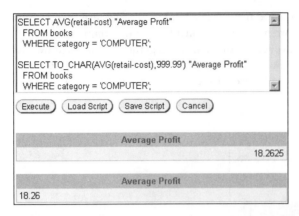

FIGURE 11-5 Using the AVG function to calculate average profit

NOTE

If you are working with the client SQL*Plus tool (not the Internet SQL*Plus tool), the entire column alias is not displayed in the output of the second query. Why? When the TO_CHAR function is included to apply a format model to the average profit, the numeric values are converted to characters for display purposes. Because the average profit is now considered to be a character string, Oracle 10*g* truncates the column alias to match the width of the displayed column.

COUNT Function

Depending on the argument used, the **COUNT** function can either (1) count the records that have non-NULL values in a specified field or (2) count the total records that meet a specific condition, including those containing NULL values. The syntax of the COUNT function is COUNT(*|[|DISTINCT|ALL] c), where c represents any type of column, numeric or nonnumeric.

The query in Figure 11-6 tells Oracle 10g to use the COUNT function to return the number of distinct categories represented by the titles currently stored in the BOOKS table.

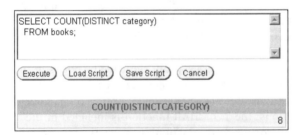

FIGURE 11-6 Using the COUNT function with a DISTINCT operation

Notice that in Figure 11-6, the DISTINCT keyword precedes the column name in the argument of the COUNT function, rather than appearing directly after the SELECT keyword in the SELECT clause. This instructs Oracle 10g to count each different value found in the Category column. If the DISTINCT keyword were entered directly after SELECT, it would apply to the entire COUNT function and would have been interpreted to mean that only duplicate rows, not duplicate category values, should be suppressed. In this case, the COUNT function returns only one value (or row), so the DISTINCT keyword would have no effect on the results.

As shown in Figure 11-7, if the DISTINCT keyword is included as part of the SELECT clause, Oracle 10g returns an actual count of how many rows are in the BOOKS table. Why? Because the DISTINCT keyword applies to the results of the COUNT function—after all the rows containing a value in the Category column have been counted.

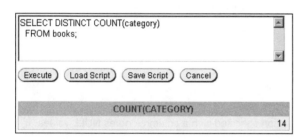

```
SELECT DISTINCT COUNT(category)
  FROM books;
```

Execute   Load Script   Save Script   Cancel

COUNT(CATEGORY)

14

FIGURE 11-7 Flawed query: the COUNT function now counts all rows with a category value

However, in this case, management wants to know how many different categories are represented by the books in the BOOKS table. Because the DISTINCT keyword should apply to the contents of the Category column, the keyword must be listed immediately before the column name, inside the argument of the COUNT function. As was shown in Figure 11-6, the 14 titles in the BOOKS table represent eight different categories.

Suppose that management asks how many orders are currently outstanding; that is, they have not been shipped to customers. One solution is to print a list of all orders that have a NULL value for the date shipped. However, you would still need to count the records returned in the results. There is a simpler solution, as shown in Figure 11-8.

```
SELECT COUNT(*)
  FROM orders
  WHERE shipdate IS NULL;
```

Execute   Load Script   Save Script   Cancel

COUNT(*)

6

FIGURE 11-8 Using the COUNT(*) function to include NULL values

When the argument supplied in the COUNT function is an asterisk (*), the existence of the entire record is counted. By counting the entire record, a NULL value will not be discarded by the COUNT function. As shown by the query in Figure 11-8, the WHERE clause restricts the rows that should be counted to only those that do not have a value stored in the Shipdate column. By contrast, look at the flawed query in Figure 11-9.

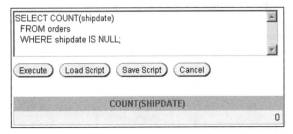

FIGURE 11-9   Flawed query: the COUNT function specified with a column ignores NULL values

The query in Figure 11-9 is a modification of the example shown in Figure 11-8 and illustrates a common query error—the asterisk has been replaced with the Shipdate column in the COUNT argument. Because the WHERE clause restricts the records to only those having a NULL value in the Shipdate column, the function returned a count of zero. Basically, because the specified column contained no value, there was nothing to count. Therefore, *whenever NULL values may affect the COUNT function, you should use an asterisk rather than a column name as the argument.*

MAX Function

The **MAX** function returns the largest value stored in the specified column. The syntax for the MAX function is MAX([DISTINCT|ALL] c), where c can represent any numeric, character, or date field. The query shown in Figure 11-10 retrieves the maximum profit generated by a book.

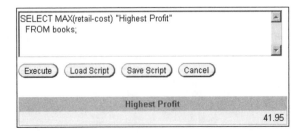

FIGURE 11-10   Using the MAX function on numeric data

As shown in the results, the largest profit earned by a single book is $41.95.

The MAX function can also be used with nonnumeric data. With nonnumeric data, the output shows the first value that occurs when a column is sorted in descending order. Similarly, if a column contains dates, the most recent date is considered to have the highest value (based on its Julian date, as discussed in Chapter 10). If the MAX function is applied to a character column, the letter *Z* would have a higher, or larger, "value" than the letter *A*.

For illustrative purposes, suppose that you needed to find the book title that appears at the end of a list of book titles sorted alphabetically in ascending order (A-Z). In other words, the book title has the highest value of all books in the BOOKS table. As shown in Figure 11-11, when the MAX function is applied to the Title column, the book title *The Wok Way to Cook* is displayed.

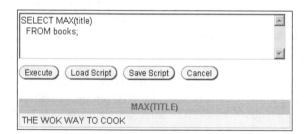

FIGURE 11-11   Using the MAX function on character data

MIN Function

In contrast to the MAX function, the **MIN** function returns the smallest value in a specified column. As with the MAX function, the MIN function works with any numeric, character, or date column. The syntax for the MIN function is MIN([DISTINCT|ALL] c), where c represents a character, numeric, or date column. Figure 11-12 shows how the MIN function can be used to find the book with the earliest publication date in the BOOKS table.

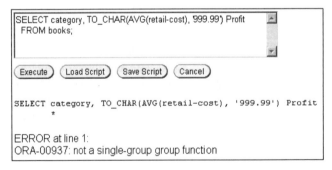

(Figure shows a SQL query interface:)

```
SELECT MIN(pubdate)
  FROM books;
```

Execute   Load Script   Save Script   Cancel

MIN(PUBDATE)
17-JUL-00

FIGURE 11-12   Using the MIN function on date data

In this case, the earliest publication date is July 17, 2000. Of course, if you wanted the most recently published book, you would substitute the MIN function with the MAX function. The MIN function uses the same logic as the MAX function for numeric and character data—except that it returns the smallest value rather than the largest value.

GROUPING DATA

In Figure 11-5, the SELECT query returned the average profit of all books in the Computer category. However, suppose that the management of JustLee Books wants to know the average profit for each category of books. One solution is to reissue the query in Figure 11-5, once for each category, using the WHERE clause to restrict the query to a specific category each time. An alternative solution is to divide the records in the BOOKS table into groups, and then calculate the average for each group—preferably all in one query. This can be done with the GROUP BY clause. The syntax for the GROUP BY clause is GROUP BY *columnname* [, *columnname*,...], where the *columnname* is the column(s) to be used to create the groups or sets of data.

What happens when you attempt to create this query by adding the Category column in the SELECT clause and without the GROUP BY clause? In Figure 11-13, the SELECT clause includes both the single column named Category and the AVG group function. Because the SELECT statement did not include a GROUP BY clause, an error message is returned.

```
SELECT category, TO_CHAR(AVG(retail-cost), '999.99') Profit
  FROM books;
```

Execute   Load Script   Save Script   Cancel

```
SELECT category, TO_CHAR(AVG(retail-cost), '999.99') Profit
       *
ERROR at line 1:
ORA-00937: not a single-group group function
```

FIGURE 11-13   Flawed query: including both aggregate and nonaggregate columns requires a
                    GROUP BY clause

To specify that groups should be created, add the GROUP BY clause to the SELECT statement. When using the GROUP BY clause, remember the following:

1. If a group function is used in the SELECT clause, then any individual columns (nonaggregate) listed in the SELECT clause must be listed in the GROUP BY clause.
2. Columns used to group data in the GROUP BY clause do not have to be listed in the SELECT clause. They are included in the SELECT clause only to have the groups identified in the output.
3. Column aliases cannot be used in the GROUP BY clause.
4. Results returned from a SELECT statement that include a GROUP BY clause will present the results in ascending order of the column(s) listed in the GROUP BY clause. To present the results in a different sort sequence, use the ORDER BY clause.

The required GROUP BY clause is added in Figure 11-14 to correct the error in the query in Figure 11-13.

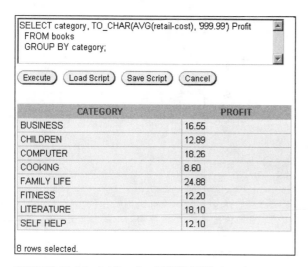

FIGURE 11-14   Adding the GROUP BY clause

In the example in Figure 11-14, the single column name listed in the SELECT clause (Category) is included in the GROUP BY clause. Then, when the query is executed, the records in the BOOKS table are first grouped by category, and then the average profit for each category is calculated. Because the Category column is listed in the SELECT clause, the name of each category is displayed with the average profit generated by each category.

Recall that in Figure 11-4, the SUM function was used to calculate the total sales for a particular day. But how can you determine how much each customer owes for an order? The SUM function is a group function and, therefore, returns one total for all rows processed; but how do you display a list of all orders and the total amount due for *each* order? This is the perfect scenario for the GROUP BY clause.

For example, if the Billing Department requests a list of the amount due from each customer for each order, use the GROUP BY function to group the rows for each order, and then use the SUM function to calculate the extended price for the items ordered and return the total amount due for each order. The SQL statement to create this list is shown in Figure 11-15.

```
SELECT customer#, order#, SUM(quantity*retail) "Order Total"
  FROM orders JOIN orderitems USING (order#)
        JOIN books USING (isbn)
  GROUP BY customer#, order#;
```

(Execute) (Load Script) (Save Script) (Cancel)

CUSTOMER#	ORDER#	Order Total
1001	1003	106.85
1001	1018	75.9
1003	1006	54.5
1003	1016	89.95
1004	1008	39.9
1005	1000	19.95
1005	1009	41.95
1007	1007	347.25
1007	1014	44
1008	1020	19.95
1010	1001	121.9
1010	1011	89.95
1011	1002	111.9
1014	1013	55.95
1015	1017	17.9
1017	1012	170.9
1018	1005	39.95
1018	1019	22
1019	1010	55.95
1020	1004	179.9
1020	1015	19.95

21 rows selected.

FIGURE 11-15 Query to calculate the total amount due for each order

When the statement is executed, Oracle 10g displays each order, the number of the customer who placed the order, and the total amount due for each order. Because the SELECT clause in Figure 11-15 includes the individual Customer# and Order# columns, these columns must also be listed in the GROUP BY clause. You might ask whether it is necessary to include the order number in the query. Suppose that a customer recently placed two orders. If the order number had not been included in the query, it is possible the SQL statement would return the total amount due from each customer, not the amount due from a customer for each order. The customer number was included to identify who placed each order.

RESTRICTING AGGREGATED OUTPUT

The **HAVING** clause is used to restrict the groups returned by a query. If you need to use a group function to restrict groups, you must use the HAVING clause because *the WHERE clause cannot contain group functions*. Although the WHERE clause restricts the records that enter the query for processing, the HAVING clause specifies which groups will be displayed in the results. Suppose you want to display categories with an average profit of more than $15.00. Figure 11-16 shows a query to accomplish this task. Notice that the HAVING clause serves as the WHERE clause for groups.

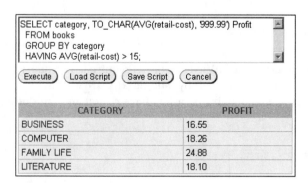

FIGURE 11-16   Using a HAVING clause to restrict groups displayed

In this case, only four categories return an average profit of more than $15.00. The syntax for the HAVING clause is the keyword HAVING followed by the group condition(s): HAVING *groupfunction comparisonoperator value*. The GROUP BY clause causes the average profit to be calculated for each category. Then, the HAVING clause checks each category average to see if it is greater than $15.00. Also, just as with the WHERE clause, the logical operators NOT, AND, and OR can be used to join group conditions in the HAVING clause, if necessary.

Keep in mind that a WHERE clause can still be used to restrict specific rows in the query. In Figure 11-17, the WHERE clause restricts the records to be processed to only those having a publication date after January 1, 2002. The GROUP BY clause groups the records that meet the publication date restriction by the category to which each book is assigned. The HAVING clause is used to restrict the group data displayed to only those with an average profit greater than $15.00. Whenever a SELECT statement includes all three clauses, the order in which they are evaluated is as follows:

1. The WHERE clause
2. The GROUP BY clause
3. The HAVING clause

In essence, the WHERE clause filters the data *before* grouping, whereas the HAVING clause filters the groups *after* the grouping occurs.

Suppose that after the Billing Department receives the list created in Figure 11-15, the department manager asks for another list of the amount due—but only for orders with a total amount due greater than $100.00. Because output is restricted (based on the results

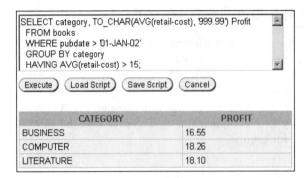

```
SELECT category, TO_CHAR(AVG(retail-cost), '999.99') Profit
  FROM books
  WHERE pubdate > '01-JAN-02'
  GROUP BY category
  HAVING AVG(retail-cost) > 15;
```

Execute   Load Script   Save Script   Cancel

CATEGORY	PROFIT
BUSINESS	16.55
COMPUTER	18.26
LITERATURE	18.10

FIGURE 11-17   Using both the WHERE and HAVING clauses

of the SUM function, which is a group function), a HAVING clause is required. The query previously shown in Figure 11-15 could be modified by adding **HAVING SUM (quantity*retail)>100**. As shown in Figure 11-18, only six orders have a total amount due greater than $100.

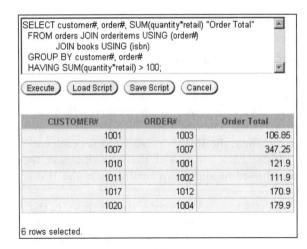

```
SELECT customer#, order#, SUM(quantity*retail) "Order Total"
  FROM orders JOIN orderitems USING (order#)
       JOIN books USING (isbn)
  GROUP BY customer#, order#
  HAVING SUM(quantity*retail) > 100;
```

Execute   Load Script   Save Script   Cancel

CUSTOMER#	ORDER#	Order Total
1001	1003	106.85
1007	1007	347.25
1010	1001	121.9
1011	1002	111.9
1017	1012	170.9
1020	1004	179.9

6 rows selected.

FIGURE 11-18   Using a HAVING clause to restrict grouped output

NESTING FUNCTIONS

When group functions are nested, *the inner function is resolved first*, just as with single-row functions. The result of the inner function is passed back as input for the outer function. Unlike single-row functions that have no restriction on how many nesting levels can occur, *group functions can be nested only to a depth of two*. As was shown in Figure 11-16, group functions can be nested inside single-row functions. In addition, single-row functions can be nested inside group functions.

Suppose you need to calculate the average sales amount per order. To do this, you first need to calculate the total sales by order and then compute the average of those amounts. In Figure 11-19, the SUM function is nested inside the AVG function to determine the average total amount for an order. The GROUP BY clause first groups all the records, based on the Order# column. Then, the total order amount is calculated for each order by the SUM function. The AVG function is used to calculate the average of the total order amounts that were calculated by the SUM function. The resulting output is the average total amount due for orders currently stored in the ORDERS table.

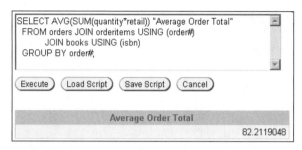

```
SELECT AVG(SUM(quantity*retail)) "Average Order Total"
   FROM orders JOIN orderitems USING (order#)
        JOIN books USING (isbn)
   GROUP BY order#;
```

Execute   Load Script   Save Script   Cancel

Average Order Total
82.2119048

FIGURE 11-19 Nesting group functions

365

HELP

Don't forget to include two closing parentheses at the end of the function in the SELECT clause. The first parenthesis closes the SUM function, and the second closes the AVG function.

STATISTICAL GROUP FUNCTIONS

Oracle 10g provides **statistical group functions** to perform calculations for data analysis. In most organizations, functional areas, such as Marketing and Accounting, need to perform data analyses to detect sales trends, price fluctuations, and so on. Oracle 10g provides functions to support basic statistical calculations such as standard deviation and variance. Although these calculations are easy to perform in Oracle 10g, most people need training in statistical analysis to interpret the results of the calculations. This chapter's discussion of the results of those calculations is intended to give you only an overview of the calculations' purpose, not to train you in statistical analysis. The statistical functions to be covered in this chapter are STDDEV and VARIANCE.

STDDEV Function

The **STDDEV** function calculates the standard deviation for a specified field. A **standard deviation** calculation is used to determine how close individual values are to the mean, or average, of a group of numbers. The syntax for the STDDEV function is STDDEV([DISTINCT|ALL] n), where n represents a numeric column.

The SELECT statement shown in Figure 11-20 displays each book category in JustLee Books' database, the average profit for each category, and the standard deviation of the profit for each category. Let's look at how to interpret these results.

```
SELECT category, AVG(retail-cost), STDDEV(retail-cost)
  FROM books
  GROUP BY category;
```

(Execute) (Load Script) (Save Script) (Cancel)

CATEGORY	AVG(RETAIL-COST)	STDDEV(RETAIL-COST)
BUSINESS	16.55	0
CHILDREN	12.89	13.0956176
COMPUTER	18.2625	11.2267074
COOKING	8.6	1.6263456
FAMILY LIFE	24.875	24.1476966
FITNESS	12.2	0
LITERATURE	18.1	0
SELF HELP	12.1	0

8 rows selected.

FIGURE 11-20 Using the STDDEV function

For the value calculated by the standard deviation to be useful, it must be compared to the calculated "average profit" for each category. For example, look at the Cooking category in Figure 11-20. The average profit for books in the Cooking category is $8.60. However, are most books close to that average, or do the majority of the books generate a small profit (perhaps of only $1), while one book generates a large profit (of perhaps $20)—which would inflate the average? The standard deviation is a statistical approximation of how many books within a certain category are within a certain range around the average.

The STDDEV function is based on the concept of normal distribution. A **normal distribution** means that if you input a large number of data values, those values tend to cluster around some average value. The basic assumption is that as you move closer to the average value, you will find the majority of data values clustered around that average value. However, some values may be extreme, and thus they may be much larger or smaller than the average value. Of course, each extreme data value affects the group's average. For example, calculating the average of 5, 6, 7, and 100 would result in a larger group average than if the 100 had not been included. When performing statistical analysis, the standard deviation is calculated to determine how closely the data match the average value for the group.

In a normal distribution, you can expect to find 68% of the books within one standard deviation (plus or minus) of the average, and 95% of the books within two standard deviations (plus or minus) of the average. In simpler terms, again using the example in Figure 11-20, 68% of the books will have a profit between $6.97 ($8.60 - $1.63) and $10.23 ($8.60 + $1.63), and 95% of the books will have a profit between $5.34 ($8.60 - $3.26) and $11.86 ($8.60 + $3.26). The standard deviation can give management a quick picture of the average profit for books in a particular category, without having to examine each book—a very time-consuming task if the category includes thousands of books.

Notice that some of the categories in Figure 11-20 have a standard deviation of zero. If the STDDEV function is processing only one record per group, the result will always be zero due to the mathematical formula used to calculate the standard deviation.

VARIANCE Function

The **VARIANCE** function is used to determine how widely data is spread in a group. The variance of a group of records is calculated based on the minimum and maximum values for a specified field. If the data values are closely clustered together, the variance will be small. However, if the data contains extreme values (unusually high or low values), the variance will be larger. The syntax for the VARIANCE function is VARIANCE([DISTINCT|ALL] *n*), where *n* represents a numeric field. Figure 11-21 presents the VARIANCE function.

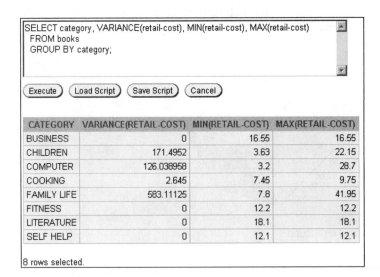

FIGURE 11-21 Using the VARIANCE function

The query in Figure 11-21 lists the categories for all books in the BOOKS table, the profit variance of each category, and (for comparison purposes) the lowest and highest profit within each category. As with the standard deviation, if a group of data consists of only one value (as is the case with the Business, Fitness, Literature, and Self Help categories), the calculated variance is zero. However, unlike standard deviation, variance is not measured with the same units (for example, dollars) as the source data used for the calculation.

To interpret the results of a VARIANCE function, you must look at how large or small the value is. For example, the Cooking category has a smaller variance than the other categories. This means that the profits for books in the Cooking category are clustered tightly together (that is, the profit does not cover a wide range). Look at the minimum and maximum profit for all books in the Cooking category, and notice that the profit range is $2.30 ($9.75 - $7.45). On the other hand, look at the Family Life category. This category has the largest variance, and if you compare the minimum and maximum profit, it has the largest profit range of all the categories presented. This should throw up a warning flag to management that some books may generate very little profit, whereas others may return a very large profit, and that considering only the average profit for books in the Family Life category should not be used as the basis for decision making.

ENHANCED AGGREGATION FOR REPORTING

Oracle provides extensions to the GROUP BY clause, which allow both aggregation across multiple dimensions or the generation of increasing levels of subtotals with a single SELECT statement. A **dimension** is a term used to describe any category used in analyzing data, such as time, geography, and product line. Each dimension could contain various levels of aggregation. For example, the time dimension may include aggregation by month, quarter, and year. Multidimensional analysis involves data aggregated across more than one dimension. A cube of data, as shown in Figure 11-22, is a common analogy used to help you visualize data with many dimensions.

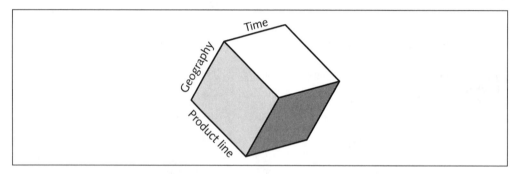

FIGURE 11-22   A multidimensional data cube

To achieve aggregated results across multiple dimensions or increasing levels of aggregation in one dimension, a series of aggregate queries joined with a UNION operator are required. However, these statements are not very efficient. The GROUPING SETS expression is a much simpler statement that achieves the same task. The GROUPING clause uses a single scan to compute all aggregates, which typically leads to improved performance.

NOTE

The advanced aggregation extensions of GROUP BY presented in this section are widely used in Online Analytical Processing (OLAP) and data warehousing.

A basic Microsoft Excel PivotTable can help demonstrate the need for multidimensional data. A pivot table allows the user to drag dimensions of data to various areas on a spreadsheet to display different aggregations. For example, Figure 11-23 displays a pivot table with two dimensions: Publisher and Category. This table analyzes the total number of books available.

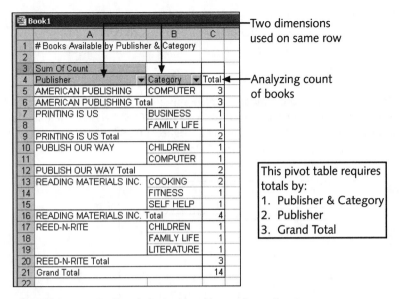

Two dimensions
used on same row

Analyzing count
of books

This pivot table requires totals by:
1. Publisher & Category
2. Publisher
3. Grand Total

FIGURE 11-23   An Excel pivot table with two dimensions on a row

This analysis requires three levels of aggregation, as listed in Figure 11-23. What happens when the user drags the Category dimension to the column area? Figure 11-24 shows the results of this action and that four levels of aggregation are now required.

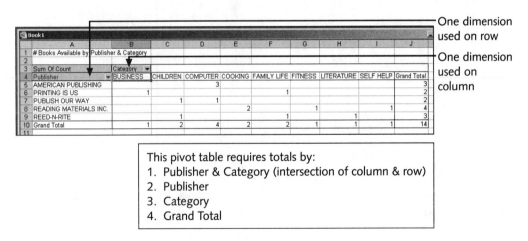

One dimension
used on row

One dimension
used on
column

This pivot table requires totals by:
1. Publisher & Category (intersection of column & row)
2. Publisher
3. Category
4. Grand Total

FIGURE 11-24   An Excel pivot table with one row and one column dimension

This type of analysis is made possible by the use of multidimensional data or a data cube. The data cube stores the data with all possible aggregations to enable quicker and more versatile analyses.

Grouping Sets

The GROUPING SETS expression is the fundamental component on which the other GROUP BY extensions of ROLLUP and CUBE are built. GROUPING SETS enables a single query statement to perform multiple GROUP BY clauses. Figure 11-25 shows the GROUPING SETS expression added to the GROUP BY clause.

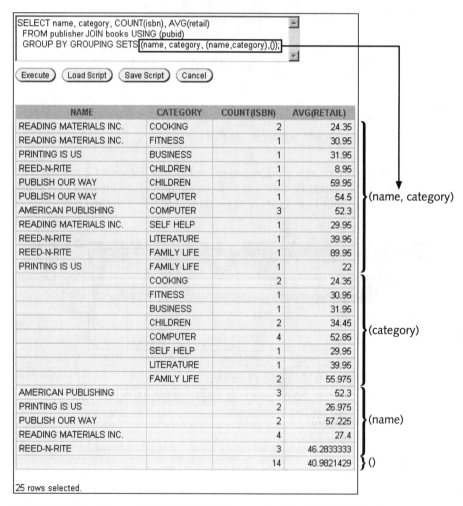

FIGURE 11-25 Using a GROUPING SETS expression in a GROUP BY clause

The column arguments in parentheses following the GROUPING SETS expression indicate the different GROUP BY operations to be performed—four different groups in this case. The () argument indicates that an overall total aggregation is desired. The blanks within the column output indicate a subtotal row. Without the GROUPING SET expression, this task would require combining results of four separate queries with the UNION operator, as shown in Figure 11-26.

```
SELECT name, category, COUNT(isbn), AVG(retail)
  FROM publisher JOIN books USING (pubid)
  GROUP BY (name, category)
UNION
SELECT NULL, category, COUNT(isbn), AVG(retail)
  FROM publisher JOIN books USING (pubid)
  GROUP BY (NULL, category)
UNION
SELECT name, NULL, COUNT(isbn), AVG(retail)
  FROM publisher JOIN books USING (pubid)
  GROUP BY (name, NULL)
UNION
SELECT NULL, NULL, COUNT(isbn), AVG(retail)
  FROM publisher JOIN books USING (pubid);

(Execute)  (Load Script)  (Save Script)  (Cancel)
```

NAME	CATEGORY	COUNT(ISBN)	AVG(RETAIL)
AMERICAN PUBLISHING	COMPUTER	3	52.3
AMERICAN PUBLISHING		3	52.3
PRINTING IS US	BUSINESS	1	31.95
PRINTING IS US	FAMILY LIFE	1	22
PRINTING IS US		2	26.975
PUBLISH OUR WAY	CHILDREN	1	59.95
PUBLISH OUR WAY	COMPUTER	1	54.5
PUBLISH OUR WAY		2	57.225
READING MATERIALS INC.	COOKING	2	24.35
READING MATERIALS INC.	FITNESS	1	30.95
READING MATERIALS INC.	SELF HELP	1	29.95
READING MATERIALS INC.		4	27.4
REED-N-RITE	CHILDREN	1	8.95
REED-N-RITE	FAMILY LIFE	1	89.95
REED-N-RITE	LITERATURE	1	39.95
REED-N-RITE		3	46.2833333
	BUSINESS	1	31.95
	CHILDREN	2	34.45
	COMPUTER	4	52.85
	COOKING	2	24.35
	FAMILY LIFE	2	55.975
	FITNESS	1	30.95
	LITERATURE	1	39.95
	SELF HELP	1	29.95
		14	40.9821429

25 rows selected.

FIGURE 11-26   Using the UNION operator to perform the same action as the GROUPING SETS expression

NOTE

The results are displayed in a different order because the UNION operation automatically performs a sorting action.

The GROUPING SETS expression allows control over the specific aggregations to be performed. Two extensions of the GROUP BY clause, CUBE and ROLLUP, perform two pre-described GROUPING SETS actions. The CUBE option performs cross-tabular type aggregations, whereas the ROLLUP option calculates subtotals.

Cube

The CUBE extension of GROUP BY instructs the system to perform aggregations for all possible combinations of the columns indicated. Figure 11-27 shows a statement that uses the CUBE option to perform all aggregate combinations on the two columns of Name (publisher name) and Category.

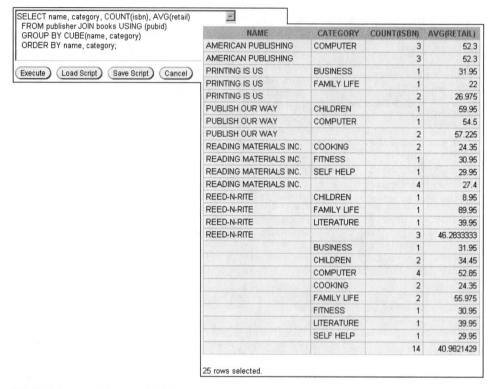

```
SELECT name, category, COUNT(isbn), AVG(retail)
  FROM publisher JOIN books USING (pubid)
  GROUP BY CUBE(name, category)
  ORDER BY name, category;
```

NAME	CATEGORY	COUNT(ISBN)	AVG(RETAIL)
AMERICAN PUBLISHING	COMPUTER	3	52.3
AMERICAN PUBLISHING		3	52.3
PRINTING IS US	BUSINESS	1	31.95
PRINTING IS US	FAMILY LIFE	1	22
PRINTING IS US		2	26.975
PUBLISH OUR WAY	CHILDREN	1	59.95
PUBLISH OUR WAY	COMPUTER	1	54.5
PUBLISH OUR WAY		2	57.225
READING MATERIALS INC.	COOKING	2	24.35
READING MATERIALS INC.	FITNESS	1	30.95
READING MATERIALS INC.	SELF HELP	1	29.95
READING MATERIALS INC.		4	27.4
REED-N-RITE	CHILDREN	1	8.95
REED-N-RITE	FAMILY LIFE	1	89.95
REED-N-RITE	LITERATURE	1	39.95
REED-N-RITE		3	46.2833333
	BUSINESS	1	31.95
	CHILDREN	2	34.45
	COMPUTER	4	52.85
	COOKING	2	24.35
	FAMILY LIFE	2	55.975
	FITNESS	1	30.95
	LITERATURE	1	39.95
	SELF HELP	1	29.95
		14	40.9821429

25 rows selected.

FIGURE 11-27   Using the CUBE extension of GROUP BY

This output matches all the aggregates performed in Figure 11-25 using the GROUPING SETS expression. If only a subset of the four aggregate levels calculated is needed, the GROUPING SETS option must be used because the CUBE option will always perform all aggregation levels.

Identifying subtotal rows is helpful in labeling, sorting, and restricting output. The GROUPING function allows the identification of subtotal rows within the results. The function returns a 1 to identify a row that displays a subtotal for a column, as shown in Figure 11-28.

FIGURE 11-28 The GROUPING function returns a 1 to identify subtotal rows

```
SELECT name, category, COUNT(isbn), AVG(retail),
        GROUPING(name), GROUPING(category)
 FROM publisher JOIN books USING (pubid)
 GROUP BY CUBE(name, category)
```

Execute   Load Script   Save Script   Cancel

NAME	CATEGORY	COUNT(ISBN)	AVG(RETAIL)	GROUPING(NAME)	GROUPING(CATEGORY)
		14	40.9821429	1	1
	COOKING	2	24.35	1	0
	FITNESS	1	30.95	1	0
	BUSINESS	1	31.95	1	0
	CHILDREN	2	34.45	1	0
	COMPUTER	4	52.85	1	0
	SELF HELP	1	29.95	1	0
	LITERATURE	1	39.95	1	0
	FAMILY LIFE	2	55.975	1	0
REED-N-RITE		3	46.2833333	0	1
REED-N-RITE	CHILDREN	1	8.95	0	0
REED-N-RITE	LITERATURE	1	39.95	0	0
REED-N-RITE	FAMILY LIFE	1	89.95	0	0
PRINTING IS US		2	26.975	0	1
PRINTING IS US	BUSINESS	1	31.95	0	0
PRINTING IS US	FAMILY LIFE	1	22	0	0
PUBLISH OUR WAY		2	57.225	0	1
PUBLISH OUR WAY	CHILDREN	1	59.95	0	0
PUBLISH OUR WAY	COMPUTER	1	54.5	0	0
AMERICAN PUBLISHING		3	52.3	0	1
AMERICAN PUBLISHING	COMPUTER	3	52.3	0	1
READING MATERIALS INC.		4	27.4	0	1
READING MATERIALS INC.	COOKING	2	24.35	0	0
READING MATERIALS INC.	FITNESS	1	30.95	0	0
READING MATERIALS INC.	SELF HELP	1	29.95	0	0

25 rows selected.

The results of the GROUPING function could be used in a DECODE operation to label subtotal rows, as shown in Figure 11-29.

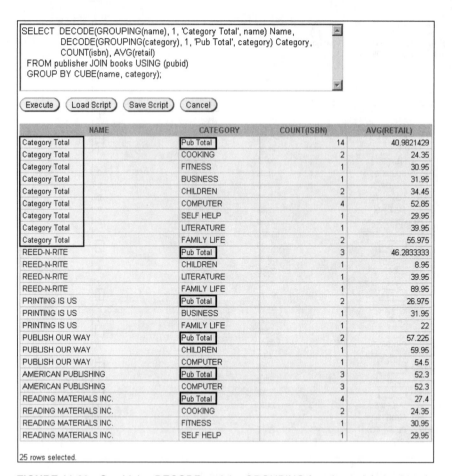

```
SELECT  DECODE(GROUPING(name), 1, 'Category Total', name) Name,
        DECODE(GROUPING(category), 1, 'Pub Total', category) Category,
        COUNT(isbn), AVG(retail)
FROM publisher JOIN books USING (pubid)
GROUP BY CUBE(name, category);
```

(Execute) (Load Script) (Save Script) (Cancel)

NAME	CATEGORY	COUNT(ISBN)	AVG(RETAIL)
Category Total	Pub Total	14	40.9821429
Category Total	COOKING	2	24.35
Category Total	FITNESS	1	30.95
Category Total	BUSINESS	1	31.95
Category Total	CHILDREN	2	34.45
Category Total	COMPUTER	4	52.85
Category Total	SELF HELP	1	29.95
Category Total	LITERATURE	1	39.95
Category Total	FAMILY LIFE	2	55.975
REED-N-RITE	Pub Total	3	46.2833333
REED-N-RITE	CHILDREN	1	8.95
REED-N-RITE	LITERATURE	1	39.95
REED-N-RITE	FAMILY LIFE	1	89.95
PRINTING IS US	Pub Total	2	26.975
PRINTING IS US	BUSINESS	1	31.95
PRINTING IS US	FAMILY LIFE	1	22
PUBLISH OUR WAY	Pub Total	2	57.225
PUBLISH OUR WAY	CHILDREN	1	59.95
PUBLISH OUR WAY	COMPUTER	1	54.5
AMERICAN PUBLISHING	Pub Total	3	52.3
AMERICAN PUBLISHING	COMPUTER	3	52.3
READING MATERIALS INC.	Pub Total	4	27.4
READING MATERIALS INC.	COOKING	2	24.35
READING MATERIALS INC.	FITNESS	1	30.95
READING MATERIALS INC.	SELF HELP	1	29.95

25 rows selected.

FIGURE 11-29 Combining DECODE and the GROUPING function to label subtotal rows

Rollup

The ROLLUP extension of GROUP BY causes cumulative subtotals to be calculated for the columns indicated. If multiple columns are indicated, subtotals are performed for each of the columns in the argument list except the far-right column. A grand total is also calculated. Figure 11-30 shows a ROLLUP operation on the same two columns used previously in the CUBE operation: Name and Category.

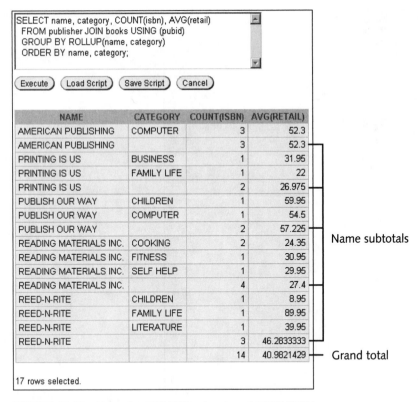

```
SELECT name, category, COUNT(isbn), AVG(retail)
  FROM publisher JOIN books USING (pubid)
  GROUP BY ROLLUP(name, category)
  ORDER BY name, category;
```

Execute   Load Script   Save Script   Cancel

NAME	CATEGORY	COUNT(ISBN)	AVG(RETAIL)
AMERICAN PUBLISHING	COMPUTER	3	52.3
AMERICAN PUBLISHING		3	52.3
PRINTING IS US	BUSINESS	1	31.95
PRINTING IS US	FAMILY LIFE	1	22
PRINTING IS US		2	26.975
PUBLISH OUR WAY	CHILDREN	1	59.95
PUBLISH OUR WAY	COMPUTER	1	54.5
PUBLISH OUR WAY		2	57.225
READING MATERIALS INC.	COOKING	2	24.35
READING MATERIALS INC.	FITNESS	1	30.95
READING MATERIALS INC.	SELF HELP	1	29.95
READING MATERIALS INC.		4	27.4
REED-N-RITE	CHILDREN	1	8.95
REED-N-RITE	FAMILY LIFE	1	89.95
REED-N-RITE	LITERATURE	1	39.95
REED-N-RITE		3	46.2833333
		14	40.9821429

Name subtotals

Grand total

17 rows selected.

FIGURE 11-30   Using the ROLLUP extension of GROUP BY

Three levels of increasing aggregation are performed: 1) combination of Name and Category columns, 2) Name subtotal, and 3) grand total. A partial ROLLUP can be accomplished by including only a subset of the columns in the GROUP BY in the ROLLUP operation. For example, if you need only subtotals by Category and each Name within each Category, the statement in Figure 11-31 uses a partial ROLLUP to accomplish this task.

```
SELECT category, name, COUNT(isbn), AVG(retail)
  FROM publisher JOIN books USING (pubid)
  GROUP BY category, ROLLUP(name)
  ORDER BY  category;
```

Execute   Load Script   Save Script   Cancel

CATEGORY	NAME	COUNT(ISBN)	AVG(RETAIL)
BUSINESS	PRINTING IS US	1	31.95
BUSINESS		1	31.95
CHILDREN	PUBLISH OUR WAY	1	59.95
CHILDREN	REED-N-RITE	1	8.95
CHILDREN		2	34.45
COMPUTER	AMERICAN PUBLISHING	3	52.3
COMPUTER	PUBLISH OUR WAY	1	54.5
COMPUTER		4	52.85
COOKING	READING MATERIALS INC.	2	24.35
COOKING		2	24.35
FAMILY LIFE	PRINTING IS US	1	22
FAMILY LIFE	REED-N-RITE	1	89.95
FAMILY LIFE		2	55.975
FITNESS	READING MATERIALS INC.	1	30.95
FITNESS		1	30.95
LITERATURE	REED-N-RITE	1	39.95
LITERATURE		1	39.95
SELF HELP	READING MATERIALS INC.	1	29.95
SELF HELP		1	29.95

19 rows selected.

FIGURE 11-31 Using a partial ROLLUP

In this case, the column outside the ROLLUP operation, Category, is considered the aggregate value. A subtotal will be calculated for the aggregate value as well as for each unique value of the ROLLUP column within the aggregate value.

Using parentheses on a collection of columns creates a composite column, which is treated as a single unit in grouping operations. First, let's review a ROLLUP operation that contains three columns, as shown in Figure 11-32.

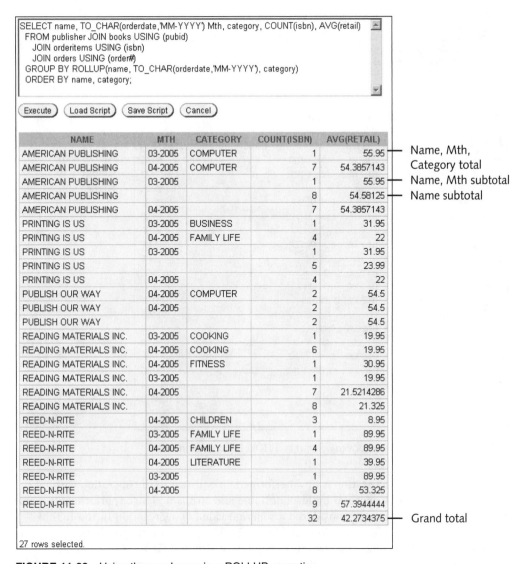

```
SELECT name, TO_CHAR(orderdate,'MM-YYYY') Mth, category, COUNT(isbn), AVG(retail)
  FROM publisher JOIN books USING (pubid)
    JOIN orderitems USING (isbn)
    JOIN orders USING (order#)
  GROUP BY ROLLUP(name, TO_CHAR(orderdate,'MM-YYYY'), category)
  ORDER BY name, category;
```

(Execute) (Load Script) (Save Script) (Cancel)

NAME	MTH	CATEGORY	COUNT(ISBN)	AVG(RETAIL)
AMERICAN PUBLISHING	03-2005	COMPUTER	1	55.95
AMERICAN PUBLISHING	04-2005	COMPUTER	7	54.3857143
AMERICAN PUBLISHING	03-2005		1	55.95
AMERICAN PUBLISHING			8	54.58125
AMERICAN PUBLISHING	04-2005		7	54.3857143
PRINTING IS US	03-2005	BUSINESS	1	31.95
PRINTING IS US	04-2005	FAMILY LIFE	4	22
PRINTING IS US	03-2005		1	31.95
PRINTING IS US			5	23.99
PRINTING IS US	04-2005		4	22
PUBLISH OUR WAY	04-2005	COMPUTER	2	54.5
PUBLISH OUR WAY	04-2005		2	54.5
PUBLISH OUR WAY			2	54.5
READING MATERIALS INC.	03-2005	COOKING	1	19.95
READING MATERIALS INC.	04-2005	COOKING	6	19.95
READING MATERIALS INC.	04-2005	FITNESS	1	30.95
READING MATERIALS INC.	03-2005		1	19.95
READING MATERIALS INC.	04-2005		7	21.5214286
READING MATERIALS INC.			8	21.325
REED-N-RITE	04-2005	CHILDREN	3	8.95
REED-N-RITE	03-2005	FAMILY LIFE	1	89.95
REED-N-RITE	04-2005	FAMILY LIFE	4	89.95
REED-N-RITE	04-2005	LITERATURE	1	39.95
REED-N-RITE	03-2005		1	89.95
REED-N-RITE	04-2005		8	53.325
REED-N-RITE			9	57.3944444
			32	42.2734375

27 rows selected.

— Name, Mth, Category total (rows 1–2)
— Name, Mth subtotal (row 3)
— Name subtotal (row 4)
— Grand total (last row)

FIGURE 11-32 Using three columns in a ROLLUP operation

The ROLLUP operation performs four levels of increasing aggregation: 1) Name, Mth, and Category, 2) Name and Mth, 3) Name, and 4) grand total. Now, we will modify this statement to combine the Mth and Category as a composite column. Figure 11-33 shows the ROLLUP with the composite column, which is created by placing parentheses around the two columns in the ROLLUP arguments.

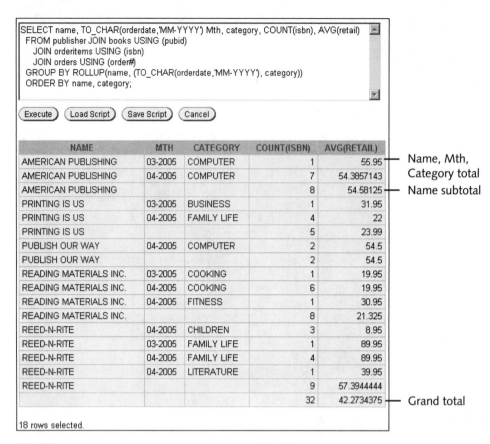

```
SELECT name, TO_CHAR(orderdate,'MM-YYYY') Mth, category, COUNT(isbn), AVG(retail)
  FROM publisher JOIN books USING (pubid)
    JOIN orderitems USING (isbn)
    JOIN orders USING (order#)
  GROUP BY ROLLUP(name, (TO_CHAR(orderdate,'MM-YYYY'), category))
  ORDER BY name, category;
```

(Execute) (Load Script) (Save Script) (Cancel)

NAME	MTH	CATEGORY	COUNT(ISBN)	AVG(RETAIL)	
AMERICAN PUBLISHING	03-2005	COMPUTER	1	55.95	— Name, Mth, Category total
AMERICAN PUBLISHING	04-2005	COMPUTER	7	54.3857143	
AMERICAN PUBLISHING			8	54.58125	— Name subtotal
PRINTING IS US	03-2005	BUSINESS	1	31.95	
PRINTING IS US	04-2005	FAMILY LIFE	4	22	
PRINTING IS US			5	23.99	
PUBLISH OUR WAY	04-2005	COMPUTER	2	54.5	
PUBLISH OUR WAY			2	54.5	
READING MATERIALS INC.	03-2005	COOKING	1	19.95	
READING MATERIALS INC.	04-2005	COOKING	6	19.95	
READING MATERIALS INC.	04-2005	FITNESS	1	30.95	
READING MATERIALS INC.			8	21.325	
REED-N-RITE	04-2005	CHILDREN	3	8.95	
REED-N-RITE	03-2005	FAMILY LIFE	1	89.95	
REED-N-RITE	04-2005	FAMILY LIFE	4	89.95	
REED-N-RITE	04-2005	LITERATURE	1	39.95	
REED-N-RITE			9	57.3944444	
			32	42.2734375	— Grand total

18 rows selected.

FIGURE 11-33   Using a composite column in a ROLLUP operation

One less level of aggregation is now performed, in contrast to the operation without the composite column. The aggregation level of Name and Mth is no longer performed because the Mth column is no longer considered a separate value of aggregation in the ROLLUP operation. The ROLLUP operation in Figure 11-33 is considered to have only two arguments: 1) Name and 2) the composite column of Mth and Category combined.

Combinations of groupings can be generated by using concatenated groupings. A concatenated grouping operation is created by listing multiple grouping sets in the grouping operation. Let's say we need accumulated aggregate totals of a book count and the average retail value of books by two different groupings: 1) Name and Category and 2) Mth and Year. Figure 11-34 shows a concatenated grouping that includes two ROLLUP operations.

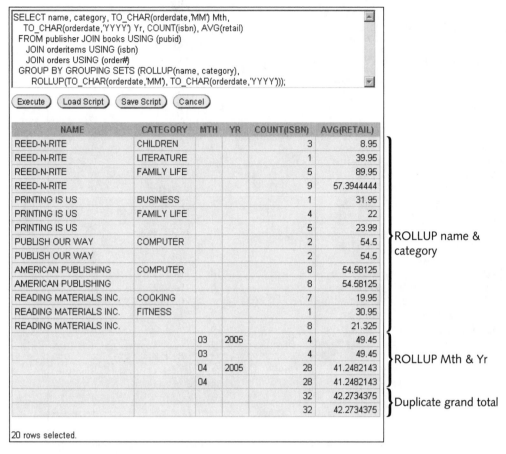

```
SELECT name, category, TO_CHAR(orderdate,'MM') Mth,
  TO_CHAR(orderdate,'YYYY') Yr, COUNT(isbn), AVG(retail)
 FROM publisher JOIN books USING (pubid)
  JOIN orderitems USING (isbn)
  JOIN orders USING (order#)
 GROUP BY GROUPING SETS (ROLLUP(name, category),
   ROLLUP(TO_CHAR(orderdate,'MM'), TO_CHAR(orderdate,'YYYY')));
```

(Execute) (Load Script) (Save Script) (Cancel)

NAME	CATEGORY	MTH	YR	COUNT(ISBN)	AVG(RETAIL)
REED-N-RITE	CHILDREN			3	8.95
REED-N-RITE	LITERATURE			1	39.95
REED-N-RITE	FAMILY LIFE			5	89.95
REED-N-RITE				9	57.3944444
PRINTING IS US	BUSINESS			1	31.95
PRINTING IS US	FAMILY LIFE			4	22
PRINTING IS US				5	23.99
PUBLISH OUR WAY	COMPUTER			2	54.5
PUBLISH OUR WAY				2	54.5
AMERICAN PUBLISHING	COMPUTER			8	54.58125
AMERICAN PUBLISHING				8	54.58125
READING MATERIALS INC.	COOKING			7	19.95
READING MATERIALS INC.	FITNESS			1	30.95
READING MATERIALS INC.				8	21.325
		03	2005	4	49.45
		03		4	49.45
		04	2005	28	41.2482143
		04		28	41.2482143
				32	42.2734375
				32	42.2734375

20 rows selected.
```

ROLLUP name & category

ROLLUP Mth & Yr

Duplicate grand total

**FIGURE 11-34**   Using concatenated groupings

As grouping operations become more complex, duplicate grouping results are sometimes generated. Duplicate grand totals are displayed at the bottom of the output in Figure 11-34. A GROUP_ID function is available for eliminating duplicate grouping results. The GROUP_ID function returns a value of 1 for duplicate output rows. Figure 11-35 adds the GROUP_ID function to the previous statement.
```

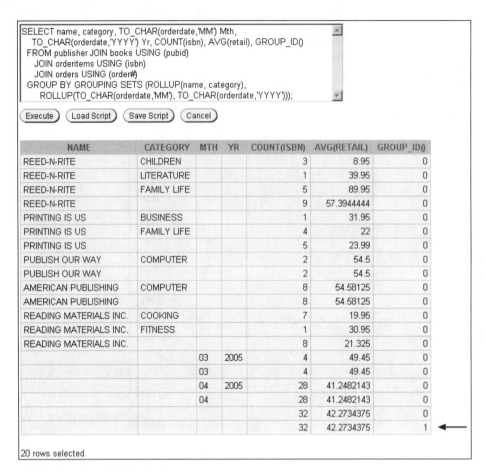

FIGURE 11-35 GROUP_ID function values

Notice that the repeated grand total row has a GROUP_ID of 1. To eliminate the duplicate row in the output, add a HAVING clause of HAVING GROUP_ID() = 0 to the query. The GROUP_ID value can be checked for any grouping level. The example in Figure 11-35 used a GROUP_ID argument of (), which represents the grand total grouping level.

NOTE

Partial groupings, composite columns, and concatenated groupings can also be performed with GROUPING SETS and CUBE operations using the same technique as shown with the ROLLUP examples.

Chapter Summary

- The AVG, SUM, STDDEV, and VARIANCE functions are used only with numeric fields.
- The COUNT, MAX, and MIN functions can be applied to any datatype.
- The AVG, SUM, MAX, MIN, STDDEV, and VARIANCE functions all ignore NULL values. COUNT(*) counts records containing NULL values. To include NULL values in other group functions, the NVL function is required.
- By default, the AVG, SUM, MAX, MIN, COUNT, STDDEV, and VARIANCE functions include duplicate values. To include only unique values, the DISTINCT keyword must be used.
- The GROUP BY clause is used to divide table data into groups.
- If a SELECT clause contains both an individual field name and a group function, the field name must also be included in a GROUP BY clause.
- The HAVING clause is used to restrict groups in a group function.
- Group functions can be nested to a depth of only two. The inner function is always solved first. The results of the nested function are used as input for the outer function.
- The functions STDDEV and VARIANCE are used to perform statistical analyses on a set of data.
- GROUPING SETS operations can be used to perform multiple GROUP BY aggregations with a single query.
- The CUBE option of the GROUP BY calculates aggregations for all possible combinations or groupings of columns included.
- The ROLLUP option of the GROUP BY calculates increasing levels of accumulated subtotals for the column list provided.

Chapter 11 Syntax Summary

The following table presents a summary of the syntax that you have learned in this chapter. You can use the table as a study guide and reference.

SYNTAX GUIDE		
Function (and Syntax)	Description	Example
Group (Multiple-Row) Functions		
SUM([DISTINCT|ALL] n)	Returns the sum or total value of the selected numeric field. Ignores NULL values.	`SELECT SUM (retail-cost) FROM books;`
AVG([DISTINCT|ALL] n)	Returns the average value of the selected numeric field. Ignores NULL values.	`SELECT AVG(cost) FROM books;`

Function (and Syntax)	Description	Example
Group (Multiple-Row) Functions		

COUNT(*|[|DISTINCT| ALL] c)	Returns the number of rows that contain a value in the identified field. Rows containing NULL values in the field will not be included in the results. To count all rows, including those with NULL values, use an * rather than a field name.	`SELECT COUNT(*)`
		`FROM books;` *or*
		`SELECT COUNT`
		`(shipdate)`
		`FROM orders;`

MAX([DISTINCT|ALL] c)	Returns the highest (maximum) value from the selected field. Ignores NULL values.	`SELECT MAX`
		`(customer#)`
		`FROM customers;`

MIN([DISTINCT|ALL] c)	Returns the lowest (minimum) value from the selected field. Ignores NULL values.	`SELECT MIN`
		`(retail-cost)`
		`FROM books;`

STDDEV([DISTINCT| ALL] n)	Returns the standard deviation of the selected numeric field. Ignores NULL values.	`SELECT STDDEV`
		`(retail)`
		`FROM books;`

VARIANCE([DISTINCT| ALL] n)	Returns the variance of the selected numeric field. Ignores NULL values.	`SELECT VARIANCE`
		`(retail)`
		`FROM books;`

Clauses		

GROUP BY columnname[,columnname, ...]	Divides data into sets or groups based on the contents of the specified column(s).	`SELECT AVG(cost)`
		`FROM books`
		`GROUP BY category;`

HAVING groupfunction Comparisonoperator value	Restricts the groups displayed in the results of a query.	`SELECT AVG(cost)`
		`FROM books`
		`GROUP BY category`
		`HAVING AVG`
		`(cost)>21;`

GROUP BY Extensions		

GROUPING SETS	Enables multiple GROUP BY clauses to be performed with a single query.	`SELECT name, category,`
		`AVG(retail) FROM publisher`
		`JOIN books USING (pubid)`
		`GROUP BY GROUPING SETS`
		`(name, category,`
		`(name,category),());`

SYNTAX GUIDE		
Function (and Syntax)	Description	Example
GROUP BY Extensions		

CUBE	Performs aggregations for all possible combinations of columns included.	`SELECT name, category,`
		`AVG(retail) FROM`
		`publisher JOIN books`
		`USING (pubid) GROUP BY`
		`CUBE(name, category)`
		`ORDER BY name, category;`

ROLLUP	Performs increasing levels of cumulative subtotals based on column list provided.	`SELECT name, category,`
		`AVG(retail) FROM`
		`publisher JOIN books`
		`USING (pubid) GROUP BY`
		`ROLLUP(name, category)`
		`ORDER BY name, category;`

Review Questions

To answer these questions, refer to the tables in Appendix A.

1. Explain the difference between single-row and group functions.

2. Which group function can be used to count NULL values?

3. Which clause can be used to restrict the groups returned by a query based on a group function?

4. Under what circumstances *must* you include a GROUP BY clause in a query?

5. In which clause should you include the condition "pubid=4" to restrict the rows processed by a query?

6. In which clause should you include the condition **MAX(cost)>39** to restrict the groups displayed in the results of a query?

7. What is the basic difference between using the ROLLUP and CUBE extensions of the GROUP BY clause?

8. What is the maximum depth allowed when nesting group functions?

9. In what order will output results be presented if a SELECT statement contains a GROUP BY clause and no ORDER BY clause?

10. Which clause is used to restrict the records retrieved from a table? Which clause restricts the groups displayed in the results of a query?

Multiple Choice

To answer these questions, refer to the tables in Appendix A.

1. Which of the following statements is true?

    a. The MIN function can be used only with numeric data.

    b. The MAX function can be used only with date values.

    c. The AVG function can be used only with numeric data.

    d. The SUM function cannot be part of a nested function.

2. Which of the following is a valid SELECT statement?

    a. SELECT AVG(retail-cost) FROM books GROUP BY category;

    b. SELECT category, AVG(retail-cost) FROM books;

    c. SELECT category, AVG(retail-cost) FROM books
       WHERE AVG(retail-cost) > 8.56
       GROUP BY category;

    d. SELECT category, AVG(retail-cost) profit
       FROM books
       GROUP BY category
       HAVING profit > 8.56;

3. Which of the following statements is correct?

    a. The WHERE clause can contain a group function only if that group function is not also listed in the SELECT clause.

    b. Group functions cannot be used in the SELECT, FROM, or WHERE clauses.

    c. The HAVING clause is always processed before the WHERE clause.

    d. The GROUP BY clause is always processed before the HAVING clause.

4. Which of the following is not a valid SQL statement?

    a. SELECT MIN(pubdate)
       FROM books
       GROUP BY category
       HAVING pubid = 4;

    b. SELECT MIN(pubdate)
       FROM books
       WHERE category = 'COOKING';

    c. SELECT COUNT(*)
       FROM orders
       WHERE customer# = 1005;

    d. SELECT MAX(COUNT(customer#))
       FROM orders
       GROUP BY customer#;

5. Which of the following statements is correct?

   a. The COUNT function can be used to determine how many rows contain a NULL value.

   b. Only distinct values are included in group functions, unless the ALL keyword is included in the SELECT clause.

   c. The HAVING clause restricts rows to be processed.

   d. The WHERE clause determines which groups will be displayed in the query results.

   e. None of the statements is correct.

6. Which of the following is a valid SQL statement?

   a. SELECT customer#, order#,
      MAX(shipdate-orderdate)
      FROM orders
      GROUP BY customer#
      WHERE customer# = 1001;

   b. SELECT customer#, COUNT(order#)
      FROM orders
      GROUP BY customer#;

   c. SELECT customer#, COUNT(order#)
      FROM orders
      GROUP BY COUNT(order#);

   d. SELECT customer#, COUNT(order#)
      FROM orders
      GROUP BY order#;

7. Which of the following SELECT statements will list only the book with the largest profit?

   a. SELECT title, MAX(retail-cost)
      FROM books
      GROUP BY title;

   b. SELECT title, MAX(retail-cost)
      FROM books
      GROUP BY title
      HAVING MAX(retail-cost);

   c. SELECT title, MAX(retail-cost)
      FROM books;

   d. none of the above

8. Which of the following is correct?

   a. A group function can be nested inside a group function.

   b. A group function can be nested inside a single-row function.

   c. A single-row function can be nested inside a group function.

   d. A and B

   e. A, B, and C

9. Which of the following functions is used to calculate the total value contained in a specified column?

    a. COUNT

    b. MIN

    c. TOTAL

    d. SUM

    e. ADD

10. Which of the following SELECT statements will list the highest retail price of all books in the Family category?

    a. SELECT MAX(retail)
       FROM books
       WHERE category = 'FAMILY';

    b. SELECT MAX(retail)
       FROM books
       HAVING category = 'FAMILY';

    c. SELECT retail
       FROM books
       WHERE category = 'FAMILY'
       HAVING MAX(retail);

    d. none of the above

11. Which of the following functions can be used to include NULL values in calculations?

    a. SUM

    b. NVL

    c. MAX

    d. MIN

12. Which of the following is not a valid statement?

    a. You must enter the ALL keyword in a function to include all nonunique values.

    b. The AVG function can be used to find the average calculated difference between two dates.

    c. The MIN and MAX functions can be used on any type of data.

    d. All of the above are valid statements.

    e. None of the above are valid statements.

13. Which of the following SQL statements will determine how many total customers were referred by other customers?

    a. SELECT customer#, SUM(referred)
       FROM customers
       GROUP BY customer#;

    b. SELECT COUNT(referred)
       FROM customers;

c. SELECT COUNT(*)
    FROM customers;

d. SELECT COUNT(*)
    FROM customers
    WHERE referred IS NULL;

Use the following SELECT statement to answer questions 14–18.

1 SELECT customer#, COUNT(*)
2 FROM customers NATURAL JOIN orders
3 NATURAL JOIN orderitems
4 WHERE orderdate > '02-APR-06'
5 GROUP BY customer#
6 HAVING COUNT(*) > 2;

14. Which line of the SELECT statement is used to restrict the number of records to be processed by the query?

    a. 1
    b. 4
    c. 5
    d. 6

15. Which line of the SELECT statement is used to restrict the groups displayed in the results of the query?

    a. 1
    b. 4
    c. 5
    d. 6

16. Which line of the SELECT statement is used to group the data contained in the database?

    a. 1
    b. 4
    c. 5
    d. 6

17. Which clause must be included for the query to execute because the SELECT clause contains the Customer# column?

    a. 1
    b. 4
    c. 5
    d. 6

18. The COUNT(*) function in the SELECT clause is used to return:
    a. the number of records in the specified tables
    b. the number of orders placed by each customer
    c. the number of NULL values in the specified tables
    d. the number of customers who have placed an order

19. Which of the following functions can be used to determine the earliest ship date for all orders recently processed by JustLee Books?
    a. COUNT function
    b. MAX function
    c. MIN function
    d. STDDEV function
    e. VARIANCE function

20. Which of the following is not a valid SELECT statement?
    a. SELECT STDDEV(retail)
       FROM books;
    b. SELECT AVG(SUM(retail))
       FROM orders NATURAL JOIN orderitems NATURAL JOIN books
       GROUP BY customer#;
    c. SELECT order#, TO_CHAR(SUM(retail, '999.99')
       FROM orderitems NATURAL JOIN books
       GROUP BY order#;
    d. SELECT title, VARIANCE(retail-cost)
       FROM books
       GROUP BY pubid;

Hands-On Assignments

To perform these activities, refer to the tables in Appendix A.

1. Determine how many books are in the Cooking category.
2. Display the number of books that have a retail price of more than $30.00.
3. Display the date of the most recently published book.
4. Determine the total profit generated by sales to customer 1017. Note: Quantity should be reflected in the total profit calculation.
5. List the least expensive book in the Computer category.
6. Determine the average profit generated by orders contained in the ORDERS table. Note: The total profit by order must be calculated before taking the average.
7. Determine how many orders have been placed by each customer in the CUSTOMERS table. Do not include in the results any customer who has not recently placed an order with JustLee Books.

8. Determine the average retail price of books by publisher name and category. Include only the categories of Children and Computer. Include only the groups that have an average retail price greater than $50.

9. List the customers living in Georgia or Florida who have recently placed an order totaling more than $80.

10. What is the retail price of the most expensive book written by Lisa White?

Advanced Challenge

To perform this activity, refer to the tables in Appendix A.

JustLee Books has a problem: Their book storage space is filling up. As a solution, management is considering limiting the inventory to only those books that return at least a 55% profit. Any book that returns less than a 55% profit would be dropped from inventory and not reordered.

This plan could, however, have a negative impact on overall sales. Management fears that if JustLee stops carrying the less-profitable books, the company might lose repeat business from its customers. As part of management's decision-making process, they want to know whether less-profitable books are frequently purchased by current customers. Therefore, management wants to know how many times these less-profitable books have been purchased recently.

Determine which books generate less than a 55% profit and how many copies of those books have been sold. Summarize your findings for management, and include a copy of the query necessary to retrieve the data from the database tables.

Case Study: *City Jail*

Note: It is assumed that the City Jail database creation script from Chapter 8 has been executed. That script makes all database objects available to complete this case study.

The city's Crimes Analysis unit has submitted the following data requests. Provide the SQL statement that would satisfy the request. Test the statements and show execution results as well.

1. The average number of crimes reported by an officer.

2. The total number of crimes by status.

3. The highest number of crimes committed by an individual.

4. The lowest fine amount assigned to an individual crime charge.

5. List of criminals (id and name) who have multiple sentences assigned.

6. The total number of crime charges successfully defended (guilty status assigned) by precinct. Include only precincts with at least seven guilty charges.

7. List the total amount of collections (fines and fees) and the total amount owed by crime classification.

Use single queries to resolve the following requests:

8. Total number of charges by crime classification and charge status. Include a grand total in the results as well.

9. Accomplish the same task as in question #8 and add the following: 1) a subtotal by each crime classification and 2) a subtotal for each charge status. Provide two different queries that will accomplish this task.

10. Accomplish the same task as in question #8 and add a subtotal by each crime classification. Provide two different queries that will accomplish this task.

CHAPTER 12

SUBQUERIES AND MERGE

LEARNING OBJECTIVES

After completing this chapter, you should be able to do the following:

- Determine when it is appropriate to use a subquery
- Identify which clauses can contain subqueries
- Distinguish between an outer query and a subquery
- Use a single-row subquery in a WHERE clause
- Use a single-row subquery in a HAVING clause
- Use a single-row subquery in a SELECT clause
- Distinguish between single-row and multiple-row comparison operators
- Use a multiple-row subquery in a WHERE clause
- Use a multiple-row subquery in a HAVING clause
- Use a multiple-column subquery in a WHERE clause
- Create an inline view using a multiple-column subquery in a FROM clause
- Compensate for NULL values in subqueries
- Distinguish between correlated and uncorrelated subqueries
- Nest a subquery inside another subquery
- Process multiple DML actions with a MERGE statement

INTRODUCTION

Suppose that the management of JustLee Books requests a list of every computer book that has a higher retail price than *Database Implementation*. In previous chapters, you would have followed this procedure: (1) Query the database to determine the retail price of *Database Implementation*, and then (2) create a second SELECT statement to find the titles of all books retailing for more than *Database Implementation*.

In this chapter, you will learn how to use an alternative approach, called a subquery, to get the same output. A **subquery** is a nested query—one complete query inside another query.

The output of the subquery can consist of a single value (a single-row subquery), several rows of values (a multiple-row subquery), or even multiple columns of data (a multiple-column subquery). This chapter addresses each of these types of subqueries. In addition, the final section of the chapter returns to the topic of DML. At that point, because the more complex queries have been covered, you can learn about the MERGE statement. With the MERGE statement, you can conditionally process multiple DML actions with a single SQL statement. Figure 12-1 provides an overview of this chapter's contents.

SUBQUERY	DESCRIPTION
Single-row subquery	Returns to the outer query one row of results that consists of one column
Multiple-row subquery	Returns to the outer query more than one row of results
Multiple-column subquery	Returns to the outer query more than one column of results
Correlated subquery	References a column in the outer query, and executes the subquery once for every row in the outer query
Uncorrelated subquery	Executes the subquery first and passes the value to the outer query
MERGE statement	Conditionally processes a series of DML statements

FIGURE 12-1 Chapter topics

NOTE

Before attempting to work through the examples provided in this chapter, run the prech12.sql file to make additions needed to the JustLee Books database. This assumes you have already executed the prech08.sql script as instructed in Chapter 8.

SUBQUERIES AND THEIR USES

Sometimes, getting an answer to a query requires a multistep operation. First, you must create a query to determine a value that you don't know but that is contained within the database. That first query is the subquery. The results of the subquery are passed as input to the **parent query**, or **outer query**. The parent query incorporates that value into its calculations to determine the final output.

Although subqueries are most commonly used in the WHERE or HAVING clause of a SELECT statement, there may be times when it is appropriate to use a subquery in the SELECT or FROM clause. When the subquery is nested in a WHERE or HAVING clause, the results returned from the subquery are used as a condition in the outer query. Any type of subquery (single-row, multiple-row, or multiple-column) can be used in the WHERE, HAVING, or FROM clause of a SELECT statement. As you will see, the only type of subquery that can be used in a SELECT clause is a single-row subquery.

NOTE

Using a subquery in a FROM clause has a specific purpose. You'll explore that in a separate section in this chapter.

Keep the following rules in mind when working with any type of subquery:

- A subquery must be *a complete query in itself*—in other words, it must have at least a SELECT and a FROM clause.
- A subquery cannot have an ORDER BY clause. If you need to present the displayed output in a specific order, list an ORDER BY clause as the last clause of the outer query.
- A subquery *must be enclosed within a set of parentheses* to separate it from the outer query.
- If you place the subquery in the WHERE or HAVING clause of an outer query, you can do so only on the *right side* of the comparison operator.

SINGLE-ROW SUBQUERIES

A **single-row subquery** is used when the results of the outer query are based on a single, unknown value. Although it is formally called "single-row," this implies that the query will return multiple columns—but only one row—of results. However, *a single-row subquery can return to the outer query only one row of results that consists of only one column*. Therefore, this text refers to the output of a single-row subquery as a **single value**.

Single-Row Subquery in a WHERE Clause

Let's compare creating multiple queries, which you've studied in previous chapters, with creating one query containing a subquery. In this chapter's introduction, management requested a list of all computer books that have a higher retail price than *Database Implementation*. As shown in Figure 12-2, the first step is to create a query to determine the retail price of that book, which is $31.40.

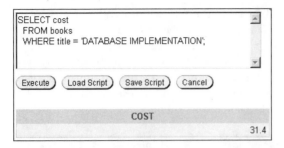

```
SELECT cost
 FROM books
 WHERE title = 'DATABASE IMPLEMENTATION';
```

COST
31.4

FIGURE 12-2   Query to determine the retail price of *Database Implementation*

To determine which computer books retail for more than $31.40, a second query must be issued that explicitly states the cost of *Database Implementation*. That second query is issued in Figure 12-3.

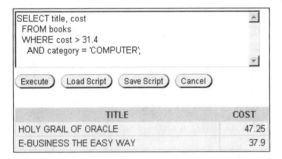

```
SELECT title, cost
 FROM books
 WHERE cost > 31.4
   AND category = 'COMPUTER';
```

TITLE	COST
HOLY GRAIL OF ORACLE	47.25
E-BUSINESS THE EASY WAY	37.9

FIGURE 12-3   Query for computer books costing more than $31.40

The WHERE clause in Figure 12-3 explicitly includes the retail price of *Database Implementation* found by the first query in Figure 12-2. The category condition restricts records to those only in the Computer Category.

However, you can obtain these same results through the use of a single-row subquery. A single-row subquery is appropriate in this example because (1) to obtain the desired results, an unknown value must be obtained, and that value is contained in the database, and (2) only one value should be returned from the inner query (that is, the retail price of *Database Implementation*).

In Figure 12-4, a single-row subquery is substituted for the Cost condition of the SELECT statement shown in Figure 12-3. The subquery is enclosed in parentheses to distinguish it from the clauses of the parent query.

In Figure 12-4, the inner query is executed first, and the result of the query, a single value of 31.4, is passed back to the outer query. The outer query is then executed, and all books having a retail price greater than $31.40 and belonging to the Computer Category are listed in the output.

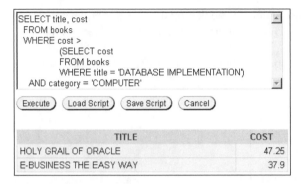

FIGURE 12-4 A single-row subquery

Operators indicate to Oracle 10*g* whether a user is creating a single-row subquery or a multiple-row subquery. The single-row operators are =, >, <, >=, <=, and <>. Although other operators, such as IN, are allowed, single-row operators instruct Oracle 10*g* that only one value is expected from the subquery. If more than one value is returned, the SELECT statement fails, and you receive an error message.

Suppose that management makes another request. They want the title of the most expensive book sold by JustLee Books. Because you learned how to use the MAX function in Chapter 11, this should be simple. You might create the query shown in Figure 12-5.

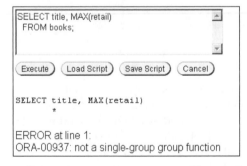

FIGURE 12-5 Flawed query: attempt to determine the book with the highest retail value

Perhaps it is not quite that simple. Remember the rule when working with group functions: If an individual field with a group function is listed in the SELECT clause, the individual field must also be listed in a GROUP BY clause. In this instance, adding a GROUP BY clause would not make sense: If a GROUP BY clause is added that contains the Title column, each book would be its own group because each title is different. In other words, the results would be the same as using **SELECT title, retail** in the query.

Thus, to retrieve the title of the most expensive book, you can use a subquery to determine the highest retail price of any book. That retail price can then be returned to an outer query and displayed in the results.

As shown in Figure 12-6, the most expensive book sold by JustLee Books is *Painless Child-Rearing*.

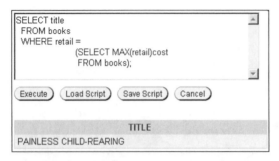

```
SELECT title
 FROM books
  WHERE retail =
              (SELECT MAX(retail)cost
               FROM books);
```

Execute   Load Script   Save Script   Cancel

TITLE

PAINLESS CHILD-REARING

FIGURE 12-6 Query to determine the title of the most expensive book

If management also requests the actual retail price of the book, you can list the Retail field in the SELECT clause of the query to have it displayed in the results.

HELP

If an error message is returned for the query in Figure 12-6, make certain that the subquery contains four parentheses—one set around the *retail* argument for the MAX function and one set around the subquery.

TIP

The query statement that will serve as the subquery should be created and executed first by itself. This allows you to verify that the query produces the expected results before embedding it into another query as a subquery.

You can include multiple subqueries in a SELECT statement. For example, suppose that management needs to know the title of all books published by the publisher of *Big Bear and Little Dove* that generate more than the average profit returned by all books sold through JustLee Books. In this case, two values are unknown: (1) the identity of the publisher of *Big Bear and Little Dove* and (2) the average profit of all books. How might you create a query that extracts those values? The SELECT statement in Figure 12-7 uses two separate subqueries in the WHERE clause to obtain the information.

```
SELECT isbn, title
 FROM books
 WHERE pubid =
         (SELECT pubid
          FROM books
          WHERE title = 'BIG BEAR AND LITTLE DOVE')
 AND retail-cost >
         (SELECT AVG(retail-cost)
          FROM books);
```

(Execute) (Load Script) (Save Script) (Cancel)

ISBN	TITLE
2491748320	PAINLESS CHILD-REARING
2147428890	SHORTEST POEMS

FIGURE 12-7   SELECT statement with two single-row subqueries

Notice that each subquery in Figure 12-7 is complete—both contain a minimum of one SELECT clause and one FROM clause. Because they are subqueries, each is enclosed in parentheses. The first subquery determines the publisher of *Big Bear and Little Dove* and returns that result to the first condition of the WHERE clause (line 3). The second subquery finds the average profit of all books sold by JustLee Books by using the AVG function, and then passes that value back to the second condition of the WHERE clause (line 7) to be compared against the profit for each book. Because the two conditions of the WHERE clause in the outer query are combined with the AND logical operator, both values returned by the subqueries must be met for a book to be listed in the output of the outer query. In this example, two books are found that are published by the publisher of *Big Bear and Little Dove* and that return more than the average profit.

Single-Row Subquery in a HAVING Clause

As previously mentioned, you can include a subquery in a HAVING clause. A HAVING clause is used when the group results of a query need to be restricted based on some condition. If the result returned from a subquery must be compared to a group function, you must nest the inner query in the outer query's HAVING clause.

For example, suppose that management needs a list of all book categories that return a higher average profit than the Literature Category does. You would follow these steps:

1. Calculate the average profit for all literature books.
2. Calculate the average profit for every category.
3. Compare the average profit for every category with the average profit for the Literature Category.

Try writing a query that accomplishes this goal, and then look at the query and output in Figure 12-8.

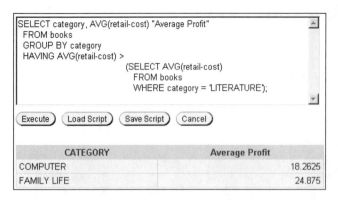

```
SELECT category, AVG(retail-cost) "Average Profit"
  FROM books
  GROUP BY category
  HAVING AVG(retail-cost) >
                    (SELECT AVG(retail-cost)
                     FROM books
                     WHERE category = 'LITERATURE');
```

Execute   Load Script   Save Script   Cancel

CATEGORY	Average Profit
COMPUTER	18.2625
FAMILY LIFE	24.875

FIGURE 12-8   Single-row subquery nested in a HAVING clause

As Figure 12-8 shows, the results are restricted to groups that have a higher average profit than the Literature Category does. Getting these results requires the HAVING clause. Because the results of the subquery are applied to groups of data, it is necessary to nest the subquery in the HAVING clause.

As shown in Figure 12-8, the subquery in a HAVING clause must follow the same guidelines as those used in WHERE clauses; that is, it must include at least a SELECT clause and a FROM clause—and it must be enclosed in parentheses.

Single-Row Subquery in a SELECT Clause

A single-row subquery can also be nested in the SELECT clause of an outer query. However, this approach is rarely used because when the subquery is listed in the SELECT clause, this means the value returned by the subquery will be displayed for every row of output generated by the parent query. For illustrative purposes, suppose that management wants to compare the price of each book in inventory against the average price of all books in inventory. One approach is to calculate the average price of all books and give management that figure along with a separate list of all books with their current retail price.

On the other hand, you could use a subquery in a SELECT clause that calculates the average retail price of all books. When a single-row subquery is included in a SELECT clause, the result of the subquery is displayed in the output of the parent query. To include a subquery in a SELECT clause, you use a comma to separate the subquery from the table columns, just as if you were listing another column. In fact, you can even give the results of the subquery a column alias. Look at the output of the query issued in Figure 12-9.

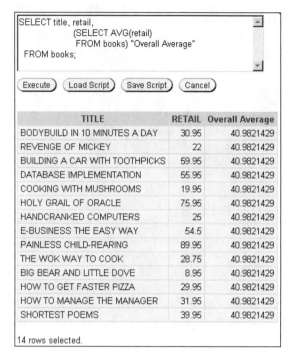

```
SELECT title, retail,
            (SELECT AVG(retail)
              FROM books) "Overall Average"
FROM books;
```

Execute   Load Script   Save Script   Cancel

TITLE	RETAIL	Overall Average
BODYBUILD IN 10 MINUTES A DAY	30.95	40.9821429
REVENGE OF MICKEY	22	40.9821429
BUILDING A CAR WITH TOOTHPICKS	59.95	40.9821429
DATABASE IMPLEMENTATION	55.95	40.9821429
COOKING WITH MUSHROOMS	19.95	40.9821429
HOLY GRAIL OF ORACLE	75.95	40.9821429
HANDCRANKED COMPUTERS	25	40.9821429
E-BUSINESS THE EASY WAY	54.5	40.9821429
PAINLESS CHILD-REARING	89.95	40.9821429
THE WOK WAY TO COOK	28.75	40.9821429
BIG BEAR AND LITTLE DOVE	8.95	40.9821429
HOW TO GET FASTER PIZZA	29.95	40.9821429
HOW TO MANAGE THE MANAGER	31.95	40.9821429
SHORTEST POEMS	39.95	40.9821429

14 rows selected.

FIGURE 12-9   Single-row subquery in a SELECT clause

As shown in Figure 12-9, to calculate the average price of all books in inventory, the SELECT clause of the outer query includes the Title and Retail columns in the column list as well as the subquery. The average calculated by the subquery is displayed for every book included in the output. The column alias, Overall Average, is assigned to the results of the subquery to indicate the contents of that column. If a column alias had not been used, the actual subquery would have been shown as the column heading—a somewhat unattractive column heading. The result of having the subquery in the SELECT clause enables management to compare each book's retail price to the average retail price for all books by looking at just one list.

Try using the subquery in Figure 12-9 to calculate the difference between the retail price and the average price. Simply restructure the SELECT clause and have retail - precede the subquery, just as if you were calculating profit using the expression (retail - cost).

MULTIPLE-ROW SUBQUERIES

Multiple-row subqueries are nested queries that can return more than one row of results to the parent query. Multiple-row subqueries are most commonly used in WHERE and HAVING clauses. The main rule to keep in mind when working with multiple-row subqueries is that *you must use multiple-row operators*. If a single-row operator is used with a subquery that returns more than one row of results, Oracle 10g returns an error message, and the SELECT statement fails. Valid multiple-row operators include IN, ALL, and ANY. Let's consider each of those operators.

IN Operator

Of the three multiple-row operators, the IN operator is the most commonly used. Figure 12-10 shows a multiple-row subquery using the IN operator. This query identifies books with a retail value that matches the highest retail value for any of the book categories.

FIGURE 12-10   Multiple-row subquery using the IN operator

The IN operator in Figure 12-10 indicates that the records processed by the outer query must match one of the values returned by the subquery (in other words, it creates an OR condition). The order of execution in Figure 12-10 is as follows:

1. The subquery determines the price of the most expensive book in each category.
2. The maximum retail price in each category is passed to the WHERE clause of the outer query.
3. The outer query compares the price of each book to the prices generated by the subquery.
4. If the retail price of a book matches one of the prices returned by the subquery, the title, retail price, and category of the book are displayed in the query's output.

ALL and ANY Operators

The ALL and ANY operators can be combined with other comparison operators to treat the results of a subquery as a set of values, rather than as individual values. Figure 12-11 summarizes the use of the ALL and ANY operators in conjunction with other comparison operators.

OPERATOR	DESCRIPTION
>ALL	More than the highest value returned by the subquery
<ALL	Less than the lowest value returned by the subquery
<ANY	Less than the highest value returned by the subquery
>ANY	More than the lowest value returned by the subquery
=ANY	Equal to any value returned by the subquery (same as IN)

FIGURE 12-11   Descriptions of ALL and ANY operator combinations

The ALL operator is fairly straightforward:

*   If the ALL operator is combined with the "greater than" symbol (>), the outer query is searching for all records with a value higher than the highest value returned by the subquery (in other words, more than ALL the values returned).
*   If the ALL operator is combined with the "less than" symbol (<), the outer query is searching for all records with a value lower than the lowest value returned by the subquery (in other words, less than ALL the values returned).

To examine the impact of using the ALL comparison operator, look at the query shown in Figure 12-12, which will be used as a subquery.

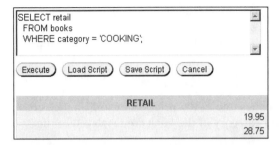

```
SELECT retail
  FROM books
  WHERE category = 'COOKING';
```

Execute   Load Script   Save Script   Cancel

RETAIL
19.95
28.75

FIGURE 12-12   Retail price of the books in the Cooking Category

The query in Figure 12-12 returns the retail prices for two books in the Cooking Category. The lowest value returned is $19.95, and the highest value is $28.75.

Suppose that you want to know the titles of all books having a retail price greater than the most expensive book in the Cooking Category. One approach is to use the MAX function in a subquery to find the highest retail price. Another approach is to use the >ALL operator, as shown in Figure 12-13.

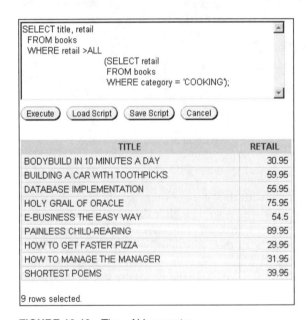

```
SELECT title, retail
  FROM books
  WHERE retail >ALL
                   (SELECT retail
                    FROM books
                    WHERE category = 'COOKING');
```

Execute   Load Script   Save Script   Cancel

TITLE	RETAIL
BODYBUILD IN 10 MINUTES A DAY	30.95
BUILDING A CAR WITH TOOTHPICKS	59.95
DATABASE IMPLEMENTATION	55.95
HOLY GRAIL OF ORACLE	75.95
E-BUSINESS THE EASY WAY	54.5
PAINLESS CHILD-REARING	89.95
HOW TO GET FASTER PIZZA	29.95
HOW TO MANAGE THE MANAGER	31.95
SHORTEST POEMS	39.95

9 rows selected.

FIGURE 12-13   The >ALL operator

This is the Oracle 10g strategy that goes into processing the SELECT command in Figure 12-13:

- The subquery shown in Figure 12-13 passes the retail prices of the two books in the Cooking Category ($19.95 and $28.75) to the outer query.
- Because the >ALL operator is used in the outer query, Oracle 10g is instructed to list all the books with a retail price higher than the largest value returned by the subquery ($28.75).
- In this example, nine books have a higher price than the most expensive book in the Cooking Category.

NOTE

You could obtain the same results shown in Figure 12-13 by using the MAX function in the subquery. With the MAX function in the subquery, a single value for the highest priced cookbook would be returned from the subquery and a multiple-row operator would not be required.

Similarly, the <ALL operator is used to determine the records that have a value less than the lowest value returned by a subquery. Thus, if you need to find books that are priced less than the least expensive book in the Cooking Category, first formulate a subquery that identifies the books in the Cooking Category. Then, you can compare the retail price of the books in the BOOKS table against the values returned by a subquery using the <ALL operator.

As in the previous query, the subquery shown in Figure 12-14 first finds the two books in the Cooking Category (Figure 12-12). The retail prices of those books ($19.95 and $28.75) are then passed to the outer query. Because $19.95 is the lowest retail price of all books in the Cooking Category, only those books having a retail price less than $19.95 are displayed in the output. In this case, Oracle 10g found only one book having a retail price lower than the least expensive book in the Cooking Category: *Big Bear and Little Dove*.

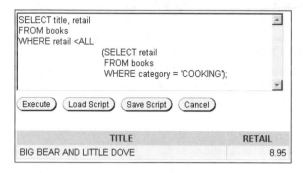

FIGURE 12-14 The <ALL operator

By contrast, the <ANY operator is used to find records that have a value less than the highest value returned by a subquery. To determine which books cost less than the most expensive book in the Cooking Category, evaluate the results of the subquery using the <ANY operator, as shown in Figure 12-15.

FIGURE 12-15 The <ANY operator

In Figure 12-15, the outer query found four books with a retail price less than the most expensive book in the Cooking Category. Notice, however, that the results also include *Cooking with Mushrooms*, which is a book found in the Cooking Category. Because the outer query compares the records to the highest value in the Cooking Category, any other book in the Cooking Category is also displayed in the query results. To eliminate any book in the Cooking Category from appearing in the output, simply add the condition **AND category <> 'COOKING'** to the WHERE clause of the outer query.

The >ANY operator is used to return records that have a value greater than the lowest value returned by the subquery. In Figure 12-16, 12 records have a retail price greater than the lowest retail price returned by the subquery ($19.95).

The =ANY operator works the same way as the IN comparison operator does. For example, in the query shown in Figure 12-17, the user is searching for the titles of books that were purchased by customers who also purchased the book with the ISBN of 0401140733. Because that book could have appeared on more than one order, and the user wants to identify all those orders, the =ANY operator is used.

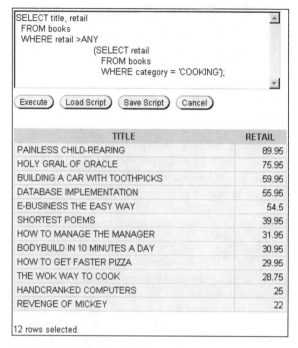

```
SELECT title, retail
  FROM books
  WHERE retail >ANY
                (SELECT retail
                   FROM books
                   WHERE category = 'COOKING');
```

Execute Load Script Save Script Cancel

TITLE	RETAIL
PAINLESS CHILD-REARING	89.95
HOLY GRAIL OF ORACLE	75.95
BUILDING A CAR WITH TOOTHPICKS	59.95
DATABASE IMPLEMENTATION	55.95
E-BUSINESS THE EASY WAY	54.5
SHORTEST POEMS	39.95
HOW TO MANAGE THE MANAGER	31.95
BODYBUILD IN 10 MINUTES A DAY	30.95
HOW TO GET FASTER PIZZA	29.95
THE WOK WAY TO COOK	28.75
HANDCRANKED COMPUTERS	25
REVENGE OF MICKEY	22

12 rows selected.

FIGURE 12-16   The >ANY operator

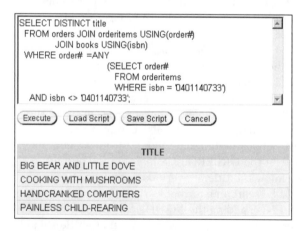

```
SELECT DISTINCT title
  FROM orders JOIN orderitems USING(order#)
        JOIN books USING(isbn)
  WHERE order# =ANY
                (SELECT order#
                   FROM orderitems
                   WHERE isbn = '0401140733')
  AND isbn <> '0401140733';
```

Execute Load Script Save Script Cancel

TITLE
BIG BEAR AND LITTLE DOVE
COOKING WITH MUSHROOMS
HANDCRANKED COMPUTERS
PAINLESS CHILD-REARING

FIGURE 12-17   The =ANY operator

The query in Figure 12-17 would have yielded the same results if the IN operator had been used instead of the =ANY operator. The DISTINCT keyword in the SELECT clause of the outer query is included because, as previously mentioned, a title could have been ordered by more than one customer and would have had multiple listings in the output.

Also notice in Figure 12-17 that the columns needed to complete the query are in three different tables: ORDERS, ORDERITEMS, and BOOKS. Because the columns necessary to perform the inner query are contained only in the ORDERITEMS table, joins among the tables are not required in the subquery. However, the columns referenced by the outer query are contained in the BOOKS table (ISBN and Title) and the ORDERITEMS table (Order#). To complete the logical join between the BOOKS and ORDERITEMS tables, the ORDERS table must also be included in the FROM clause of the outer query.

EXISTS Operator

The **EXISTS** operator is used to determine whether a condition is present in a subquery. The results of the operator are Boolean—it is TRUE if the condition exists and FALSE if it does not. If the results are TRUE, the records meeting the condition are displayed.

To better understand how the EXISTS operator works, let's look at an example. Suppose that management requests a list of all recently ordered books. Recall that the ORDERS table contains all orders for the current month and any orders from the previous month that have not yet shipped—these are considered recent orders. To provide this list, you can take one of two approaches. One approach is to join the BOOKS and ORDERITEMS tables and display the title for each book. The other approach is to use the EXISTS operator to determine which books have been ordered recently. See the query shown in Figure 12-18.

As shown in Figure 12-18, the SELECT and FROM clauses of the outer query indicate that the titles of the books in the BOOKS table should be displayed in the output. However, which titles should be displayed? The WHERE clause in the outer query uses the EXISTS operator, which can be interpreted to mean "include only those books that exist in, or are identified by, the subquery." In this example, the subquery simply identifies all book ISBNs that are contained in the ORDERITEMS table—which happen to be all the books that have been ordered lately. Because the BOOKS table also contains the ISBN column, the outer query lists the title of every book with a matching ISBN in the ORDER-ITEMS table (in other words, it also exists in the ORDERITEMS table) in the output.

However, what if you had wanted to know the title of every book that had *not* been ordered recently? In that case, you would be searching for books that do not exist in the ORDERITEMS table. As shown in Figure 12-19, adding the NOT logical operator before the EXISTS operator in the WHERE clause of the outer query specifies that a title should be displayed only if the book's ISBN is not contained in the ORDERITEMS table (in other words, the result of the EXISTS condition is FALSE).

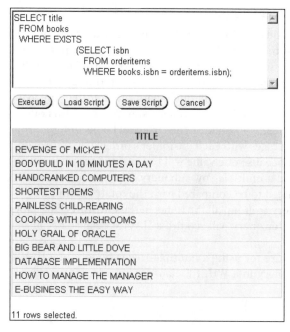

```
SELECT title
  FROM books
  WHERE EXISTS
              (SELECT isbn
                 FROM orderitems
                 WHERE books.isbn = orderitems.isbn);
```

Execute   Load Script   Save Script   Cancel

TITLE
REVENGE OF MICKEY
BODYBUILD IN 10 MINUTES A DAY
HANDCRANKED COMPUTERS
SHORTEST POEMS
PAINLESS CHILD-REARING
COOKING WITH MUSHROOMS
HOLY GRAIL OF ORACLE
BIG BEAR AND LITTLE DOVE
DATABASE IMPLEMENTATION
HOW TO MANAGE THE MANAGER
E-BUSINESS THE EASY WAY

11 rows selected.

FIGURE 12-18   Subquery using the EXISTS operator

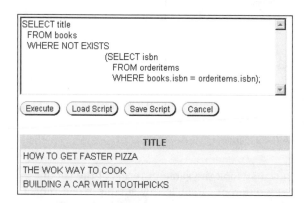

```
SELECT title
  FROM books
  WHERE NOT EXISTS
                    (SELECT isbn
                       FROM orderitems
                       WHERE books.isbn = orderitems.isbn);
```

Execute   Load Script   Save Script   Cancel

TITLE
HOW TO GET FASTER PIZZA
THE WOK WAY TO COOK
BUILDING A CAR WITH TOOTHPICKS

FIGURE 12-19   Subquery using NOT in conjunction with the operator

NOTE

Your results for some queries may be listed in a different order than shown when the query contains no sorting instructions.

Multiple-Row Subquery in a HAVING Clause

Thus far, you have seen multiple-row subqueries in a WHERE clause, but they can also be included in a HAVING clause. When the results of the subquery are being compared to grouped data in the outer query, the subquery *must* be nested in a HAVING clause in the parent query. For illustrative purposes, suppose that you needed to determine whether any customer's recently placed order has a total "amount due" that is greater than the average "amount due" for all orders originating from that customer's state. Getting this output requires that you first determine the average "amount due" from all books ordered from each state, and then compare the state averages with each customer's order. The state averages can be calculated in a subquery, but because one value will be returned for each state contained in the ORDERS table, you need a multiple-row subquery. The average "amount due" for the orders from each state will need to be compared with the total "amount due" for each order; this requires the outer query to group all the items in the ORDERITEMS table by the Order# column. Therefore, the outer query will require a HAVING clause because the comparison is based on grouped data.

As shown in Figure 12-20, the structure for using a multiple-row subquery in a HAVING clause is the same as using the subquery in a WHERE clause.

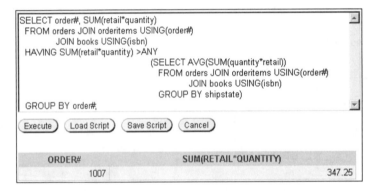

FIGURE 12-20 Multiple-row subquery in a HAVING clause

Single-row and multiple-row subqueries may look the same in terms of the subqueries themselves; however, they are distinctly different. A single-row subquery can return only *one* data value, whereas a multiple-row subquery can return *several* values. Thus, if you execute a subquery that returns more than one data value and the comparison operator is intended to be used only with single-row subqueries, you receive an error message, and the query is not executed, as shown in Figure 12-21.

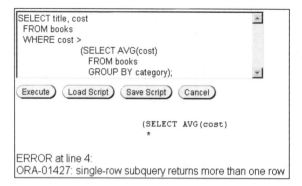

```
SELECT title, cost
  FROM books
  WHERE cost >
                (SELECT AVG(cost)
                   FROM books
                   GROUP BY category);
```

(Execute) (Load Script) (Save Script) (Cancel)

```
                (SELECT AVG(cost)
                        *
```

ERROR at line 4:
ORA-01427: single-row subquery returns more than one row

FIGURE 12-21   Flawed query: using a single-row operator for a multiple-row subquery

MULTIPLE-COLUMN SUBQUERIES

Now that you've examined multiple-row subqueries, let's look at multiple-column subqueries. A **multiple-column subquery** returns more than one column to the outer query. A multiple-column subquery can be listed in the FROM, WHERE, or HAVING clause of a query. If the multiple-column subquery is included in the FROM clause of the outer query, the subquery actually generates a temporary table that can be referenced by other clauses of the outer query.

Multiple-Column Subquery in a FROM Clause

When a multiple-column subquery is used in the FROM clause of an outer query, it creates a temporary table that can be referenced by other clauses of the outer query. This temporary table is more formally called an **inline view**. If the temporary table generated by the subquery contains grouped data, this data can be referenced or used just like individual data values.

Suppose that you need a list of all books in the BOOKS table that have a higher-than-average selling price as compared to books in the same category. You need to display each book's title, retail price, category, and the average selling price of books in that category. Because the average selling price is based on grouped data, this presents a problem. How might you solve it? Look at Figure 12-22.

In Figure 12-22, a multiple-column subquery is nested in the FROM clause of the outer table. The subquery creates a temporary table. The subquery determines the categories that exist in the BOOKS table and the average selling price of every book in that particular category. However, how do you display the title of each book in the BOOKS table, its retail price, the category of the book, and the average price of all books in that same category? The BOOKS table contains the individual data for each book, and the subquery has created a temporary table that stores the grouped data. Notice that on line 4 of the query in Figure 12-22, a table alias has been assigned to the results of the subquery, so the columns contained in the subquery (Category and Cataverage) can be referenced by other clauses in the outer SELECT statement.

In essence, the query in Figure 12-22 is referencing, or obtaining, data from two different tables. The tables have been joined using the traditional approach—through the

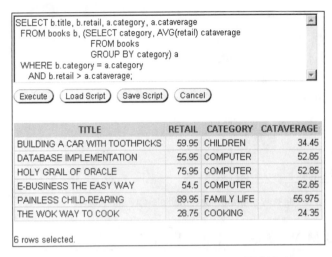

FIGURE 12-22 Multiple-column subquery in a FROM clause

WHERE clause of the outer query. The problem with using the traditional approach is that both tables contain a column called Category—this will create a problem with ambiguity if the Category column is referenced anywhere in the outer query. To avoid ambiguity, the Category column needs a column qualifier to identify which table contains the category data to be displayed. Therefore, table aliases are used in the SELECT and WHERE clauses to identify the table containing the column being referenced.

As shown in Figure 12-23, the query could have also been created using an ANSI join operation supported by Oracle 10g.

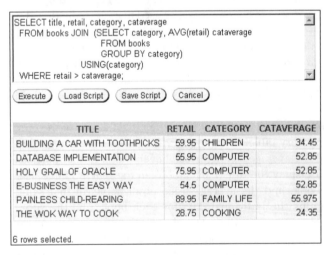

FIGURE 12-23 Using a join with a multiple-column subquery in the FROM clause

Because both tables (BOOKS and the temporary table created by the subquery) in Figure 12-23 contain a column named Category, the tables are linked in the FROM clause with a JOIN using the Category field. Because column qualifiers are not allowed with the JOIN method, the temporary table created by the subquery is not assigned a table alias.

Multiple-Column Subquery in a WHERE Clause

When a multiple-column subquery is included in the WHERE or HAVING clause of the outer query, the IN operator is used by the outer query to evaluate the results of the subquery. The results of the subquery consist of more than one column of results.

The syntax for the outer WHERE clause is WHERE (*columnname, columnname,...*) IN *subquery*. Keep these rules in mind:

- Because the WHERE clause contains more than one column name, the column list must be enclosed within parentheses.
- Column names listed in the WHERE clause must be in the same order as they are listed in the SELECT clause of the subquery.

NOTE

Double-check that the field list presented in the WHERE clause of the outer query is enclosed in parentheses and is in the same order as the list of fields given in the SELECT clause of the subquery.

Previously, in Figure 12-10, the subquery returned the price of the most expensive book in each category, and the outer query generated a list of the title, retail price, and category of books matching the retail price returned by the subquery. The overall result of the outer query was to display the title, retail price, and category for the most expensive book in each category. However, that query will work only if two books do not have the same retail price. For example, suppose that books in two different categories both have the same price as one of the values returned by the subquery. Then, one (or possibly both) would not be the most expensive book in its category, and you would receive erroneous results. To create a query specifically to create a list of the most expensive books in each category, a multiple-column subquery is more appropriate. Look at the example in Figure 12-24.

In Figure 12-24, the subquery finds the highest retail value in each category and passes both the category names and the retail prices back to the outer query.

NOTE

Although a multiple-column subquery can be used in the HAVING clause of an outer query, it is usually used only when analyzing extremely large sets of numeric data that have been grouped and is generally presented in more advanced courses focusing upon quantitative methods.

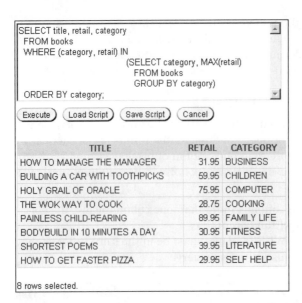

```
SELECT title, retail, category
  FROM books
  WHERE (category, retail) IN
                      (SELECT category, MAX(retail)
                          FROM books
                          GROUP BY category)
  ORDER BY category;
```

| Execute | Load Script | Save Script | Cancel |

TITLE	RETAIL	CATEGORY
HOW TO MANAGE THE MANAGER	31.95	BUSINESS
BUILDING A CAR WITH TOOTHPICKS	59.95	CHILDREN
HOLY GRAIL OF ORACLE	75.95	COMPUTER
THE WOK WAY TO COOK	28.75	COOKING
PAINLESS CHILD-REARING	89.95	FAMILY LIFE
BODYBUILD IN 10 MINUTES A DAY	30.95	FITNESS
SHORTEST POEMS	39.95	LITERATURE
HOW TO GET FASTER PIZZA	29.95	SELF HELP

8 rows selected.

FIGURE 12-24   Multiple-column subquery in a WHERE clause

NULL VALUES

As with everything else, NULL values present a problem when using subqueries. Because a NULL value is the same as the absence of data, a NULL cannot be returned to an outer query for comparison purposes—it is not equal to anything, not even another NULL. Therefore, if a NULL value is passed from a subquery, the results of the outer query will be "no rows selected." Although the statement will not fail (it will not generate an Oracle 10g error message), you will not get the expected results, as you can see from the example in Figure 12-25.

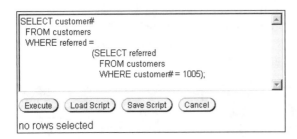

```
SELECT customer#
  FROM customers
  WHERE referred =
                (SELECT referred
                    FROM customers
                    WHERE customer# = 1005);
```

| Execute | Load Script | Save Script | Cancel |

no rows selected

FIGURE 12-25   Flawed query: NULL results from a subquery

In Figure 12-25, the user is trying to determine whether the customer who referred customer 1005 has referred any other customers to JustLee Books. The problem is that no rows are listed as output from the outer query. The question is this: Are no rows listed because the customer who referred customer 1005 has not referred any other customers, or because customer 1005 was not originally referred to JustLee Books (in which case the Referred column is NULL)? If no one referred customer 1005, should the output of the outer query be a list of all customers who were not referred by other customers?

In this case, customer 1005 was not referred by any other customer; thus, the Referred column is NULL. Because a NULL value is passed to the outer query, no matches are found because the condition is "WHERE referred = NULL". The "IS NULL" operation is required to identify NULL values in a conditional clause.

However, what if customer 1005 was not referred by another customer and you wanted a list of all customers who were also not referred by other customers? As always, it is the NVL function to the rescue.

NVL in Subqueries

If it is possible for a subquery to return a NULL value to the outer query for comparison, the NVL function should be used to substitute an actual value for the NULL. However, you must keep two things in mind:

1. The substitution of the NULL value must occur for the NULL value both in the subquery and in the outer query.
2. The value substituted for the NULL value must be one that could not possibly exist anywhere else in that column.

Figure 12-26 uses the same premise as Figure 12-25 and provides an example of these two rules.

In the query presented in Figure 12-26, the NVL function is included whenever the Referred column is referenced—in both the subquery and the outer query. In this example, a zero is substituted for the NULL value. Because the value contained in the Referred column is actually the customer number of a customer in the CUSTOMERS table, and because no customer has the customer number of zero, substituting a zero for the NULL value will not accidentally make a NULL record equivalent to a non-NULL record.

Whenever you substitute a value for a NULL, make certain no other record contains the substituted value. For example, use ZZZ for a customer name; in the case of a date field, use a date that absolutely would not exist in the database.

IS NULL in Subqueries

Although problems exist when passing a NULL value from a subquery to an outer query, searches for NULL values are allowed in a subquery. As with regular queries, you can still search for NULL values using the IS NULL comparison operator.

For example, suppose that you need to find the title of all books that have been ordered but have not yet shipped to the customers. The subquery presented in Figure 12-27 identifies the orders that have not yet shipped—the ship date is NULL.

As shown in Figure 12-27, the order number for each order is returned to the outer query, and the title for each book is displayed. The DISTINCT keyword is used to suppress duplicate titles from being listed. Although the subquery searches for records containing NULL values, it is the Order# column that is passed back to the outer query. The Order# column is the primary key for the ORDERS table, and no NULL values can exist in that field. Therefore, there is no need to use the NVL function in this example.

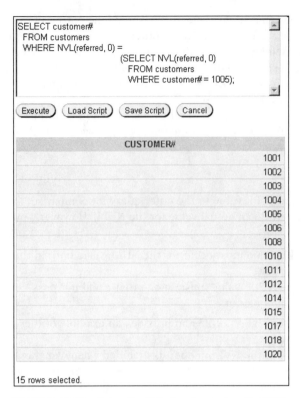

FIGURE 12-26 Using the NVL function to handle NULL values

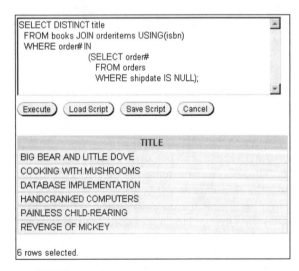

FIGURE 12-27 Using IS NULL in a subquery

Correlated Subqueries

Thus far you have, for the most part, studied **uncorrelated subqueries**: The subquery is executed first, the results of the subquery are passed to the outer query, and then the outer query is executed. By contrast, in a **correlated subquery**, Oracle 10g uses a different procedure to execute a query. The query using the EXISTS operator from Figure 12-18 is a correlated subquery and uses that different procedure. The query is displayed again in Figure 12-28 for easy reference.

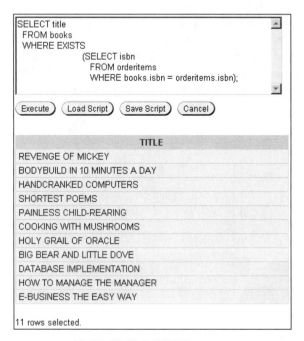

```
SELECT title
  FROM books
 WHERE EXISTS
             (SELECT isbn
                FROM orderitems
               WHERE books.isbn = orderitems.isbn);
```

Execute   Load Script   Save Script   Cancel

TITLE
REVENGE OF MICKEY
BODYBUILD IN 10 MINUTES A DAY
HANDCRANKED COMPUTERS
SHORTEST POEMS
PAINLESS CHILD-REARING
COOKING WITH MUSHROOMS
HOLY GRAIL OF ORACLE
BIG BEAR AND LITTLE DOVE
DATABASE IMPLEMENTATION
HOW TO MANAGE THE MANAGER
E-BUSINESS THE EASY WAY

11 rows selected.

FIGURE 12-28   Correlated subquery

Although the query in Figure 12-28 is a multiple-row subquery, the execution of the entire query requires that each row in the BOOKS table be processed to determine whether it also exists in the ORDERITEMS table. In other words, Oracle 10g executes the outer query first, and when it encounters the WHERE clause of the outer query, it is evaluated to determine whether that row is TRUE (whether it exists in the ORDERITEMS table). If it is TRUE, the book's title is displayed in the results. The outer query is executed again for the next book in the BOOKS table and compared to the contents of the ORDERITEMS table, and so on, for each row of the BOOKS table. In other words, *a correlated subquery is a subquery that is processed, or executed, once for each row in the outer query.*

How does Oracle 10g distinguish between an uncorrelated and a correlated subquery? Simply speaking, *if a subquery references a column from the outer query, then it is a correlated subquery*. Notice that in the subquery in Figure 12-28, the WHERE clause specifies the ISBN column of the BOOKS table. Because the BOOKS table is not included in the FROM clause of the subquery, it is forced to use data that is processed by the outer query (the ISBN of the books being processed during that execution of the outer query).

If the comparison of the ISBNs from each table had been performed in the outer query (in other words, if the two tables had been joined in the outer clause and not in the subquery), the results would have been completely different. Because the subquery would no longer reference a column contained in the outer query, it would have been considered an uncorrelated subquery. With an uncorrelated subquery, the subquery would have been executed first, and then the results would have been passed back to the outer query. Because the subquery is used to identify every ISBN contained in the ORDERITEMS table, each ISBN listed in the table would have been returned to the outer query. Then, the outer query would find the title for that book and display that title.

Notice that in Figure 12-29, the subquery no longer specifies the column ISBN from the BOOKS table—the join is created in the outer query. Therefore, the subquery is executed first, and for each value returned, it is matched to the contents of the BOOKS table. As shown in Figure 12-29, several titles are listed more than once. Why? Because their ISBN is listed more than once in the ORDERITEMS table and is returned from the subquery each time it is encountered.

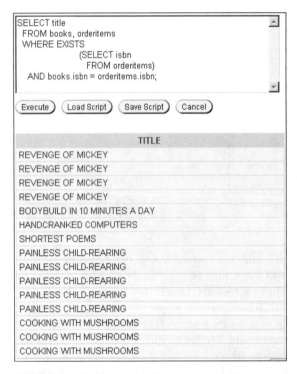

FIGURE 12-29 Uncorrelated subquery (partial output shown)

NOTE

Your results for some queries may be listed in a different order than shown when the query contains no sorting instructions.

NESTED SUBQUERIES

You can nest subqueries inside the FROM, WHERE, or HAVING clauses of other subqueries. In Oracle 10g, subqueries in a WHERE clause can be nested to a depth of 255 subqueries, and there is no depth limit when the subqueries are nested in a FROM clause. When nesting subqueries, you may want to use the following strategy:

- Determine exactly what you are trying to find. This is the goal of the query.
- Write the innermost subquery first.
- After you write the innermost query, look at the value you are able to pass back to the outer query. If this is not the value needed by the outer query (for example, it references the wrong column), analyze how the data needs to be converted to get the correct rows, and if necessary, use another subquery between the outer query and the nested subquery. In some cases, you might need to create several layers of subqueries to link the value returned by the innermost subquery to the value needed by the outer query.

The most common reason for nesting subqueries is to create a chain of data. For example, suppose that you need to find the name of the customer who has ordered the most books from JustLee Books (not including multiple quantities of the same book) on *one* order. Figure 12-30 shows a query that provides these results.

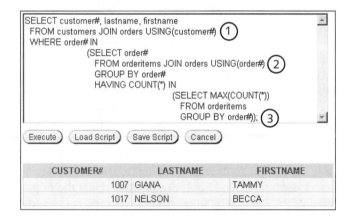

FIGURE 12-30   Nested subqueries

Here are the steps required to create the query shown in Figure 12-30:

1. The goal of the query is to count the number of items placed on each order and identify the order—or orders, in case of a tie—with the most items. The nested subquery identified by ③ in Figure 12-30 finds the order that has the most books.
2. The value of the highest count of the items ordered is then passed back to the outer subquery, ②.

3. The outer subquery, ②, is then used to identify which order(s) has the same number of items as the highest number of items previously found by the nested subquery.

4. After the order number(s) has been identified, it is then passed back to the outer query, ①, which determines the customer number and name of the person who placed the order. In this case, two customers tied for placing an order that has the most items.

If the outer (first) subquery, ②, had not been included, the inner (second) subquery would have tried to return the value from the MAX(COUNT(*)) functions directly to the outer query for comparison with the Order# column. By including the first subquery, ②, Oracle 10g is able to determine which orders have the same COUNT(*) value as the value returned by the second subquery, and can then pass the order number back to the outer query to determine the customer's information. In this case, there are two customers who meet the criteria—fortunately, the user included the IN operator in the query, thus avoiding an error message.

> **NOTE**
>
> Don't forget to include the extra set of parentheses for the nested function on line 8, or you will receive an error message.

MERGE STATEMENT

With a MERGE statement, a series of DML actions can occur with a single SQL statement. The DML statements INSERT, UPDATE, and DELETE were covered in Chapter 5. However, conditionally updating one data source based on another was not addressed. Now that you have an understanding of more complex SQL statements, this topic is appropriate.

In a data warehousing environment, one often needs to conditionally update one table based on another table. For example, a BOOKS table may exist in the JustLee Books production system for recording orders. Any book price changes, category changes, and new book additions would be entered into this table. Another copy of the BOOKS table may be kept for querying and reporting. Many organizations do not want to slow down the production system and, therefore, copies of tables are maintained on separate servers to handle query and reporting requests. In this scenario, the tables used for reporting need periodic updating. The MERGE statement assists in this task, as it can compare two data sources or tables and determine which rows need updating and which need inserting.

Let's look at an example involving two BOOKS tables. A table named BOOKS_1 will serve as the reporting table and BOOKS_2 will serve as the production table. In this case, the BOOKS_2 table will be the input source and BOOKS_1 will be the target. If the book exists in both tables, an UPDATE will be needed to capture any changes in retail price of category assignments. If a book is in the BOOKS_2 table but not in the BOOKS_1 table, an INSERT will be needed to add the book to BOOKS_1. First, query both tables to review existing data, as shown in Figure 12-31.

Rows 1–3 of the BOOKS_2 table, as numbered in Figure 12-31, will be used to update the existing rows of the BOOKS_1 table, because these three match based on ISBN.

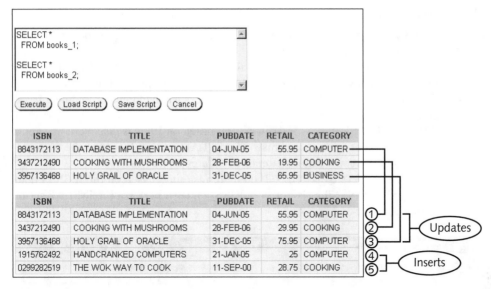

FIGURE 12-31 Current contents of the BOOKS_1 and BOOKS_2 tables

Rows 4–5 of BOOKS_2 will be added or INSERTED into the BOOKS_1 table, as these books do not currently exist in this table. Figure 12-32 shows a MERGE statement that will conditionally perform the UPDATES and INSERTS.

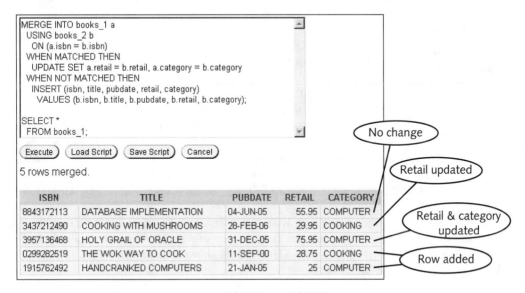

FIGURE 12-32 MERGE statement with UPDATE and INSERT

The following explains each part of this MERGE statement:

- MERGE INTO books_1 a: The BOOKS_1 table is to be changed and a table alias of "a" is assigned to this table.
- USING books_2 b: The BOOKS_2 table will provide the data to update and/or insert into BOOKS_1 and a table alias of "b" is assigned to this table.
- ON (a.isbn = b.isbn): The rows of the two tables will be joined or matched based on isbn.
- WHEN MATCHED THEN: If a row match based on ISBN is discovered, execute the UPDATE action in this clause. The UPDATE action instructs the system to modify only two columns (Retail and Category).
- WHEN NOT MATCHED THEN: If no match is found based on the ISBN (a book exists in BOOKS_2 that is not in BOOKS_1), then perform the INSERT action in this clause.

NOTE

A MERGE statement containing an UPDATE and an INSERT clause are also called UPSERT statements.

Including both the WHEN MATCHED and WHEN NOT MATCHED is not required. If only a particular DML operation is needed, only the appropriate clause needs to be included.

Execute a ROLLBACK statement so the BOOKS_1 data is set to the original three rows before performing the next example.

You can also include a WHERE condition in the matching clauses of a MERGE statement to conditionally perform the DML action based on a data value. Let's return to the previous example, but add a condition to UPDATE or INSERT only rows that have a category of COMPUTER assigned in the BOOKS_2 table. Figure 12-33 shows WHERE clauses added to the previous MERGE statement.

Recall that the BOOKS_2 table contains two rows with books in the Cooking Category. These two rows are no longer processed due to the added WHERE conditions. Therefore, the retail price of the book titled *Cooking with Mushrooms* is not updated and the book titled *The Wok Way to Cook* is not inserted. Also, the book titled *Holy Grail of Oracle* is originally assigned a category of BUSINESS in the BOOKS_1 table and COMPUTER in the BOOKS_2 table. The WHERE clause condition is checking the category data in the BOOKS_2 table, which is COMPUTER, so this row is updated in the BOOKS_1 table by the MERGE statement.

Execute a ROLLBACK statement so the BOOKS_1 data is set to the original three rows before performing the next example.

When a match is found during a MERGE statement, a DELETE can also be conditionally processed. For example, let's assume that the reporting table requires data only for books with a retail price of at least $50. Figure 12-34 shows a conditional DELETE action added to the WHEN MATCHED clause.

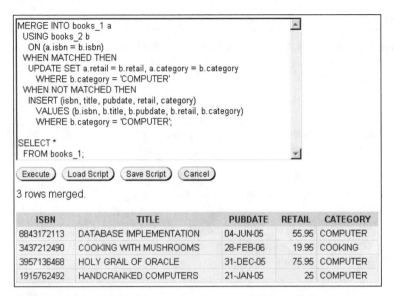

```
MERGE INTO books_1 a
  USING books_2 b
  ON (a.isbn = b.isbn)
  WHEN MATCHED THEN
    UPDATE SET a.retail = b.retail, a.category = b.category
      WHERE b.category = 'COMPUTER'
  WHEN NOT MATCHED THEN
    INSERT (isbn, title, pubdate, retail, category)
      VALUES (b.isbn, b.title, b.pubdate, b.retail, b.category)
      WHERE b.category = 'COMPUTER';

SELECT *
  FROM books_1;
```

(Execute) (Load Script) (Save Script) (Cancel)

3 rows merged.

ISBN	TITLE	PUBDATE	RETAIL	CATEGORY
8843172113	DATABASE IMPLEMENTATION	04-JUN-05	55.95	COMPUTER
3437212490	COOKING WITH MUSHROOMS	28-FEB-06	19.95	COOKING
3957136468	HOLY GRAIL OF ORACLE	31-DEC-05	75.95	COMPUTER
1915762492	HANDCRANKED COMPUTERS	21-JAN-05	25	COMPUTER

FIGURE 12-33   Using WHERE conditions in a MERGE statement

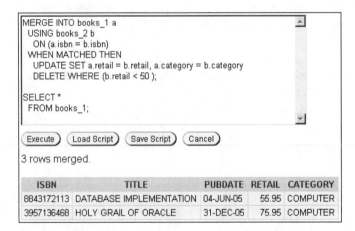

```
MERGE INTO books_1 a
  USING books_2 b
  ON (a.isbn = b.isbn)
  WHEN MATCHED THEN
    UPDATE SET a.retail = b.retail, a.category = b.category
    DELETE WHERE (b.retail < 50 );

SELECT *
  FROM books_1;
```

(Execute) (Load Script) (Save Script) (Cancel)

3 rows merged.

ISBN	TITLE	PUBDATE	RETAIL	CATEGORY
8843172113	DATABASE IMPLEMENTATION	04-JUN-05	55.95	COMPUTER
3957136468	HOLY GRAIL OF ORACLE	31-DEC-05	75.95	COMPUTER

FIGURE 12-34   Conditional DELETE in a MERGE statement

The BOOKS_1 table contains only two rows rather than three rows because the DELETE action removed the book titled *Cooking with Mushrooms*. This row is deleted because the retail amount is $29.95 in the BOOKS_2 table, and this amount meets the delete condition of being below $50. This MERGE statement only processes matched rows, so the other two rows in the BOOKS_2 table with retail amounts below $50 are not processed.

Chapter Summary

- A subquery is a complete query nested in the SELECT, FROM, HAVING, or WHERE clause of another query. The subquery must be enclosed in parentheses and have a SELECT and a FROM clause, at a minimum.

- Subqueries are completed first. The result of the subquery is used as input for the outer query.

- A single-row subquery can return a maximum of one value.

- Single-row operators include =, >, <, >=, <=, and <>.

- Multiple-row subqueries return more than one row of results.

- Operators that can be used with multiple-row subqueries include IN, ALL, ANY, and EXISTS.

- Multiple-column subqueries return more than one column to the outer query. The columns of data are passed back to the outer query in the same order in which they are listed in the SELECT clause of the subquery.

- NULL values returned by a multiple-row or multiple-column subquery will not present a problem if the IN or =ANY operator is used. The NVL function can be used to substitute a value for a NULL value when working with subqueries.

- Correlated subqueries reference a column contained in the outer query. When using correlated subqueries, the subquery is executed once for each row processed by the outer query.

- Subqueries can be nested to a maximum depth of 255 subqueries in the WHERE clause of the parent query. The depth is unlimited for subqueries nested in the FROM clause of the parent query.

- With nested subqueries, the innermost subquery is executed first, then the next highest level subquery is executed, and so on, until the outermost query is reached.

- A MERGE statement allows multiple DML actions to be conditionally performed while comparing data of two tables.

Chapter 12 Syntax Summary

The following tables present a summary of the syntax that you have learned in this chapter. You can use the tables as a study guide and reference.

SYNTAX GUIDE	
Subquery Types	
Subquery Processing	**Example**

Correlated Subquery: References a column in the outer query. Executes the subquery once for every row in the outer query.	`SELECT title`
	`FROM books b`
	`WHERE b.isbn IN`
	`    (SELECT isbn`
	`     FROM orderitems o`
	`     WHERE b.isbn = o.isbn);`

Uncorrelated Subquery: Executes the subquery first and passes the value to the outer query.	`SELECT title`
	`FROM books b, orderitems o`
	`WHERE books isbn IN`
	`    (SELECT isbn`
	`     FROM orderitems)`
	`AND b.isbn = o.isbn;`

MULTIPLE-ROW COMPARISON OPERATORS	
Operator	**Description**
>ALL	More than the highest value returned by the subquery
<ALL	Less than the lowest value returned by the subquery
<ANY	Less than the highest value returned by the subquery
>ANY	More than the lowest value returned by the subquery
=ANY	Equal to any value returned by the subquery (same as IN)
[NOT] EXISTS	Row must match a value in the subquery
DML ACTIONS WITH MERGE	
Merge Statement	**Example**

MERGE Statement: Conditionally performs a series of DML actions	`MERGE INTO books_1 a`
	`USING books_2 b`
	`ON (a.isbn = b.isbn)`
	`WHEN MATCHED THEN`
	`UPDATE SET a.retail = b.retail,`
	`a.category = b.category`
	`WHEN NOT MATCHED THEN`
	`INSERT (isbn, title, pubdate, retail, category)`
	`VALUES (b.isbn, b.title, b.pubdate, b.retail,`
	`b.category);`

Review Questions

1. What is the difference between a single-row subquery and a multiple-row subquery?

2. What comparison operators are required for multiple-row subqueries?

3. What happens if a single-row subquery returns more than one row of results?

4. Which clause(s) cannot be used in a subquery?

5. If a subquery is used in the FROM clause of an outer query, how are the results of the subquery referenced in other clauses of the outer query?

6. Why might a MERGE statement be used?

7. How can Oracle 10*g* determine whether clauses of a SELECT statement belong to an outer query or a subquery?

8. When should a subquery be nested in a HAVING clause?

9. What is the difference between correlated and uncorrelated subqueries?

10. What type of situation requires the use of a subquery?

Multiple Choice

To answer these questions, refer to the tables in Appendix A.

1. Which query will identify the customers living in the same state as the customer named Leila Smith?

   a. SELECT customer# FROM customers WHERE state =
      (SELECT state FROM customers WHERE lastname = 'SMITH');

   b. SELECT customer# FROM customers WHERE state =
      (SELECT state FROM customers WHERE lastname = 'SMITH' OR
      firstname = 'LEILA');

   c. SELECT customer# FROM customers WHERE state =
      (SELECT state FROM customers
      WHERE lastname = 'SMITH' AND firstname = 'LEILA'
      ORDER BY customer);

   d. SELECT customer# FROM customers WHERE state =
      (SELECT state FROM customers
      WHERE lastname = 'SMITH' AND firstname = 'LEILA');

2. Which of the following is a valid SELECT statement?

   a. SELECT order# FROM orders WHERE shipdate =
      SELECT shipdate FROM orders WHERE order# = 1010;

   b. SELECT order# FROM orders WHERE shipdate =
      (SELECT shipdate FROM orders)
      AND order# = 1010;

   c. SELECT order# FROM orders WHERE shipdate =
      (SELECT shipdate FROM orders WHERE order# = 1010);

   d. SELECT order# FROM orders HAVING shipdate =
      (SELECT shipdate FROM orders WHERE order# = 1010);

3. Which of the following operators is considered a single-row operator?

    a.   IN

    b.   ALL

    c.   <>

    d.   <>ALL

4. Which of the following queries will determine which customers have ordered the same books as customer 1017?

    a.   SELECT order# FROM orders WHERE customer# = 1017;

    b.   SELECT customer# FROM orders JOIN orderitems USING(order#)
        WHERE isbn = (SELECT isbn FROM orderitems
        WHERE customer# = 1017);

    c.   SELECT customer# FROM orders WHERE order# =
        (SELECT order# FROM orderitems WHERE customer# = 1017);

    d.   SELECT customer# FROM orders JOIN orderitems (order#)
        WHERE isbn IN (SELECT isbn FROM orderitems
        JOIN orders USING(order#)
        WHERE customer# = 1017);

5. Which of the following statements is valid?

    a.   SELECT title FROM books WHERE retail <
        (SELECT cost FROM books WHERE isbn = '9959789321');

    b.   SELECT title FROM books WHERE retail =
        (SELECT cost FROM books WHERE isbn = '9959789321'
        ORDER BY cost);

    c.   SELECT title FROM books WHERE category IN
        (SELECT cost FROM orderitems WHERE isbn = '9959789321');

    d.   none of the above statements

6. Which of the following statements is correct?

    a.   If a subquery is used in the FROM clause of the outer query, the data contained in the temporary table cannot be referenced by clauses used in the outer query.

    b.   The temporary table created by a subquery in the FROM clause of the outer query must be assigned a table alias or it cannot be joined with another table using the JOIN keyword.

    c.   If a temporary table is created through a subquery in the FROM clause of an outer query, the data in the temporary table can be referenced by another clause in the outer query.

    d.   none of the above

7. Which of the following queries identifies other customers who were referred to JustLee Books by the same individual that referred Jorge Perez?

   a. SELECT customer# FROM customers WHERE referred =
      (SELECT referred FROM customers
      WHERE firstname = 'JORGE' AND lastname = 'PEREZ');

   b. SELECT referred FROM customers
      WHERE (customer#, referred) =
      (SELECT customer# FROM customers
      WHERE firstname = 'JORGE' AND lastname = 'PEREZ');

   c. SELECT referred FROM customers
      WHERE (customer#, referred) IN
      (SELECT customer# FROM customers
      WHERE firstname = 'JORGE' AND lastname = 'PEREZ');

   d. SELECT customer# FROM customers
      WHERE customer# =
      (SELECT customer# FROM customers
      WHERE firstname = 'JORGE' AND lastname = 'PEREZ');

8. In which of the following situations would it be appropriate to use a subquery?

   a. when you need to find all customers living in a particular region of the country

   b. when you need to find all publishers who have toll-free telephone numbers

   c. when you need to find the name of all books that were shipped on the same date as an order placed by a particular customer

   d. when you need to find all books published by Publisher 4

9. Which of the following queries will identify customers who have ordered the same books as customers 1001 and 1005?

   a. SELECT customer# FROM orders JOIN books USING(isbn)
      WHERE isbn =
      (SELECT isbn FROM orderitems JOIN books USING(isbn)
      WHERE customer# = 1001 OR customer# = 1005);

   b. SELECT customer# FROM orders JOIN books USING(isbn)
      WHERE isbn <ANY
      (SELECT isbn FROM orderitems JOIN books USING(isbn)
      WHERE customer# = 1001 OR customer# = 1005);

   c. SELECT customer# FROM orders JOIN books USING(isbn)
      WHERE isbn =
      (SELECT isbn FROM orderitems JOIN books USING(isbn)
      WHERE customer# = 1001 OR 1005);

   d. SELECT customer# FROM orders JOIN orderitems USING(order#)
      WHERE isbn IN
      (SELECT isbn FROM orders JOIN orderitems USING(order#)
      JOIN books USING(isbn)
      WHERE customer# = 1001 OR customer# = 1005);

10. Which of the following operators is used to find all values that are greater than the highest value returned by a subquery?

    a. >ALL

    b. <ALL

    c. >ANY

    d. <ANY

    e. IN

11. Which query will determine the customer who has ordered the most books from JustLee Books?

    a. SELECT customer# FROM orders JOIN orderitems USING(order#)
       JOIN customers USING(customer#)
       HAVING SUM(quantity) =
       (SELECT MAX(SUM(quantity)) FROM orders JOIN orderitems USING(order#)
       GROUP BY customer#)
       GROUP BY customer#;

    b. SELECT customer# FROM orders JOIN orderitems USING(order#)
       WHERE SUM(quantity) =
       (SELECT MAX(SUM(quantity)) FROM orderitems GROUP BY
       customer#);

    c. SELECT customer# FROM orders WHERE MAX(SUM(quantity)) =
       (SELECT MAX(SUM(quantity) FROM orderitems GROUP BY order#);

    d. SELECT customer# FROM orders WHERE quantity =
       (SELECT MAX(SUM(quantity)) FROM orderitems GROUP BY
       customer#);

12. Which of the following statements is correct?

    a. The IN comparison operator cannot be used with a subquery that returns only one row of results.

    b. The equals (=) comparison operator cannot be used with a subquery that returns more than one row of results.

    c. In an uncorrelated subquery, the statements in the outer query are executed first, and then the statements in the subquery are executed.

    d. A subquery can be nested only in the SELECT clause of an outer query.

13. What is the purpose of the following query?

    SELECT isbn, title
    FROM books
    WHERE (pubid, category) IN
    (SELECT pubid, category
    FROM books
    WHERE title LIKE '%ORACLE%');

    a. It determines which publisher published a book belonging to the Oracle Category and then lists all other books published by that same publisher.

    b. It lists all the publishers and categories containing the value ORACLE.

    c. It lists the ISBN and title of all books belonging to the same category and having the same publisher as any book with the letters ORACLE in its title.

    d. None of the above—the query contains a multiple-row operator, and because the inner query returns only one value, the SELECT statement will fail and return an error message.

14. A subquery must be placed in the HAVING clause of the outer query if:

    a. The inner query needs to reference the value returned to the outer query.

    b. The value returned by the inner query is to be compared to grouped data in the outer query.

    c. The subquery returns more than one value to the outer query.

    d. None of the above—subqueries cannot be used in the HAVING clause of an outer query.

15. Which of the following SQL statements will list all books written by the author who wrote *The Wok Way to Cook*?

    a. SELECT title FROM books WHERE isbn IN
    (SELECT isbn FROM bookauthor
    HAVING authorid IN 'THE WOK WAY TO COOK);

    b. SELECT isbn FROM bookauthor WHERE authorid IN
    (SELECT authorid FROM books JOIN bookauthor USING(isbn)
    WHERE title = 'THE WOK WAY TO COOK');

    c. SELECT title FROM bookauthor WHERE authorid IN
    (SELECT authorid FROM books JOIN bookauthor USING(isbn)
    WHERE title = 'THE WOK WAY TO COOK);

    d. SELECT isbn FROM bookauthor HAVING authorid =
    SELECT authorid FROM books JOIN bookauthor USING(isbn)
    WHERE title = 'THE WOK WAY TO COOK';

16. Which of the following statements is correct?

    a. If the subquery returns only a NULL value, the only records returned by an outer query are those that contain an equivalent NULL value.

    b. A multiple-column subquery can be used only in the FROM clause of an outer query.

    c. A subquery can contain only one condition in its WHERE clause.

    d. The order of the columns listed in the SELECT clause of a multiple-column subquery must be in the same order as the corresponding columns listed in the WHERE clause of the outer query.

17. In a MERGE statement, within which conditional clause would an INSERT be placed?

    a. USING

    b. WHEN MATCHED

    c. WHEN NOT MATCHED

    d. INSERTS are not allowed in a MERGE statement.

18. Given the following query, which statement is correct?

    SELECT order# FROM orders WHERE order# IN
    (SELECT order# FROM orderitems
    WHERE isbn = '9959789321');

    a. The statement will not execute because the subquery and the outer query do not reference the same table.

    b. The outer query is not necessary because it has no effect on the results displayed.

    c. The query will fail if only one result is returned to the outer query because the WHERE clause of the outer query uses the IN comparison operator.

    d. No rows will be displayed because the ISBN in the WHERE clause is enclosed in single quotation marks.

19. Given the following SQL statement, which statement is most accurate?

    SELECT customer#
    FROM customers JOIN orders USING(customer#)
    WHERE shipdate-orderdate IN
    (SELECT MAX(shipdate-orderdate)
    FROM orders
    WHERE shipdate IS NULL);

    a. The SELECT statement will fail and return an Oracle error message.

    b. The outer query will display no rows in its results because the subquery passes a NULL value to the outer query.

    c. The customer number will be displayed for those customers whose orders have not yet shipped.

    d. The customer number of all customers who have not placed an order will be displayed.

20. Which of the following statements is correct?

    a. In a correlated subquery, the outer query is executed once for every row processed by the inner query.

    b. In an uncorrelated subquery, the outer query is executed once for every row processed by the inner query.

    c. In a correlated subquery, the inner query is executed once for every row processed by the outer query.

    d. In an uncorrelated subquery, the inner query is executed once for every row returned by the outer query.

Hands-On Assignments

To perform these activities, refer to the tables in Appendix A. Use a subquery to accomplish each of the tasks. First, execute the query you will use as the subquery to verify the results.

1. Determine which books have a retail price that is less than the average retail price of all books sold by JustLee Books.

2. Determine which books cost less than the average cost of other books in the same category.

3. Determine which orders were shipped to the same state as order 1014.

4. Determine which orders had a higher total amount due than order 1008.

5. Determine which author or authors wrote the book(s) most frequently purchased by customers of JustLee Books.

6. List the title of all books in the same category as books previously purchased by customer 1007. Do not include books already purchased by this customer.

7. List the city and state for the customer(s) who experienced the longest shipping delay.

8. Determine which customers placed orders for the least expensive book (in terms of retail price) carried by JustLee Books.

9. Determine how many different customers have placed an order for books written or cowritten by James Austin.

10. Determine which books were published by the publisher of *The Wok Way to Cook.*

Advanced Challenge

To perform this activity, refer to the tables in Appendix A.

Currently, JustLee Books bills customers for orders by enclosing an invoice with each order when it is shipped. A customer then has 10 days to send in the payment. Of course, this has resulted in the company having to list some debts as "uncollectible." By contrast, most other online booksellers receive payment through a customer's credit card at the time of purchase. Although payment would be deposited within 24 hours into JustLee Books' bank account by accepting credit cards, there is a downside to this alternative. When a merchant accepts credit cards for payment, the company that processes the credit card sales (usually called a "credit card clearinghouse") deducts a 1.5 percent processing fee from the total amount of the credit card sale.

The management of JustLee Books is trying to determine whether the surcharge for credit card processing is more than the amount usually deemed "uncollectible" when customers are sent an invoice. Historically, the average amount that has been lost by JustLee Books is about 4 percent of the total amount due from orders that have a higher-than-average amount due. In other words, it is usually customers who have an order with a larger-than-average invoice total who default on their payment.

To determine how much money would be lost or gained, by accepting credit card payments, management has requested that you do the following:

1. Determine how much the surcharge would be for all recently placed orders if payment had been made by a credit card.

2. Determine the total amount that can be expected to be written off as "uncollectible" based on recently placed orders having an invoice total that is more than the average of all recently placed orders.

Based on the results of these two calculations, you should determine whether the company will lose money by accepting payment via credit card. State your findings in a memo to management. Include the SQL statement(s) necessary to calculate the expected surcharge and the expected amount of "uncollectible" payments.

431

Case Study: *City Jail*

It is assumed that the City Jail database creation script from Chapter 8 has been executed. That script makes all database objects available for the completion of this case study.

The city's Crimes Analysis unit has submitted the following data requests. Provide the SQL statements using subqueries that would satisfy the requests. Test the statements and show execution results as well.

1. List the name of each officer who has reported more than the average number of crimes reported by officers.

2. List the criminal names for all criminals who have a less than average number of crimes and are not listed as violent offenders.

3. List appeal information for each appeal that has a less than average number of days between the filing and hearing dates.

4. List the names of the probation officers who have had a less than average number of criminals assigned.

5. List each crime that has had the highest number of appeals recorded.

6. List the information on crime charges for each charge that has had a fine that is above average and sum paid that is below average.

7. List all the names of all criminals who have had any of the crime code charges that were involved in crime id 10089.

8. Use a correlated subquery to determine which criminals have had at least one probation period assigned.

9. List the names of the officers who have booked the highest number of crimes. Note that more than one officer may be listed.

    Use a MERGE statement to resolve the following request.

10. The criminal data warehouse contains a copy of the CRIMINALS table that needs to be periodically updated from the production CRIMINALS table. The data warehouse table is named CRIMINALS_DW. Use a single SQL statement to update the data warehouse table to reflect any data changes for existing criminals and to add new criminals.

VIEWS

INTRODUCTION

Views are database objects that store a SELECT statement and allow the results of the query to be used as a table. Views have two purposes:

- Simplify issuing complex SQL queries

- Restrict users' access to sensitive data

Although views are database objects, they do not actually store data. A view stores a query and is used to access data in the underlying base tables. You can think of a view as the result of a stored query: The results are given a name that allows the results to be used as the source for queries just as you would use a table. In fact, a view can be referenced in the FROM clause of a SELECT statement, just like you reference any table. Views can include one column or as many as all the columns in specified tables.

Figure 13-1 depicts the basic processing of a view. When a query references a view, the query contained in the view is processed and the results are treated as a virtual or temporary table. In this case, any query referencing the BOOK_VU view would be able to examine only the books in the Cooking Category.

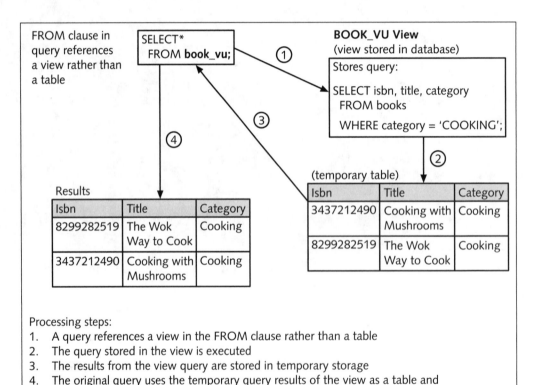

Processing steps:
1. A query references a view in the FROM clause rather than a table
2. The query stored in the view is executed
3. The results from the view query are stored in temporary storage
4. The original query uses the temporary query results of the view as a table and completes execution

FIGURE 13-1 View processing

Let's examine how views can help simplify complex queries. First, many nontechnical users may not be familiar with SQL coding. Data retrieval from JustLee Books' tables can require fairly complex queries. For example, suppose that an employee needs to find a customer's total balance due for an order. Just to determine that amount, the employee must simultaneously query the CUSTOMERS, ORDERS, ORDERITEMS, and BOOKS tables. The average employee probably lacks the training needed to create a query that joins multiple tables and performs calculations needed to perform this seemingly simple task—to determine the total balance due for an order. To simplify the user's task, one option is to create a view that contains all the information the user needs. Thus, rather than teaching users how to create queries with joins and calculations, you can instead show them how to perform simple queries on the virtual table that includes customer names, order numbers, order dates, book titles, quantity of books ordered, retail prices, and the calculated extended price.

Application development can also be simplified by using views. For example, a developer may have several screens that require customer order details which involve a complex query. Rather than programming the query in multiple places in the application, the developer could create a view that stores the complex query. Then, the developer need only perform a simple query on the view within the application code.

A view can be used not only to simplify complex queries, but also to restrict access to what management may consider "sensitive data." For example, the BOOKS table contains both the cost and retail price of each book in inventory. What happens if management decides that the cost of each book should not be accessible by every employee in the company? Do you delete the column from the BOOKS table? If so, then how would you calculate the profit for each book sold? Rather than providing users with access to the actual table storing all the data for the books, you can give them access to data via a view that includes only the data appropriate for the user, based on her job duties.

Views are most commonly used to query data. Some developers, however, may want to perform a DML activity on data accessed via a view. This chapter presents the commands and guidelines regulating DML operations for data accessed by views. The last section of the chapter introduces materialized views, which are views that permanently store data. Figure 13-2 provides an overview of this chapter's contents.

VIEW TYPE	DESCRIPTION
Simple view	A view based upon a subquery that only references one table and does not include any group functions, expressions, or a GROUP BY clause
Complex view	A view based upon a subquery that retrieves or derives data from one or more tables—and may also contain functions or grouped data
Inline view	A subquery used in the FROM clause of a SELECT statement to create a "temporary" table that can be referenced by the SELECT and WHERE clauses of the outer statement
Materialized view	A view that replicates data by physically storing the results of the view query

COMMAND	COMMAND SYNTAX
Create a view	`CREATE [OR REPLACE] [FORCE|NOFORCE] VIEW viewname (columnname, ...)AS subquery[WITH CHECK OPTION [CONSTRAINT constraintname]][WITH READ ONLY];`
Drop a view	`DROP VIEW viewname`
Create an inline view	`SELECT columnname, ...FROM (subquery)WHERE ROWNUM<=N;`

Create a materialized view	`CREATE MATERIALIZED VIEW custbal_mv`
	`  REFRESH COMPLETE`
	`  START WITH SYSDATE NEXT SYSDATE+7`
	`  AS SELECT customer#, city, state, order#,`
	`SUM(quantity*retail) Amtdue`
	`    FROM customers JOIN orders`
	`USING(customer#)`
	`         JOIN orderitems USING(order#)`
	`         JOIN books USING(isbn)`
	`    GROUP BY customer#, city, state, order#;`

FIGURE 13-2   Overview of view concepts

NOTE

Before attempting to work through the examples provided in this chapter, run the prech08.sql script as instructed in Chapter 8.

CREATING A VIEW

A view is created with the **CREATE VIEW** command. The syntax for the CREATE VIEW command is shown in Figure 13-3.

Let's get an overview of the syntax elements shown in Figure 13-3.

- *CREATE VIEW/CREATE OR REPLACE VIEW*: You use the **CREATE VIEW** keywords to create a view, choosing a name that is not being used by another database object in the current schema. *There is no way to modify or change an existing view,* so if you need to change a view, you must use the **CREATE OR REPLACE VIEW** keywords. The OR REPLACE option instructs Oracle 10g

```
CREATE [OR REPLACE] [FORCE|NOFORCE] VIEW
    viewname (columnname, ...)
AS select statement
[WITH CHECK OPTION [CONSTRAINT constraintname]]
[WITH READ ONLY];
```

FIGURE 13-3   Syntax of the CREATE VIEW command

that a view with the same name may already exist, and if it does, the option instructs Oracle to replace the previous version of the view with the one defined in the new command.

- *FORCE/NOFORCE*: If you attempt to create a view based upon a table(s) that does not yet exist or is currently unavailable (for example, offline), Oracle 10g returns an error message, and the view is not created. However, if you include the **FORCE** keyword in the CREATE clause, Oracle 10g creates the view in spite of the absence of any referenced tables. For example, **NOFORCE** is the default mode for the CREATE VIEW command, and that means all the tables and columns must be valid, or the view will not be created. This approach is commonly used when a new database is being developed and the data has not yet been loaded, or entered, into the database objects.

- *View name*: As previously mentioned, you should give each view a name that is not already assigned to another database object in the same schema.

- *Column names*: If you want to assign new names for the columns that are displayed by the view, the new column names can be listed after the VIEW keyword, within a set of parentheses. *The number of names listed must match the number of columns returned by the SELECT statement.* An alternative is to use column aliases in the query. If column aliases are given in the query, Oracle 10g uses the aliases as the column names in the view that is created.

- *AS clause*: The query listed after the AS keyword must be a complete SELECT statement (both SELECT and FROM clauses are required) and can reference more than one table. The query can also include single-row and group functions, WHERE and GROUP BY clauses, nested subqueries, and so on. However, as with subqueries, the query cannot include the ORDER BY clause. The results of the query will be the content of the view being created.

- *WITH CHECK OPTION*: The WITH CHECK OPTION constraint ensures that any DML operations performed on the view (such as adding rows or changing data) do not prevent the row from being accessed by the view because it no longer meets the condition in the WHERE clause. For example, if a view consists of books only in the Cooking Category, and the user attempts to change the category of one of the books included in the view to the Family Life Category, the change will not be allowed if the WITH CHECK OPTION constraint was included when the view was created. Why? Because the change would mean that the book will no longer be listed in the view because it consists of books only in the Cooking Category. If the WITH CHECK OPTION constraint is omitted when the view is created, any valid DML operation is allowed, even if the result is that the row(s) being changed would no longer be included in the view. However, if a view is being created with the sole purpose of displaying data, the WITH READ ONLY option can be used instead to ensure that the data cannot be changed.

- *WITH READ ONLY OPTION*: This option prevents a user from issuing any DML actions on the view. This option is heavily used in situations in which it's important that users can only query data and not make any changes to it.

Next, let's look at these operations in more depth. First, you'll see how to create a simple view, then how to change a simple view, and finally how to create a complex view.

Creating a Simple View

You create a simple view from a subquery that references only one table and does not include a group function, expression (such as retail-cost), or GROUP BY clause. For example, when JustLee Books' service representatives assist customers with orders, they need to access the ISBN, title, and retail price of every book in JustLee's inventory. However, management does not want representatives to view the books' actual cost. The solution: Create a simple view to allow service representatives to access only the data needed to assist customers—and not access irrelevant columns, such as Publisher id, Cost, and so on.

The CREATE VIEW command to create the view needed by customer service representatives is shown in Figure 13-4.

```
CREATE VIEW inventory
 AS SELECT isbn, title, retail price
     FROM books
 WITH READ ONLY;
```

FIGURE 13-4 Command to create the INVENTORY view

As indicated in the CREATE VIEW clause, the name of the new view will be INVENTORY. Note these other elements in Figure 13-4:

1. Because another view with this name does not exist, the OR REPLACE clause is not necessary.
2. The only columns included in the view are the ISBN, Title, and Retail columns from the BOOKS table. Notice that the Retail column has been assigned the column alias of Price. Thus, whenever the Retail column is referenced through a query on the INVENTORY view, it must be called Price in the query.
3. The WITH READ ONLY clause is used so that no customer service representative can accidentally change the ISBN, title, or price of a book. Any changes made to the data contained in a simple view created without the WITH READ ONLY option automatically update the underlying BOOKS table.

As shown in Figure 13-5, after the INVENTORY view is created, users can reference it in the FROM clause of a SELECT statement in the same manner as they would any table.

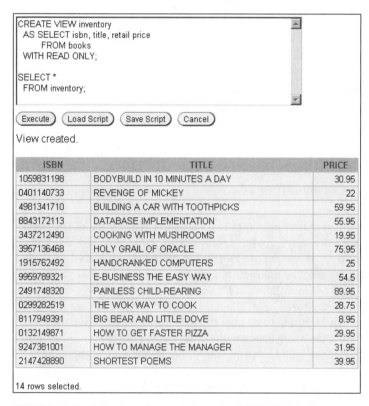
NOTE

If the command in Figure 13-5 returns an error message, make certain that a view with the same name does not already exist. If such a view does exist, add the keywords OR REPLACE to the CREATE VIEW clause and then execute the command again.

```
CREATE VIEW inventory
   AS SELECT isbn, title, retail price
        FROM books
   WITH READ ONLY;

SELECT *
   FROM inventory;
```

(Execute) (Load Script) (Save Script) (Cancel)

View created.

ISBN	TITLE	PRICE
1059831198	BODYBUILD IN 10 MINUTES A DAY	30.95
0401140733	REVENGE OF MICKEY	22
4981341710	BUILDING A CAR WITH TOOTHPICKS	59.95
8843172113	DATABASE IMPLEMENTATION	55.95
3437212490	COOKING WITH MUSHROOMS	19.95
3957136468	HOLY GRAIL OF ORACLE	75.95
1915762492	HANDCRANKED COMPUTERS	25
9959789321	E-BUSINESS THE EASY WAY	54.5
2491748320	PAINLESS CHILD-REARING	89.95
0299282519	THE WOK WAY TO COOK	28.75
8117949391	BIG BEAR AND LITTLE DOVE	8.95
0132149871	HOW TO GET FASTER PIZZA	29.95
9247381001	HOW TO MANAGE THE MANAGER	31.95
2147428890	SHORTEST POEMS	39.95

14 rows selected.

FIGURE 13-5   Selecting all records from the INVENTORY view

Although the INVENTORY view can be retrieved with a SELECT statement as if it were a regular table, the view in Figure 13-5 is created with a **WITH READ ONLY** option. This prevents any DML operations from being performed on the data. In Figure 13-6, the user is unsuccessfully attempting to update the data contained in the Price and Title columns of the INVENTORY view.

439

Views

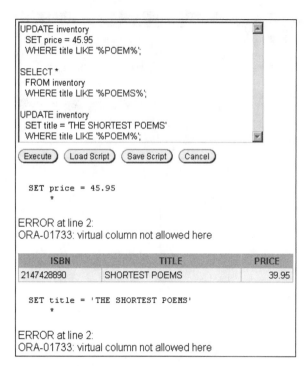

```
UPDATE inventory
  SET price = 45.95
  WHERE title LIKE '%POEM%';

SELECT *
  FROM inventory
  WHERE title LIKE '%POEMS%';

UPDATE inventory
  SET title = 'THE SHORTEST POEMS'
  WHERE title LIKE '%POEM%';
```

[Execute] [Load Script] [Save Script] [Cancel]

```
    SET price = 45.95
        *
```

ERROR at line 2:
ORA-01733: virtual column not allowed here

ISBN	TITLE	PRICE
2147428890	SHORTEST POEMS	39.95

```
    SET title = 'THE SHORTEST POEMS'
        *
```

ERROR at line 2:
ORA-01733: virtual column not allowed here

FIGURE 13-6   Failed updates on the INVENTORY view

Figure 13-6 shows that the user first attempts to change the retail price of the book called *Shortest Poems*. Because the user could not remember the exact title of the book, a search pattern is used in the WHERE clause to identify the book being updated. The per-cent signs (%) indicate that there may be characters appearing before and after the word *Poems*, but the book's title must contain the word *Poems*. However, Oracle 10*g* returns the error message "virtual column not allowed here." Because the error message does not seem to indicate the problem, the user attempts to change the title for the book to see whether the problem is due to a column alias being used for the column that is actually called Retail in the BOOKS table. However, these failed attempts at changing the data in the view are not due to any column references. Rather, it is because the view was created with the WITH READ ONLY option, and no DML operations are allowed at all. As you can see, some-times Oracle 10*g* error messages can be obscure.

DML Operations on a Simple View

If management later decides that the INVENTORY view should be used to alter the retail prices of the books currently in inventory, it can be re-created using the CREATE OR REPLACE VIEW command—without the WITH READ ONLY option. Because the INVEN-TORY view already exists in the database, the OR REPLACE keywords must be included, or Oracle 10*g* returns an error message indicating that the view already exists.

In Figure 13-7, the INVENTORY view has been re-created—without a WITH READ ONLY option, so updates are allowed.

```
CREATE OR REPLACE VIEW inventory
  AS SELECT isbn, title, retail price
    FROM books;

SELECT *
  FROM inventory
  WHERE title LIKE '%BODYBUILD%';
```

(Execute) (Load Script) (Save Script) (Cancel)

View created.

ISBN	TITLE	PRICE
1059831198	BODYBUILD IN 10 MINUTES A DAY	30.95

FIGURE 13-7 Re-create the view to allow DML activity

After the view shown in Figure 13-7 is re-created, anyone with access to the view can change the ISBN, title, or retail price of any book in the BOOKS table. As shown, the original retail price of *Bodybuild in 10 Minutes a Day* was $30.95. Figure 13-8 shows an UPDATE command issued on the view to change the retail price of the book to $49.95. The previous version of the view would not have allowed the data to be changed.

```
UPDATE inventory
  SET price = 49.95
  WHERE title LIKE '%BODYBUILD%';

SELECT *
  FROM inventory
  WHERE title LIKE '%BODYBUILD%';

SELECT *
  FROM books
  WHERE title LIKE '%BODYBUILD%';
```

(Execute) (Load Script) (Save Script) (Cancel)

1 row updated.

ISBN	TITLE	PRICE
1059831198	BODYBUILD IN 10 MINUTES A DAY	49.95

ISBN	TITLE	PUBDATE	PUBID	COST	RETAIL	CATEGORY
1059831198	BODYBUILD IN 10 MINUTES A DAY	21-JAN-01	4	18.75	49.95	FITNESS

FIGURE 13-8 Issue DML on a simple view

As shown in Figure 13-8, the view now accepts the change and, as a result, the price value in the base table (the BOOKS table) is altered.

The basic rule for DML operations on a simple view (or a complex view) is this: As long as the view was not created with the WITH READ ONLY option, any DML operation is allowed on a simple view if it does not violate an existing constraint on the underlying base table. In essence, you can add, modify, and even delete data in an underlying base table as long as the operation is not prevented by one of the following constraints:

- PRIMARY KEY
- NOT NULL
- UNIQUE
- FOREIGN KEY
- WITH CHECK OPTION

NOTE

If the SELECT statements in Figures 13-7 or 13-8 return an error message or do not select any rows, make certain the book title in the WHERE clause is enclosed in single quotation marks and includes a percent sign (%) at the end of the search pattern for BODYBUILD.

Let's look at another example. As shown in Figure 13-9, the OUTSTANDING view has been created to display all the orders contained in the ORDERS table that have not yet been shipped to the customer.

FIGURE 13-9 Create a view with the WITH CHECK OPTION constraint

You should note that because a WITH CHECK OPTION constraint was included in the CREATE VIEW command in Figure 13-9, the OUTSTANDING view cannot be used to update the shipping date of any of the six records that have not yet been shipped because changing this would remove the record(s) from the view. Any attempt to change the ship date of an order to a non-NULL value returns an error message, and the update fails, as shown in Figure 13-10.

If the purpose of the OUTSTANDING view is to allow users to enter the ship date of an order when it is shipped, it should be re-created *without* the WITH CHECK OPTION constraint, as shown in Figure 13-11.

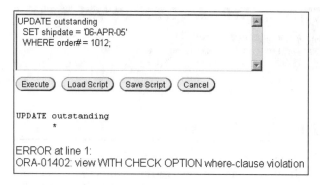

```
UPDATE outstanding
  SET shipdate = '06-APR-05'
  WHERE order# = 1012;



Execute   Load Script   Save Script   Cancel


UPDATE outstanding
       *

ERROR at line 1:
ORA-01402: view WITH CHECK OPTION where-clause violation
```

FIGURE 13-10   Error returned on update that violates the WITH CHECK OPTION

```
CREATE OR REPLACE VIEW outstanding
  AS SELECT customer#, order#, orderdate, shipdate
     FROM orders
       WHERE shipdate IS NULL;

UPDATE outstanding
  SET shipdate = '06-APR-05'
  WHERE order# = 1012;

SELECT *
FROM outstanding;

Execute   Load Script   Save Script   Cancel

View created.

1 row updated.
```

CUSTOMER#	ORDER#	ORDERDATE	SHIPDATE
1020	1015	04-APR-05	
1003	1016	04-APR-05	
1001	1018	05-APR-05	
1018	1019	05-APR-05	
1008	1020	05-APR-05	

FIGURE 13-11   Update succeeds on view created without the WITH CHECK OPTION

In Figure 13-11, the view has been re-created and a command has been issued to update the ship date of order 1012 to April 6, 2005. Because the WITH CHECK OPTION constraint no longer exists on the OUTSTANDING view, the update is allowed. However, the record will no longer be included in the view after the change occurs because the Shipdate is no longer NULL.

443

CREATING A COMPLEX VIEW

You create a complex view using the same CREATE VIEW command as a simple view. However, the SELECT statement contained in a complex view retrieves or derives data from one or more tables—and may contain functions or grouped data. *The main difference in functionality between simple and complex views is that certain DML operations are not permitted with complex views.* To discuss how complex views react to various DML operations, this section presents three different views.

1. The first complex view presented is based on one table, but it uses an expression for one of the columns.
2. The second complex view presented is based on two tables, and it also uses an expression for one of the columns.
3. The third complex view presented is derived from four tables, and it includes a group function and a GROUP BY clause.

DML Operations on a Complex View with an Arithmetic Expression

As you will see, different factors affect the type of DML operations that are allowed on complex views. For example, if a view contains a column that is the result of an arithmetic expression or grouped data, or if it is based upon multiple tables and it is difficult to determine exactly which table should be modified, then certain DML operations will not work. Look at the complex view in Figure 13-12, which shows creating and updating a view called PRICES.

The complex view created in Figure 13-12 seems like a simple view, except that it uses the expression "retail-cost" to calculate, or derive, the Profit column. Also shown is an UPDATE command using **(SET retail = 29.95)** that changes the retail price of *Revenge of Mickey* from $22.00 to $29.95. Again, the view acts like a simple view because the UPDATE command worked to make the change. But what about removing rows in the view? Look at Figure 13-13.

In Figure 13-13, the DELETE command is used to remove the book called *Revenge of Mickey* from the PRICES view, which actually removes the book from the BOOKS table. So far, so good—the DELETE command works for this view. Now what about adding a record to the view?

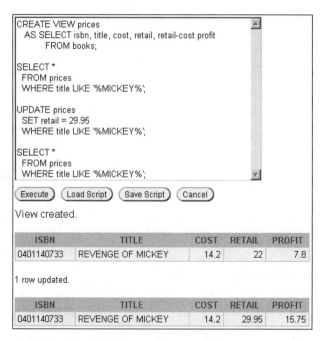

```
CREATE VIEW prices
    AS SELECT isbn, title, cost, retail, retail-cost profit
        FROM books;

SELECT *
 FROM prices
 WHERE title LIKE '%MICKEY%';

UPDATE prices
 SET retail = 29.95
 WHERE title LIKE '%MICKEY%';

SELECT *
 FROM prices
 WHERE title LIKE '%MICKEY%';
```

(Execute) (Load Script) (Save Script) (Cancel)

View created.

ISBN	TITLE	COST	RETAIL	PROFIT
0401140733	REVENGE OF MICKEY	14.2	22	7.8

1 row updated.

ISBN	TITLE	COST	RETAIL	PROFIT
0401140733	REVENGE OF MICKEY	14.2	29.95	15.75

FIGURE 13-12   Create and update a complex view named PRICES

```
DELETE FROM prices
    WHERE title LIKE '%MICKEY%';
```

(Execute) (Load Script) (Save Script) (Cancel)

1 row deleted.

FIGURE 13-13   Deleting a book via the PRICES view

Before attempting to add a record to the view, you need to remember that the Profit column is based upon the expression "retail-cost" to derive the dollar profit generated by the book. So, when you add a new book to the PRICES view, do you enter the actual profit generated, or should you let Oracle 10g calculate the profit? We will attempt an INSERT first in which we include a Profit amount and then in which we exclude it. The INSERTS in Figure 13-14 shows the outcome of these two attempts.

The first attempt includes the amount of profit that would be generated by the new book being added to the view. However, Oracle 10g does not accept the calculated profit entered by the user. The error message "virtual column not allowed here" is one that may appear often until you learn the rules for DML operations on complex views. In this case, the error message means that because one of the columns in the view is based upon an arithmetic expression, a value cannot be inserted into that column of the view. In other words, a Profit column does not exist in the base BOOKS table, so this value has no place to be stored in the database. The second INSERT attempt in Figure 13-14 tries to add a new record to the PRICES view excluding the Profit value. This returns the error message "not enough

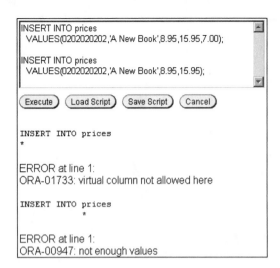

```
INSERT INTO prices
   VALUES(0202020202,'A New Book',8.95,15.95,7.00);

INSERT INTO prices
   VALUES(0202020202,'A New Book',8.95,15.95);
```

(Execute) (Load Script) (Save Script) (Cancel)

```
INSERT INTO prices
       *
ERROR at line 1:
ORA-01733: virtual column not allowed here

INSERT INTO prices
              *
ERROR at line 1:
ORA-00947: not enough values
```

FIGURE 13-14   Failed attempts to add a new book via the PRICES view

values." That seems simple enough; it appears to indicate that the view contains five values, yet we indicate only four values in the INSERT. Therefore, the only way to add a new book to the PRICES view is to use a column list in the INSERT, as shown in Figure 13-15.

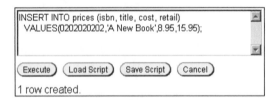

```
INSERT INTO prices (isbn, title, cost, retail)
   VALUES(0202020202,'A New Book',8.95,15.95);
```

(Execute) (Load Script) (Save Script) (Cancel)

1 row created.

FIGURE 13-15   Successful attempt to add a new book via the PRICES view

Thus, you have discovered one of the rules governing DML operations for complex views:

Values cannot be inserted into columns that are based on arithmetic expressions.

Another consideration is the NOT NULL constraints that exist on the base table. What if the BOOKS table contains a NOT NULL constraint on the Pubid column? Would the INSERT via the PRICES view still work? No! The PRICES view does not contain the Pubid column and, therefore, does not allow a value to be indicated for this column in an INSERT. Figure 13-16 demonstrates the addition of a NOT NULL constraint on Pubid column and then another attempt of the previously successful INSERT. First, the previous row added must be deleted because this does not contain a Pubid value and would violate the NOT NULL constraint.

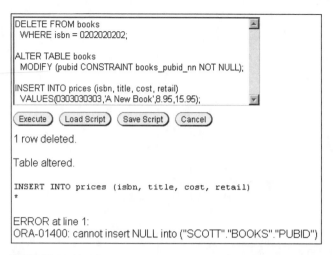

```
DELETE FROM books
  WHERE isbn = 0202020202;

ALTER TABLE books
  MODIFY (pubid CONSTRAINT books_pubid_nn NOT NULL);

INSERT INTO prices (isbn, title, cost, retail)
  VALUES(0303030303,'A New Book',8.95,15.95);
```

(Execute) (Load Script) (Save Script) (Cancel)

1 row deleted.

Table altered.

```
INSERT INTO prices (isbn, title, cost, retail)
*
```

ERROR at line 1:
ORA-01400: cannot insert NULL into ("SCOTT"."BOOKS"."PUBID")

FIGURE 13-16 Constraint violation with INSERT via the PRICES view

DML Operations on a Complex View Containing Data from Multiple Tables

To add a little more complexity to the complex view called PRICES, the view has been re-created in Figure 13-17 to include the name of the publishers from the PUBLISHER table.

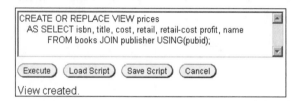

```
CREATE OR REPLACE VIEW prices
  AS SELECT isbn, title, cost, retail, retail-cost profit, name
    FROM books JOIN publisher USING(pubid);
```

(Execute) (Load Script) (Save Script) (Cancel)

View created.

FIGURE 13-17 PRICES view with a table join

Now let's try to perform the same type of DML operations that were performed on the previous version of the PRICES view. First, we attempt to update the price of one of the books, as shown in Figure 13-18.

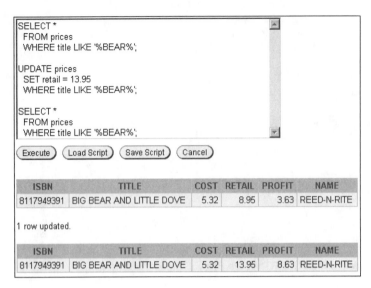

```
SELECT *
 FROM prices
 WHERE title LIKE '%BEAR%';

UPDATE prices
 SET retail = 13.95
 WHERE title LIKE '%BEAR%';

SELECT *
 FROM prices
 WHERE title LIKE '%BEAR%';
```

Execute   Load Script   Save Script   Cancel

ISBN	TITLE	COST	RETAIL	PROFIT	NAME
8117949391	BIG BEAR AND LITTLE DOVE	5.32	8.95	3.63	REED-N-RITE

1 row updated.

ISBN	TITLE	COST	RETAIL	PROFIT	NAME
8117949391	BIG BEAR AND LITTLE DOVE	5.32	13.95	8.63	REED-N-RITE

FIGURE 13-18   Updating the Retail column via the PRICES view

In Figure 13-18, the retail price of the book titled *Big Bear and Little Dove* has been changed from $8.95 to $13.95. As with the previous version of the PRICES view, the DML operation to modify a record worked. However, the change made to the view was that the name of the publisher was included from the PUBLISHER table, so what happens if the Name column is updated? Look at Figure 13-19.

```
UPDATE prices
 SET name = 'PRINT IS US'
 WHERE title LIKE '%BEAR%';
```

Execute   Load Script   Save Script   Cancel

```
   SET name = 'PRINT IS US'
        *
ERROR at line 2:
ORA-01779: cannot modify a column which maps to a non key-preserved table
```

FIGURE 13-19   Failed attempt to UPDATE the publisher name via the PRICES view

When Oracle 10g attempted to update the name of the publisher of *Big Bear and Little Dove*, the UPDATE command failed, and you received the error message "cannot modify a column which maps to a non key-preserved table." Perhaps taking a step back and analyzing the underlying tables will help you to understand the meaning of the error message—and what caused the error message to occur.

The PRICES view was built using columns from the BOOKS and PUBLISHER tables. When a view includes columns from more than one table, updates can be applied only to *one* table. The table that can be updated is the one that includes the primary key of an underlying table and is basically being used as the primary key for the view. The PRICES view includes the primary key for the BOOKS table, so basically any UPDATE command can be performed on the columns from the BOOKS table—if the change does not violate any constraints from that table (for example, you cannot change the primary key if it is used as a reference for a FOREIGN KEY constraint or if it would no longer be unique). In the PRICES view, the BOOKS table is known as the **key-preserved table**. In essence, a key-preserved table is the table that contains the primary key that the view is using to uniquely identify each record displayed by the view. By contrast, the Name column is from the PUBLISHER table. The primary key for the PUBLISHER table is the Pubid column, and it is not included in the view. However, even if the Pubid column had been included, Oracle 10g would not have considered the column as the primary key for the PRICES view because it could have appeared more than once in the contents of the view. Therefore, Oracle 10g treats the data from the PUBLISHER table as coming from a **non key-preserved table** because it does not uniquely identify the records in the PRICES view.

One way to make sense of this problem is to consider that the BOOKS table actually stores the publishers' ID numbers. Thus, if Oracle 10g were to change the name of a publisher, as instructed by the UPDATE command shown in Figure 13-19, does that mean you want the Pubid column updated in the BOOKS table as well? Or would you change the name of the publisher in the PUBLISHER table? If the name is changed in the PUBLISHER table, then every book that had the same Pubid as *Big Bear and Little Dove* would now be published by the publisher Print Is Us, and that was not the intention of the command. Now you have discovered a second rule that applies to complex views:

`DML operations cannot be performed on a non key-preserved table.`

The rule that no DML operations can be performed on a non key-preserved table may be a little broad, considering that you have not tried to delete a record from the PRICES view. Therefore, let's attempt a DELETE, as shown in Figure 13-20.

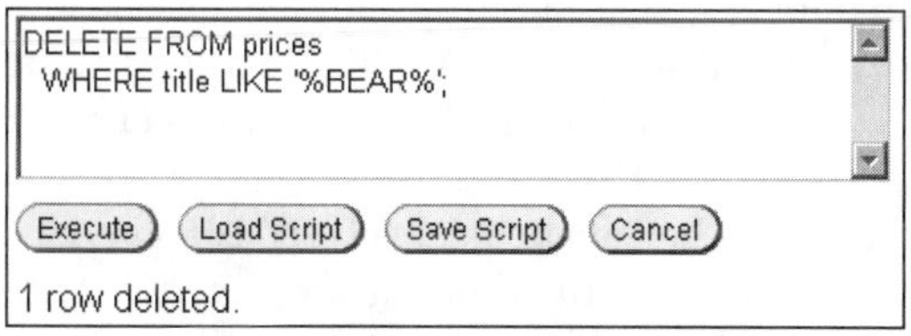

FIGURE 13-20   Deletion of a book via the PRICES view

Given the results of the DELETE command in Figure 13-20, it might appear that there is a problem with the second rule. You were not allowed to update the publisher's name for the book, but you were allowed to delete the row for the book. However, what actually occurred is that the book *Big Bear and Little Dove* was deleted from the BOOKS table, which is the key-preserved table. No change occurred in the PUBLISHER table so, technically, the command did not perform a DML operation on the non key-preserved table; therefore, the rule was not violated.

DML Operations on a Complex View Containing Functions or Grouped Data

Views are also considered to be complex if they contain a function or a GROUP BY clause. To determine what effect they have on DML commands, let's create a view that displays the total balance due for each order placed by a customer, as shown in Figure 13-21.

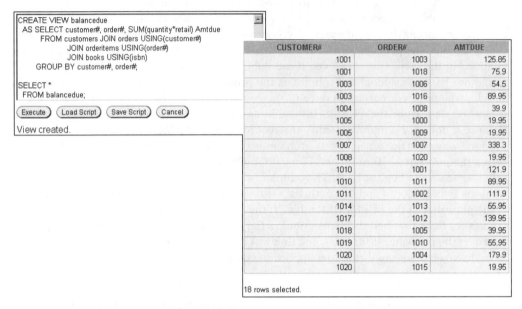

CREATE VIEW balancedue
  AS SELECT customer#, order#, SUM(quantity*retail) Amtdue
    FROM customers JOIN orders USING(customer#)
      JOIN orderitems USING(order#)
      JOIN books USING(isbn)
  GROUP BY customer#, order#;

SELECT *
  FROM balancedue;

(Execute) (Load Script) (Save Script) (Cancel)

View created.

CUSTOMER#	ORDER#	AMTDUE
1001	1003	125.85
1001	1018	75.9
1003	1006	54.5
1003	1016	89.95
1004	1008	39.9
1005	1000	19.95
1005	1009	19.95
1007	1007	338.3
1008	1020	19.95
1010	1001	121.9
1010	1011	89.95
1011	1002	111.9
1014	1013	55.95
1017	1012	139.95
1018	1005	39.95
1019	1010	55.95
1020	1004	179.9
1020	1015	19.95

18 rows selected.

FIGURE 13-21   View including aggregated data

The BALANCEDUE view created in Figure 13-21 first groups the items on each customer's order, and then it calculates the total amount due, based on the number of books ordered and each book's retail price. Adding a record to the BALANCEDUE view is not allowed because the Amtdue column is derived from a function (it is treated the same way as an expression). Even without the SUM function, Oracle 10g would still not allow a record to be added because of the GROUP BY clause: The data being displayed is grouped; it is not possible to add an individual record to the view. In addition, *the function and GROUP BY clause prevents any of the data displayed from being changed, because each record displayed may represent more than one row in the underlying key-preserved table.* (Try a little experiment on your own and see whether you can add or modify a record, but don't be too disappointed if you receive an error message.) Thus, the question becomes, "Will Oracle 10g allow a row to be deleted from the view?" Figure 13-22 shows an attempt.

As shown in Figure 13-22, apparently the answer is NO! The rationale behind not allowing a record to be deleted from the view is that the data is grouped, so it is hard to clarify exactly what would be deleted. Rather than allow a user to mistakenly delete what could possibly be several rows in the underlying key-preserved table, the operation is simply not allowed. Therefore, if you really want to delete a particular customer's orders, you would

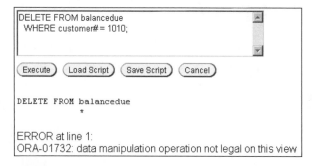

FIGURE 13-22 Failed DELETE on view with grouped data

have to delete them directly from the ORDERS table. (Of course, if that deletion violates any existing constraints between the ORDERS and ORDERITEMS tables, then the deletion may not be possible without including an ON CASCADE DELETE option.) And now a third rule has been identified:

DML operations are not permitted if the view includes a group function or a GROUP BY clause.

DML Operations on a Complex View Containing DISTINCT or ROWNUM

Before we summarize the lessons you have learned about complex views and DML operations, you should look at how to use the DISTINCT keyword during view creation and examine the impact of ROWNUM on DML operations in a view. Let's look at those next.

When using the DISTINCT keyword in a subquery to create a view, remember that in a SELECT clause, the keyword instructs Oracle 10g to suppress duplication. In other words, if more than one row of a table contains the same data, Oracle displays the row only once in the view. If you consider each unique row as one group, the DISTINCT keyword acts almost like a GROUP BY clause. Therefore, a fourth rule has emerged:

DML operations on a view that is created with the DISTINCT keyword are not permitted.

> **NOTE**
>
> If the rationale behind not permitting DML operations doesn't seem valid, try to create a "distinct" view of the different ISBNs in the ORDERITEMS table and attempt some DML operations.

The second point you should examine applies to the concept of **ROWNUM**. ROWNUM is a pseudocolumn that applies to every table, even though it is not displayed through a SELECT * command, or even the DESCRIBE command. Oracle 10g assigns every record in a table a row number to indicate the row's position within the table.

The query issued in Figure 13-23 instructs Oracle 10g to list the last name of each customer and the record's position in the CUSTOMERS table. However, the query also requires that the last names be presented in alphabetical order.

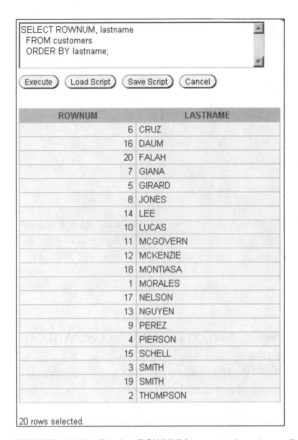

```
SELECT ROWNUM, lastname
  FROM customers
  ORDER BY lastname;
```

Execute   Load Script   Save Script   Cancel

ROWNUM	LASTNAME
6	CRUZ
16	DAUM
20	FALAH
7	GIANA
5	GIRARD
8	JONES
14	LEE
10	LUCAS
11	MCGOVERN
12	MCKENZIE
18	MONTIASA
1	MORALES
17	NELSON
13	NGUYEN
9	PEREZ
4	PIERSON
15	SCHELL
3	SMITH
19	SMITH
2	THOMPSON

20 rows selected.

FIGURE 13-23   Display ROWNUM on sorted customer list

As shown in the query results in Figure 13-23, the customer with the last name of Cruz is listed first because the list is sorted alphabetically. However, the customer is actually stored in the sixth row of the CUSTOMERS table. In fact, based on customer numbers, the customer named Morales occupies the first record in the CUSTOMERS table.

NOTE

If the SELECT command in Figure 13-23 did not include an ORDER BY clause, the rows would have been displayed in ROWNUM order, as this reflects the physical order of the rows in the table.

So, what does ROWNUM have to do with DML operations on a complex view? If the query of the complex view listed ROWNUM as one of the columns to be included in the view, no DML operation would be allowed on the view. Because ROWNUM is a pseudo-column that Oracle 10g uses to assign a value to each row, Oracle 10g does not allow any additions, deletions, or modifications on the data displayed in the view. This results in the final rule:

DML operations are not allowed on views that include the pseudocolumn ROWNUM.

Summary Guidelines of DML Operations on a Complex View

A summary of the guidelines regulating DML operations on complex views follows.

- DML operations that violate a constraint are not permitted.
- A value cannot be added to a column that contains an arithmetic expression.
- DML operations are not permitted on non key-preserved tables.
- DML operations are not permitted on views that include group functions, a GROUP BY clause, the DISTINCT keyword, or the ROWNUM pseudocolumn.

DROPPING A VIEW

A view can be dropped or deleted using the DROP VIEW command. The syntax for the command is shown in Figure 13-24.

453

```
DROP VIEW viewname;
```

FIGURE 13-24   Syntax for the DROP VIEW command

The command to drop the PRICES view we used earlier is shown in Figure 13-25.

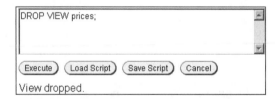

FIGURE 13-25   Command to drop the PRICES view

After the command is successfully executed, Oracle 10g returns the message that the view has been dropped, as shown in Figure 13-25. However, the data that was displayed by the view is still available in the underlying tables originally used to create the view. All that has been deleted is the database object named PRICES that pointed to the data contained in the underlying tables.

CREATING AN INLINE VIEW

In Chapter 12, we used a subquery in the FROM clause of a SELECT statement to create a "temporary" table that could be referenced by the SELECT and WHERE clauses of the statement. It was considered temporary because nowhere in the database did it actually store a copy of the data returned by the subquery. That temporary table is very similar to what Oracle 10g calls an **inline view**. The main difference between an inline view and the views we've discussed so far in this chapter is that *an inline view exists only while the command is being executed.* It is not a permanent database object and cannot be referenced again by a subsequent query. It is most often used to provide a temporary data source while a command is being executed. Perhaps the most common usage for an inline view is to perform a "TOP-N" analysis.

"TOP-N" Analysis

Suppose that you want to find the five books that generate the most profit. In Chapter 6, we used the group function MAX to find the most profitable book. However, using that function yields only the highest value in a column. How would you find the five highest values? This is where **"TOP-N" analysis** is used. In a "TOP-N" analysis, the concepts of an inline view and the pseudocolumn ROWNUM are merged together to create a temporary list of records in a sorted order, and then the top "N," or number of records, are retrieved. An inline view must be used to perform the analysis because an ORDER BY clause must be used by the subquery to put the records in the correct order before the subquery is passed to the outer query—and ORDER BY clauses are not allowed in the CREATE VIEW command.

The syntax to perform a "TOP-N" analysis is shown in Figure 13-26.

```
SELECT columnname, ...
  FROM (subquery)
  WHERE ROWNUM <= N;
```

FIGURE 13-26   Syntax for "TOP-N" analysis

To determine the five books that generate the most profit, the subquery needs to calculate the profit for each book and then sort the results in descending order by profit before passing the values to the outer query. The subquery is shown in Figure 13-27.

To perform the analysis, the subquery must be nested in the FROM clause of a SELECT statement to create the inline view. When the sorted results are passed from the subquery, each row is assigned a row number to identify its position. Then, a WHERE clause is added to the outer query to select only those books that have a ROWNUM that is less than or equal to the desired "N." In this case, "N" is five because you are looking for the five most profitable books.

The command to determine the five most profitable books is shown in Figure 13-28.

When the command shown in Figure 13-28 is executed, Oracle 10g displays only the books that have a row number less than or equal to five. Because the data is received by the outer query in descending order based upon profit, the top five most profitable books are assigned the row numbers one through five, as shown in Figure 13-28.

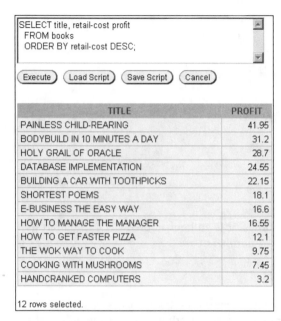

FIGURE 13-27   Subquery needed to perform "TOP-N" analysis

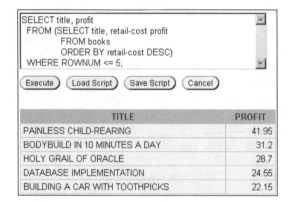

FIGURE 13-28   "TOP-N" analysis to identify the five most profitable books

As shown, the five most profitable books have a profit range between $22.15 and $41.95. But what if management wants to know the titles of the three least profitable books? The simplest solution is to use a subquery to sort the data, so the book that is least profitable receives the first row number, the second least profitable book receives the second row number, and so on. This query and output are shown in Figure 13-29.

The subquery in Figure 13-29 has been modified so the data is sorted in ascending order before it is passed to the outer query. The WHERE clause has also been changed to select only the first three books that are shown in the results. In this case, the lowest profit generated is $3.20, the next lowest profit is $7.45, and the third lowest profit is $9.75. Although

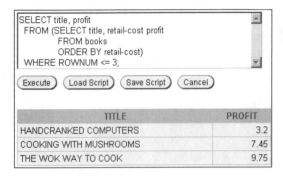

```
SELECT title, profit
   FROM (SELECT title, retail-cost profit
         FROM books
         ORDER BY retail-cost)
   WHERE ROWNUM <= 3;
```

Execute   Load Script   Save Script   Cancel

TITLE	PROFIT
HANDCRANKED COMPUTERS	3.2
COOKING WITH MUSHROOMS	7.45
THE WOK WAY TO COOK	9.75

FIGURE 13-29 The three least profitable books

the query actually returns the lowest "N" values, it is still considered "TOP-N" analysis because the results consist of the top "N" row numbers.

MATERIALIZED VIEWS

A **materialized view** allows you to store the data retrieved by the view query, and lets you reuse this data without executing the view query again. In other words, a materialized view allows the replication of data. These are often referred to as snapshots, as they take a picture or capture a set of data at a specific point in time. The data already exists in the base tables, so why would we want to replicate data? Several business needs may make materialized views attractive.

- Complex queries or queries on large databases can require a significant amount of processing, which can affect the operational users on the system. Most businesses want to maintain optimal performance for transactional processing. Replicating data for reporting and analysis allows system resources to be dedicated to transactional processes.
- Remote users could significantly improve query performance by replicating data to a local database. Rather than transferring data across states or nations, a local copy could be used for satellite offices.
- Data analysis needs may require data to be frozen for a specific time for comparison purposes.

Materialized views also have some disadvantages. First, additional storage space is required to house the copied data. Second, if modifications are made to data via the materialized view, these changes need to be synchronized with the base tables. Finally, if the data provided in the materialized view needs to be constantly updated, the reduction in processing may be minimal. A materialized view not only retrieves the data with a query, but it also must physically store the results.

Creating a materialized view is similar to creating the views we worked with previously in this chapter. A number of additional options exist, such as defining the data refresh schedule and determining whether the data in the view should be read-only. The example in Figure 13-30 demonstrates the creation of a materialized view.

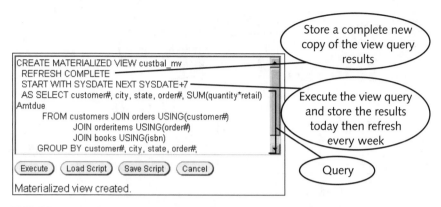

FIGURE 13-30 Create a materialized view named CUSTBAL_MV

As shown in Figure 13-31, you can query the materialized view in the same way you query views with no physical properties.

```
SELECT *
  FROM custbal_mv;
```

(Execute) (Load Script) (Save Script) (Cancel)

CUSTOMER#	CITY	STATE	ORDER#	AMTDUE
1001	EASTPOINT	FL	1003	125.85
1001	EASTPOINT	FL	1018	75.9
1003	TALLAHASSEE	FL	1006	54.5
1003	TALLAHASSEE	FL	1016	89.95
1004	BOISE	ID	1008	39.9
1005	SEATTLE	WA	1000	19.95
1005	SEATTLE	WA	1009	19.95
1007	AUSTIN	TX	1007	338.3
1008	CHEYENNE	WY	1020	19.95
1010	ATLANTA	GA	1001	121.9
1010	ATLANTA	GA	1011	89.95
1011	CHICAGO	IL	1002	111.9
1014	CODY	WY	1013	55.95
1017	KALMAZOO	MI	1012	139.95
1018	MACON	GA	1005	39.95
1019	MORRISTOWN	NJ	1010	55.95
1020	TRENTON	NJ	1004	179.9
1020	TRENTON	NJ	1015	19.95

18 rows selected.

FIGURE 13-31 Query a materialized view

You remove a materialized view with a DROP command containing the keyword MATERIALIZED, as shown in Figure 13-32.

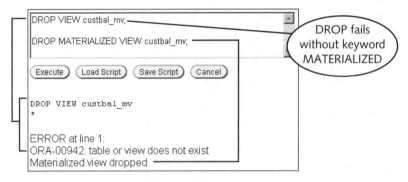

FIGURE 13-32 DROP a materialized view

Chapter Summary

- A view is a pseudo or virtual table that is used to retrieve data that exists in the underlying database tables.
- The view query must be executed each time the view is used.
- A view can be used to simplify queries or to restrict access to sensitive data.
- A view is created with the CREATE VIEW command.
- A view cannot be modified. To change a view, it must be dropped and then re-created, or the CREATE OR REPLACE VIEW command must be used.
- Any DML operation can be performed on a simple query if it does not violate a constraint.
- A view that contains expressions or functions, or that joins multiple tables, is considered a complex view.
- A complex view can be used to update only one table. The table must be a key-preserved table.
- Data cannot be added to a view column that contains an expression.
- DML operations are not permitted on non key-preserved tables.
- DML operations are not permitted on views that include group functions, a GROUP BY clause, the ROWNUM pseudocolumn, or the DISTINCT keyword.
- Oracle 10g assigns a row number to every row in a table to indicate its position in the table. The row number can be referenced by the keyword ROWNUM.
- A view can be dropped with the DROP VIEW command. The data is not affected, because it exists in the original tables.
- An inline view can be used only by the current statement and can include an ORDER BY clause.
- "TOP-N" analysis uses the row number of sorted data to determine a range of top values.
- Materialized views physically store view query results.

Chapter 13 Syntax Summary

The following table presents a summary of the syntax that you have learned in this chapter. You can use the table as a study guide and reference.

SYNTAX GUIDE		
Element	**Command Syntax**	**Example**
Command to create a new view	CREATE [FORCE|NOFORCE]VIEW viewname(columnname, ...)]AS subquery[WITH CHECK OPTION [CONSTRAINT constraintname]][WITH READ ONLY];	CREATE VIEW inventory AS SELECT isbn, title, retail price FROM books WITH READ ONLY;

SYNTAX GUIDE

Element	Command Syntax	Example
Command to replace an existing view	CREATE OR REPLACE [FORCE|NOFORCE] VIEW *viewname (columnname, ...)]* AS subquery[WITH CHECK OPTION [CONSTRAINT *constraintname]][WITH READ ONLY];*	CREATE OR REPLACE VIEW inventory AS SELECT isbn, title, retail price FROM books;
Command to drop a view	DROP VIEW *viewname;*	DROP VIEW inventory;
Command to create an inline view	SELECT columnname, ...FROM *(subquery)*WHERE ROWNUM<=*N;*	SELECT title, profit FROM (SELECT title, retail-cost profit FROM books ORDER BY retail-cost DESC)WHERE ROWNUM <=5;
Command to create a materialized view	CREATE MATERIALIZED VIEW viewname [REFRESH option] [START WITH date] AS subquery;	CREATE MATERIALIZED VIEW custbal_mv REFRESH COMPLETE START WITH SYSDATE NEXT SYSDATE+7 AS SELECT customer#, city, state, order#, SUM(quantity*retail) Amtdue FROM customers JOIN orders USING(customer#) JOIN orderitems USING(order#) JOIN books USING(isbn) GROUP BY customer#, city, state, order#;

Review Questions

To answer the following questions, refer to the tables in Appendix A.

1. How is a simple view different from a complex view?
2. Under what circumstances is a DML operation not allowed on a simple view?
3. When should the FORCE keyword be used in the CREATE VIEW command?
4. What is the purpose of the WITH CHECK OPTION constraint?
5. List the guidelines regarding DML operations on complex views.
6. How can you ensure that no user will be able to change the data displayed by a view?
7. What is the difference between a key-preserved and non key-preserved table?
8. What command can be used to modify a view?
9. What is unique about materialized views as compared to other views?
10. What happens to the data displayed by a view when the view is deleted?

Multiple Choice

To answer the following questions, refer to the tables in Appendix A.

Questions 1–7 are based upon successful execution of the following statement:

```
CREATE VIEW changeaddress
AS SELECT customer#, lastname, firstname, order#, shipstreet,
        shipcity, shipstate, shipzip
  FROM customers NATURAL JOIN orders
  WHERE shipdate IS NULL
WITH CHECK OPTION;
```

1. Which of the following statements is correct?

   a. No DML operations can be performed on the CHANGEADDRESS view.

   b. The CHANGEADDRESS view is a simple view.

   c. The CHANGEADDRESS view is a complex view.

   d. The CHANGEADDRESS view is an inline view.

2. Assuming that there is only a primary key and that foreign key constraints exist on the underlying tables, which of the following commands will return an error message?

   a. UPDATE changeaddress SET shipstreet = '958 ELM ROAD'
      WHERE customer# = 1020;

   b. INSERT INTO changeaddress VALUES (9999, 'LAST', 'FIRST',
      9999, '123 HERE AVE', 'MYTOWN', 'AA', 99999);

   c. DELETE FROM changeaddress WHERE customer# = 1020;

   d. all of the above

   e. only a and b

   f. only a and c

   g. none of the above

3. Which of the following is the key-preserved table for the CHANGEADDRESS view?

   a. CUSTOMERS table

   b. ORDERS table

   c. Both tables together serve as a composite key-preserved table.

   d. none of the above

4. Which of the following columns serves as the primary key for the CHANGEADDRESS view?

   a. Customer#

   b. Lastname

   c. Firstname

   d. Order#

   e. Shipstreet

5. If a record for a customer is deleted from the CHANGEADDRESS view, the customer information is then deleted from which underlying table?

    a. CUSTOMERS

    b. ORDERS

    c. CUSTOMERS and ORDERS

    d. Neither—the DELETE command cannot be used on the CHANGEADDRESS view.

6. Which of the following is correct?

    a. ROWNUM cannot be used with the view because it is not included in the results returned from the subquery.

    b. The view is a simple view because it does not include a group function or a GROUP BY clause.

    c. The data in the view cannot be presented in descending order by customer number because an ORDER BY clause is not allowed when working with views.

    d. all of the above

    e. none of the above

7. Assuming one of the orders has shipped, which of the following is true?

    a. The CHANGEADDRESS view cannot be used to update the shipping date of the order due to the WITH CHECK OPTION constraint.

    b. The CHANGEADDRESS view cannot be used to update the shipping date of the order because the ship date column is not included in the view.

    c. The CHANGEADDRESS view cannot be used to update the shipping date of the order because the ORDERS table is not the key-preserved table.

    d. The CHANGEADDRESS view cannot be used to update the shipping date of the order because the UPDATE command cannot be used on the data in the view.

Questions 8–12 are based upon successful execution of the following command:

```
CREATE VIEW changename
AS SELECT customer#, lastname, firstname
   FROM customers
WITH CHECK OPTION;
```

Assume that the only constraint on the CUSTOMERS table is a PRIMARY KEY constraint.

8. Which of the following is a correct statement?

    a. No DML operations can be performed on the CHANGENAME view.

    b. The CHANGENAME view is a simple view.

    c. The CHANGENAME view is a complex view.

    d. The CHANGENAME view is an inline view.

9. Which of the following columns serves as the primary key for the CHANGENAME view?

    a. Customer#

    b. Lastname

    c. Firstname

    d. The view does not have or need a primary key.

10. Which of the following DML operations could never be used on the CHANGENAME view?

    a. INSERT

    b. UPDATE

    c. DELETE

    d. All of the above are valid DML operations for the CHANGENAME view.

11. The INSERT command cannot be used with the CHANGENAME view because:

    a. A key-preserved table is not included in the view.

    b. The view was created with the WITH CHECK OPTION constraint.

    c. The inserted record would not be accessible by the view.

    d. None of the above—an INSERT command can be used on the table as long as the primary key constraint is not violated.

12. If the CHANGENAME view needs to include the customer's zip code as a means of verifying the change (that is, to authenticate the user), then which of the following is true?

    a. The CREATE OR REPLACE VIEW command can be used to re-create the view with the necessary column included in the new view.

    b. The ALTER VIEW...ADD COLUMN command can be used to add the necessary column to the existing view.

    c. The CHANGENAME view can be dropped and then the CREATE VIEW command can be used to re-create the view with the necessary column included in the new view.

    d. All of the above can be performed to include the customer's zip code in the view.

    e. Only a and b will include the customer's zip code in the view.

    f. Only a and c will include the customer's zip code in the view.

    g. None of the above will include the customer's zip code in the view.

13. Which of the following DML operations cannot be performed on a view that contains a group function?

    a. INSERT

    b. UPDATE

    c. DELETE

    d. All of the above can be performed on a view that contains a group function.

    e. None of the above can be performed on a view that contains a group function.

14. A user cannot perform any DML operations on which of the following?

    a. views that are created with the WITH READ ONLY option

    b. views that include the DISTINCT keyword

    c. views that include a GROUP BY clause

    d. All of the above allow DML operations.

    e. None of the above allows DML operations.

15. A "TOP-N" analysis is performed by determining the rows with:

    a. the highest ROWNUM values

    b. a ROWNUM value greater than or equal to N

    c. the lowest ROWNUM values

    d. a ROWNUM value less than or equal to N

16. To assign names to the columns in a view, you can do which of the following?

    a. Assign aliases in the subquery, and the aliases will be used for the column names.

    b. Use the ALTER VIEW command to change the column names.

    c. Assign names for up to three columns in the CREATE VIEW clause before the subquery is listed in the AS clause.

    d. None of the above—columns cannot be assigned names for a view; they must keep their original names.

17. Which of the following is correct?

    a. The ORDER BY clause cannot be used in the subquery of a CREATE VIEW command.

    b. The ORDER BY clause cannot be used in an inline view.

    c. The DISTINCT keyword cannot be used in an inline view.

    d. The WITH READ ONLY option must be used with an inline view.

18. If you try to add a row to a complex view that includes a GROUP BY clause, you will receive which of the following error messages?

    a. virtual column not allowed here

    b. data manipulation operation not legal on this view

    c. cannot map to a column in a non key-preserved table

    d. None of the above—no error message will be returned.

19. A simple view can contain which of the following?

    a. data from one or more tables

    b. an expression

    c. a GROUP BY clause for data retrieved from one table

    d. five columns from one table

    e. all of the above

    f. none of the above

20. A complex view can contain which of the following?

    a. data from one or more tables

    b. an expression

    c. a GROUP BY clause for data retrieved from one table

    d. five columns from one table

    e. all of the above

    f. none of the above

Hands-On Assignments

To perform the following activities, refer to the tables in Appendix A.

1. Create a view that will list the name of the contact person at each publisher and the person's phone number. Do not include the publisher's ID in the view. Name the view CONTACT.

2. Change the CONTACT view so that no users can accidentally perform DML operations on the view.

3. Create a view called HOMEWORK13 that will include the columns named Col1 and Col2 from the FIRSTATTEMPT table. Make certain the view will be created even if the FIRSTATTEMPT table does not exist.

4. Attempt to view the structure of the HOMEWORK13 view.

5. Create a view that will list the ISBN and title for each book in inventory along with the name and telephone number of the individual to contact in the event the book needs to be reordered. Name the view REORDERINFO.

6. Try to change the name of one of the individuals in the REORDERINFO view to your name. Was there an error message displayed when performing this step? If so, what was the cause of the error message?

7. Select one of the books in the REORDERINFO view and try to change the ISBN of the book. Was there an error message displayed when performing this step? If so, what was the cause of the error message?

8. Delete the record in the REORDERINFO view that contains your name (if that step was not performed successfully, then delete one of the contacts already listed in the table). Was there an error message displayed when performing this step? If so, what was the cause of the error message?

9. Issue a rollback command to undo any changes made with any previous DML operations.

10. Delete the REORDERINFO view.

Advanced Challenge

To perform the following activity, refer to the tables in Appendix A.

The Marketing Department of JustLee Books is about to begin an aggressive marketing campaign to generate sales to repeat customers. Their strategy will be to look at existing customers' previous purchases, and then based on the categories from which those customers have made purchases, JustLee Books will send promotional information about other books in the same category that are highly profitable books for the company.

The Marketing Department has requested that you identify the five most frequently purchased books and the percentage of profit generated by each book. The percentage of profit can be calculated by using the formula ((retail-cost)/cost*100). The employees in the Marketing Department will use the potential profitability of the marketing campaign to determine how much money to budget for the campaign.

In a memo, provide management with a list of the five most frequently purchased books and the percentage of profit generated by each book.

Case Study: *City Jail*

Note: It is assumed that the City Jail database creation script from Chapter 8 has been executed. That script makes all database objects available to complete this case study.

The City Jail Technologies Department is constructing an application to allow users in the Crimes Analysis Unit to query data more easily. This system requires the creation of a number of views as listed. Provide the SQL statement to accomplish each task and test your views with a query.

1. Create a statement that always returns the names of the three criminals with the highest number of crimes committed.

2. Create a view that includes details for all crimes, including criminal ID, criminal name, criminal parole status, crime id, date of crime charge, crime status, charge id, crime code, charge status, pay due date, and amount due. This view should not allow any DML operations to be performed. Each time the view is used in the application, the data should be queried from the database (for example, each use of the view should reflect the most current data in the database).

3. Create a view that includes all data for officers, including the total number of crimes in which they participated in filing charges. To speed up the officer queries, store this view data and schedule the data to be updated every two weeks.

CHAPTER 14

FORMATTING READABLE OUTPUT

LEARNING OBJECTIVES

After completing this chapter, you should be able to do the following:

- Add a column heading with a line break to a report
- Format the appearance of numeric data in a column
- Specify the width of a column
- Substitute a text string for a NULL value in a report
- Add a multiple-line header to a report
- Display a page number in a report
- Add a title and a footer to a report
- Change the setting of an environment variable
- Suppress duplicate report data
- Clear changes made by the COLUMN and BREAK commands
- Perform calculations in a report with the COMPUTE command
- Save report output to a file
- Save report output in HTML format in client SQL*Plus

INTRODUCTION

In all the queries you've issued up to now, the results have been displayed as a list. Although you have manipulated columns presented in output and have used column aliases as column headings in query results, you have not yet used SQL*Plus formatting features to create a report. Some of the formatting features available in SQL*Plus include report headers and footers, formatting models for column data, suppression of duplicate data for group reports, and summary calculations. Although many formatting options are available, this chapter presents only the most commonly used features. It is helpful to understand these formatting features to quickly format output in SQL*Plus; however, most complex reports are created using various software programs and report generators such as Oracle Reports or Crystal Reports.

The basic steps for creating a report in SQL*Plus are to (1) enter the SQL*Plus format options settings for the report, and (2) enter a query to retrieve the desired rows for the report. The effect of various SQL*Plus format options varies depending on the output format being used. For example, the default output of the client SQL*Plus tool is pure text. The default output format for the Internet SQL*Plus tool, however, is an HTML table. Both the text and HTML output will be displayed for each of the examples in this chapter to illustrate the effect of the formatting options in both output formats. Figure 14-1 provides a listing of SQL*Plus formatting options covered in this chapter.

Figure 14-1 provides an overview of this chapter's contents.

FORMATTING OUTPUT	
Command	Description
START *or* @	Executes a script file
COLUMN	Defines the appearance of column headings and the format of the column data
TTITLE	Adds a header to the top of each report page
BTITLE	Adds a footer to the bottom of each report page
BREAK	Suppresses duplicated data for a specific column(s) when presented in a sorted order
COMPUTE	Performs calculations in a report based on the AVG, SUM, COUNT, MIN, or MAX statistical functions
SPOOL	Redirects output to a text file
COLUMN Options	Description
HEADING	Adds a column heading to a specified column
FORMAT	Defines the width of columns and applies specific formats to columns containing numeric data

FIGURE 14-1 List of common formatting options

FORMATTING OUTPUT	
COLUMN Options	Description
NULL	Indicates text to be substituted for NULL values in a specified column
SQL*Plus Environment Variables	Description
UNDERLINE	Specifies the symbol to be used to separate a column heading from the contents of the column
LINESIZE	Establishes the maximum number of characters that can appear on a single line of output
PAGESIZE	Establishes the maximum number of lines that can appear on one page of output
MARKUP HTML	Enables output from client SQL*Plus to be saved as HTML

FIGURE 14-1   List of common formatting options (continued)

If you have executed the prech08.sql file as instructed in Chapter 8, you will be able to work through the queries shown in this chapter. Review the prech08.sql script to be sure you still have all the objects previously created. If not, rerun the script.

COLUMN COMMAND

You can use the **COLUMN** command to format both a column heading and the data being displayed within the column, depending on the option(s) you use. The basic syntax for the COLUMN command is shown in Figure 14-2.

```
COLUMN [columnname|columnalias] [option]
```

FIGURE 14-2   Syntax of the COLUMN command

Let's examine the elements of the syntax example displayed in Figure 14-2:

- The column referenced is identified immediately after the COLUMN command.
- A column can be assigned a column alias in a SQL statement; however, if you include a column alias, you must use it to identify the column rather than the column name whenever the column is referenced in a SQL*Plus formatting option.
- The option(s) given in the COLUMN command specifies how the display will be affected.

In this section, you will learn how to use the options shown in Figure 14-3.

OPTION	DESCRIPTION
FORMAT *formatmodel*	Applies a specified format to the column data
HEADING *columnheading*	Indicates the heading to be used for a specified column
NULL *textmessage*	Identifies the text message to be displayed in the place of NULL values

FIGURE 14-3 Options available for the COLUMN command

FORMAT Option

The **FORMAT** option allows you to apply a format model to the data displayed within a column. For example, recall that in previous chapters, a book's retail price of $54.50 was displayed as 54.5. With the FORMAT option, you can display the insignificant zero in the output. You can also include the dollar sign, so the retail price would be displayed as $54. 50. Some of the formatting options available with the COLUMN command are shown in Figure 14-4.

FORMAT CODE	DESCRIPTION	EXAMPLE	OUTPUT
9	Identifies the position of numeric data (leading zeros are suppressed)	99999	54
$	Includes a floating dollar sign in output	$9999	$54
,	Indicates where to include a comma when needed (traditionally for the thousands and millions position)	9,999	54 *or* 1,212
.	Indicates the number of decimal positions to be displayed	9999.99	54.50
A*n*	Identifies the width of a column in a report	A32	Assigns a width of 32 spaces to a column

FIGURE 14-4 Codes for column format models

Suppose that the first report you need to create is one that displays the ISBN, title, cost, and retail price for books having a retail price greater than $50. To ensure that the cost and retail price for each book are displayed with two decimal places, you can create a model to specify exactly how data in the Cost and Retail columns should appear. Look at the script and output shown in Figure 14-5, and then we will examine each of the commands.

Let's examine the elements of Figure 14-5.

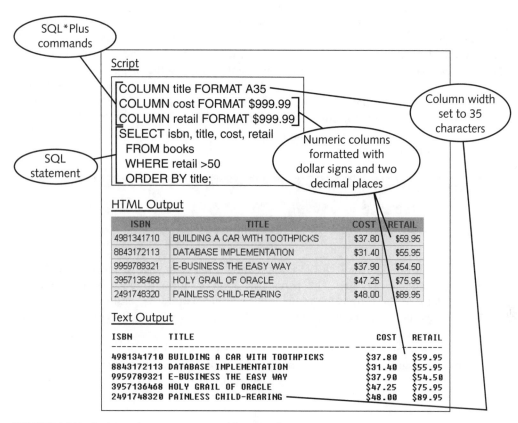

FIGURE 14-5 Script and output to create a formatted report

- The first line formats the Title column to a width of 35 spaces. Because the BOOKS table defines the column as having a width of only 30 spaces, the larger width setting will create extra blank space between the report's Title and Cost columns. The width setting affects only text output; it does *not* affect HTML output. By contrast, because the ISBN column did not have a width format applied, there is a default of one space after the data displayed in the ISBN column (and before the Title column).

- The second and third lines apply a format model to the numeric data displayed in the report's Cost and Retail columns. The format model uses a dollar sign ($) and a series of nines. The dollar sign instructs Oracle 10g to include the symbol with each value displayed in the columns. The two nines that appear after the decimal indicate that two decimal positions should be included. This forces the inclusion of insignificant zeros, which are suppressed by default in query results.

- After the Title, Cost, and Retail columns have been defined, the query to retrieve the rows to be displayed in the report is provided.

- Notice that the column commands do not end with a semicolon. A semicolon is not required for SQL*Plus commands. However, the SQL statement must still end with a semicolon as shown.

Formatting Readable Output

At this point you might be thinking, "Can't we use the TO_CHAR function in the SQL statement to accomplish the numeric formatting tasks?" Yes, the TO_CHAR function could be used to add the dollar symbol and to force the display of two decimal places. Furthermore, concatenation could be used to add additional width to the display of columns. So, why use the COLUMN FORMAT command? It is somewhat a matter of preference. However, various output formats may be desired for a given query. In this situation, the SQL*Plus COLUMN command might be useful because you could reuse the SQL statement, and the various SQL*Plus format options could be used to vary the output display depending on the situation.

NOTE

The COLUMN command does not have a specific format model for a DATE datatype column, other than increasing the column width. Therefore, the TO_CHAR function must be used to format date values.

Even though some of the COLUMN command options can also be accomplished easily with SQL statement features, this is not true of all the formatting features provided by SQL*Plus. Many of the formatting options of SQL*Plus cannot be easily accomplished within SQL statements or achieved at all within SQL. You will discover this as we proceed through the chapter.

If the set of format commands will be used repeatedly with the SQL statement to produce a standard report, you can save the script to a file and call it, rather than reentering the entire script each time the report is needed. If you are using the client SQL*Plus tool, the script shown in Figure 14-5 can be entered into a file, using either of the following approaches:

1. Enter the commands to be saved and then type ed or edit at the SQL> prompt in SQL*Plus to open a text editor such as Notepad. Select **File, Save As** from the menu, and name the script. Use the .sql file extension to indicate that the file is an SQL script file.
2. The second approach is to open a text editor (such as Notepad) outside of SQL*Plus and enter the commands directly into the text editor. Then save the file as outlined in the previous procedure to create the script file. If you use a word-processing program rather than a plain text editor, realize that many of these programs will embed hidden codes in the document, and the script file will not execute properly. If you must use such a word-processing program, make certain that when you save the file, you change the file type to either "All Files" or "ANSI file" (depending on the options available); otherwise, hidden codes will be included in the script.

After the script file has been created, you can execute it in SQL*Plus by entering the SQL*Plus START command shown in Figure 14-6.

```
START drive:\filename
```

FIGURE 14-6 Command to execute a script file in client SQL*Plus

NOTE

If you save the script file with an .sql extension, the extension does not have to be included when referencing the file in the START command. If another file extension is used, such as .txt, you must include the file extension with the filename in the START command.

In Figure 14-6, the **START** keyword is used to indicate that a script should be executed. The ***drive:*** specifies the drive and path name where the file is stored; *filename* represents the name of the script file to be executed.

NOTE

The START keyword can be substituted with the symbol commonly referred to as the "at" sign (@) to execute a script. For example: `@d:\filename`

If you are using the Internet SQL*Plus interface, you can enter the script shown in Figure 14-5 into a file using either of the following approaches:

1. Enter the script into the iSQL*Plus workspace and click the **Save Script** button to save the script as a file.
2. Open a text editor (such as Notepad) outside iSQL*Plus and enter the commands directly into the text editor, saving the file with the .sql file extension.

After the script file has been created, you can execute it in iSQL*Plus by using either of the following approaches:

1. Click the **Load Script** button and locate the file to run. The load action will bring the contents of the file into the workspace area, and you can then execute the script.
2. The script files can be saved and called from a Web site location, which can be useful if the scripts are to be shared with a number of users. The script can be called using the START command, as shown in Figure 14-7.

```
START http://www.website.com/filename
```

FIGURE 14-7 Command to execute a script file from iSQL*Plus

Let's return to the report generated in Figure 14-5. Suppose that after reviewing the results, you want to make improvements. First, you want to add additional space between the ISBN and Title columns. Second, you want to remove the dollar signs from the Cost and Retail columns. Repeating a dollar sign for every currency value in a column can be distracting and adds unnecessary clutter.

Let's modify the original script as shown in Figure 14-8 to make these changes.

Notice that the distance between the ISBN and Title columns has been increased. Because the ISBN column is defined in the BOOKS table as a VARCHAR2 datatype, the data is displayed left-aligned within the column. This results in excess space appearing on the

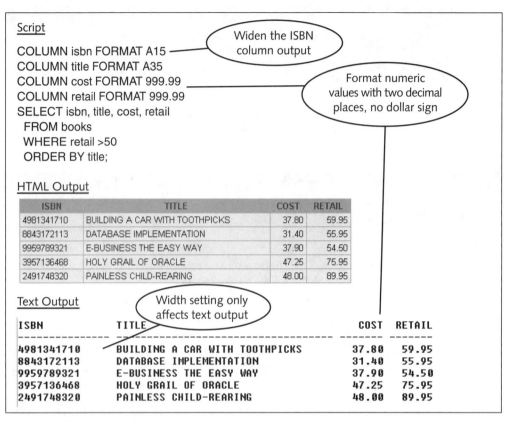

Script

COLUMN isbn FORMAT A15 ⟵ Widen the ISBN column output
COLUMN title FORMAT A35
COLUMN cost FORMAT 999.99 ⟵ Format numeric values with two decimal places, no dollar sign
COLUMN retail FORMAT 999.99
SELECT isbn, title, cost, retail
  FROM books
  WHERE retail >50
  ORDER BY title;

HTML Output

ISBN	TITLE	COST	RETAIL
4981341710	BUILDING A CAR WITH TOOTHPICKS	37.80	59.95
8843172113	DATABASE IMPLEMENTATION	31.40	55.95
9959789321	E-BUSINESS THE EASY WAY	37.90	54.50
3957136468	HOLY GRAIL OF ORACLE	47.25	75.95
2491748320	PAINLESS CHILD-REARING	48.00	89.95

Text Output — Width setting only affects text output

```
ISBN            TITLE                                 COST    RETAIL
--------------- ----------------------------------- ------- -------
4981341710      BUILDING A CAR WITH TOOTHPICKS        37.80   59.95
8843172113      DATABASE IMPLEMENTATION               31.40   55.95
9959789321      E-BUSINESS THE EASY WAY               37.90   54.50
3957136468      HOLY GRAIL OF ORACLE                  47.25   75.95
2491748320      PAINLESS CHILD-REARING                48.00   89.95
```

FIGURE 14-8 Revised script

right side of the ISBN column and creates a greater distance between the last digit of a book's ISBN and its corresponding title. On the other hand, columns defined as the NUMBER datatype display their contents right-aligned. Therefore, any excess space that has been assigned would appear in front of the column.

To increase the space between the Cost and Retail columns, increase the number of nines listed in each column's format model. For example, there are four blank spaces between the last digit in the Cost column and the first digit in the Retail column. To add an additional six spaces between the columns, simply add 6 nines to the format model for the Retail column, as shown in Figure 14-9. This width change will affect only text output, just as we have seen with character columns.

Because no book has a retail price of more than $99.99, any excess nines in the format model for the Retail column will be represented by a blank space. In essence, the series of nines establishes the width of that column. Although the space between the Cost and Retail columns has increased, the number of dashes separating the column headings and their contents has also increased. The additional dashes in the Retail column are not balanced in appearance with the dashes in the Cost column. There is, however, a simple solution to this problem, presented in the next section.

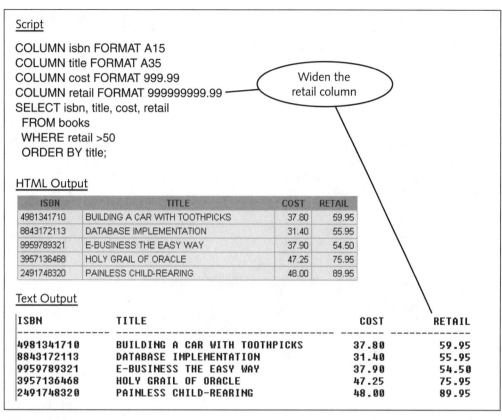

FIGURE 14-9 Modifying the width of numeric column output

HEADING Option

The **HEADING** option of the COLUMN command is used to specify a column heading for a particular column. By default, the report's column headings are simply the BOOKS table's column names. The HEADING option is similar to the use of a column alias: It provides a substitute heading for the display of the output. There are, however, two significant differences:

1. A column name assigned by the HEADING option of the COLUMN command cannot be referenced in a SELECT statement.
2. A column name assigned by the HEADING option can contain line breaks. In other words, the column name can be displayed on more than one line.

Let's work through an example. If you had assigned the column alias "Retail Price" to the Retail column of a SELECT statement, the column would have been 12 spaces wide in the output (one space for each letter and one for the space between the two words). If you used the HEADING option, however, you could indicate that the word "Retail" should appear on one line and the word "Price" should appear on a second line. This would give the report a more professional appearance and would not waste space. A single vertical bar (|) indicates where a line break should occur in a column heading. For example, 'Retail|Price' causes the first word, Retail, to appear on one line, and the second word, Price, to appear on the next line. (Note the use of single quotation marks to create a literal string.)

Modify the previous script to add a HEADING option on the Retail column, as shown in Figure 14-10.

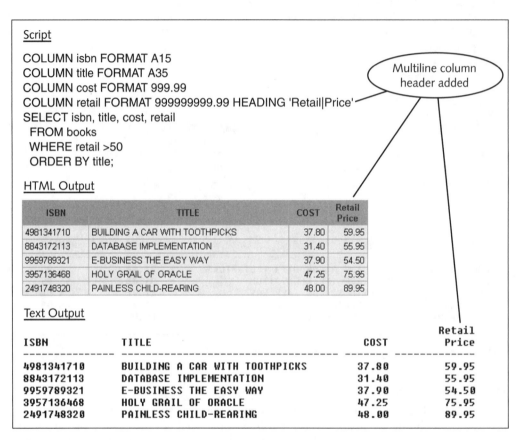

FIGURE 14-10 Heading inserted for the Retail column

Of course, now the problem is that the Title and Cost columns both have uppercase column headings, but the heading for the Retail column retains the case used in the HEADING option. To change the case of the heading of the Title and Cost columns, the simplest approach is to include a heading option for each of those columns and indicate the correct case, as shown in Figure 14-11.

As shown in Figure 14-11, the headings for the Title, Cost, and Retail Price columns are now displayed with the first letter of each word capitalized and the remaining letters in lowercase. However, the problem with the dashes separating the column headings from the column data in the text output still remains.

Using dashes to separate the headings from the data in text output is defined by the SQL*Plus environment variable **UNDERLINE**. An environment variable, also called a system variable, is used to determine how specific elements will be displayed in SQL*Plus.

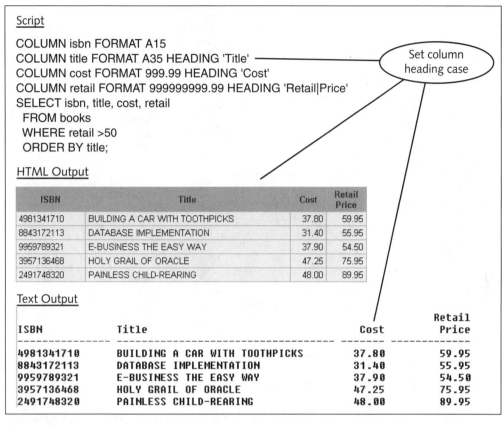

Script

```
COLUMN isbn FORMAT A15
COLUMN title FORMAT A35 HEADING 'Title'
COLUMN cost FORMAT 999.99 HEADING 'Cost'
COLUMN retail FORMAT 999999999.99 HEADING 'Retail|Price'
SELECT isbn, title, cost, retail
  FROM books
  WHERE retail >50
  ORDER BY title;
```

Set column heading case

HTML Output

ISBN	Title	Cost	Retail Price
4981341710	BUILDING A CAR WITH TOOTHPICKS	37.80	59.95
8843172113	DATABASE IMPLEMENTATION	31.40	55.95
9959789321	E-BUSINESS THE EASY WAY	37.90	54.50
3957136468	HOLY GRAIL OF ORACLE	47.25	75.95
2491748320	PAINLESS CHILD-REARING	48.00	89.95

Text Output

```
                                                         Retail
ISBN             Title                            Cost    Price
---------------  ----------------------------   -------  --------
4981341710       BUILDING A CAR WITH TOOTHPICKS   37.80    59.95
8843172113       DATABASE IMPLEMENTATION          31.40    55.95
9959789321       E-BUSINESS THE EASY WAY          37.90    54.50
3957136468       HOLY GRAIL OF ORACLE             47.25    75.95
2491748320       PAINLESS CHILD-REARING           48.00    89.95
```

FIGURE 14-11   Using the HEADING option to indicate the case of column headings

The **SET** command is used to change the value assigned to an environment variable. The UNDERLINE variable setting determines whether a header separator is displayed and what character will be used.

In our text report, we will create the illusion that the assigned width for a column is less than it is and still leave adequate spacing between the columns. This formatting issue does not exist with the HTML output, so this task addresses only the text report. This is a two-step process. First, you need to change the UNDERLINE variable so no dashes (or any other symbol) will be displayed. After the default dashes are eliminated, you can then add the desired number of dashes to the HEADING option for each column. Follow these steps, as shown in Figure 14-12:

1. At the SQL> prompt, enter SET UNDERLINE off and then press the **Enter** key. (This command says that no underlining should occur; that is, nothing should be used to separate the column heading from the column data in the output.)

2. Add a HEADING option for each column to include the number of underline dashes desired. Typically the number of dashes added match the character width setting for the column output.

Notice that the COLUMN command for the Title and Retail columns is now displayed on two lines. When a SQL*Plus command needs to be entered on more than one line, a

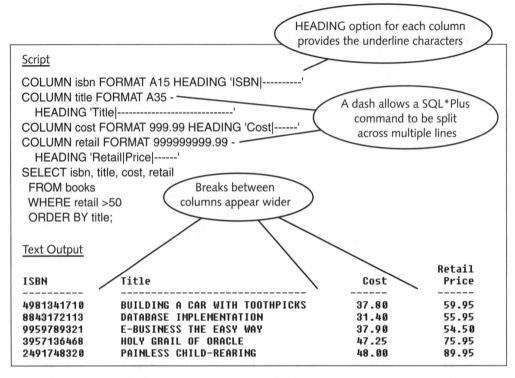

Script

COLUMN isbn FORMAT A15 HEADING 'ISBN|---------'
COLUMN title FORMAT A35 -
    HEADING 'Title|----------------------------'
COLUMN cost FORMAT 999.99 HEADING 'Cost|------'
COLUMN retail FORMAT 999999999.99 -
    HEADING 'Retail|Price|------'
SELECT isbn, title, cost, retail
  FROM books
  WHERE retail >50
  ORDER BY title;

Text Output

```
                                                                  Retail
ISBN             Title                                    Cost     Price
---------        -----------------------------------      ------   ------
4981341710       BUILDING A CAR WITH TOOTHPICKS            37.80    59.95
8843172113       DATABASE IMPLEMENTATION                  31.40    55.95
9959789321       E-BUSINESS THE EASY WAY                  37.90    54.50
3957136468       HOLY GRAIL OF ORACLE                     47.25    75.95
2491748320       PAINLESS CHILD-REARING                   48.00    89.95
```

FIGURE 14-12   Manually controlling the underline characters

dash is used at the end of the first line to indicate that the command is not complete and continues on the next line. In this case, the COLUMN command for both columns was broken just before the HEADING option and then continued on the next line. The second line for each command is also indented. This is not a requirement; however, it does improve the readability of the script.

NOTE

The underline character can be set to something other than the default, which is a single dash. For example, SET UNDERLINE '*' would cause an asterisk (*) to be used as the underline character.

NULL Option

Although this report does not contain NULL values, you will probably work with other reports that will contain NULL values. In some instances, blank spaces might be appropriate for those NULL values. For example, consider the orders JustLee Books ships to its customers. If management wants a report to view the lag time for shipping current orders, you might want to substitute the words "Not Shipped" for any NULL values that occur in the Shipdate column. With the NULL option of the COLUMN command, this is easy to accomplish, as shown in Figure 14-13. The first line turns back on the underline formatting setting.

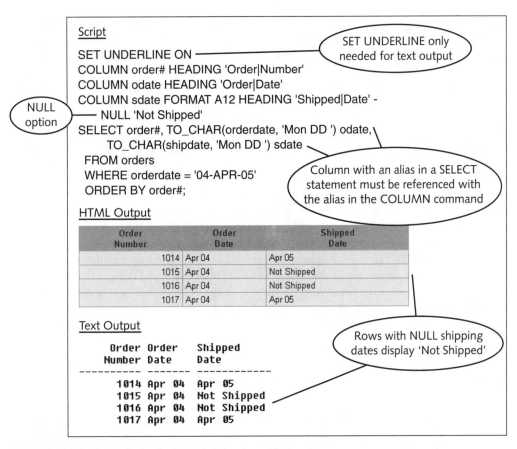

Script

```
SET UNDERLINE ON
COLUMN order# HEADING 'Order|Number'
COLUMN odate HEADING 'Order|Date'
COLUMN sdate FORMAT A12 HEADING 'Shipped|Date' -
       NULL 'Not Shipped'
SELECT order#, TO_CHAR(orderdate, 'Mon DD ') odate,
       TO_CHAR(shipdate, 'Mon DD ') sdate
FROM orders
WHERE orderdate = '04-APR-05'
ORDER BY order#;
```

SET UNDERLINE only needed for text output

NULL option

Column with an alias in a SELECT statement must be referenced with the alias in the COLUMN command

HTML Output

Order Number	Order Date	Shipped Date
1014	Apr 04	Apr 05
1015	Apr 04	Not Shipped
1016	Apr 04	Not Shipped
1017	Apr 04	Apr 05

Text Output

```
Order   Order   Shipped
Number  Date    Date
------- ------- -----------
 1014   Apr 04  Apr 05
 1015   Apr 04  Not Shipped
 1016   Apr 04  Not Shipped
 1017   Apr 04  Apr 05
```

Rows with NULL shipping dates display 'Not Shipped'

FIGURE 14-13 Script that substitutes a value for a NULL value

> **NOTE**
>
> Recall that you can also use the NVL2 function in SQL to substitute a value for a NULL value. However, NVL2 would be included in the SQL statement, whereas the NULL option is used in a SQL*Plus command.

Notice that the COLUMN command for the Shipdate column includes the NULL option. This instructs Oracle 10g that the text message should be displayed in place of the NULL value when the report is executed. Because the text message is a literal string, it will appear in the case given in the NULL option.

In the SELECT statement in Figure 14-13, the data in the Orderdate and Shipdate columns is retrieved by using the TO_CHAR function. The TO_CHAR function allows a format model to be applied. However, when the TO_CHAR function is used, the individual column names Orderdate and Shipdate cannot be referenced by the COLUMN command. Recall that when a function is applied to column data, the function is displayed as the column name in the output. The same thing occurs when a function is applied to a SELECT statement for a report. Therefore, a column alias must be assigned when a function is used, so the COLUMN command can reference the column through the assigned alias. In this case, odate and sdate are assigned as the column aliases and are then used in the HEADING option of the COLUMN command to provide a column heading for each of the date columns.

If you are using the Internet SQL*Plus interface, a number of the settings discussed in this chapter can be set via menus. If you go to the Preferences area and select Script Formatting, you will see an alphabetical list of numerous format settings that can be modified by making choices on this screen rather than entering SQL*Plus line commands. For example, you will find an entry for NULL Text on this screen along with a text box for entering the value you want to substitute or display for a NULL value in the output.

HEADERS AND FOOTERS

To create a more polished look for reports, headers and footers need to be applied. Technically, a header serves as the title for a report. The header and footer of a report are set with the TTITLE and BTITLE commands, respectively. The TTITLE command indicates the text or variables (such as page number) to be displayed at the top of the report. The BTITLE command defines what is to be printed at the bottom of the report. The syntax for both the TTITLE and BTITLE commands is shown in Figure 14-14.

```
TTITLE|BTITLE [option [text|variable]…] [ON|OFF]
```

FIGURE 14-14 Syntax of the TTITLE and BTITLE commands

Let's look at the individual elements of the command in Figure 14-14. You can use options to specify the formats to be applied to the title or footer data, where the data should appear, and on which line. Some of the options available with TTITLE and BTITLE are shown in Figure 14-15.

OPTION	DESCRIPTION
CENTER	Centers data to be displayed between the left and right margins of a report
FORMAT	Applies a format model to data to be displayed (uses same format model elements as the COLUMN command)
LEFT	Aligns data to be displayed to the left margin of the report
RIGHT	Aligns data to be displayed to the right margin of the report
SKIP n	Indicates the number of lines to skip before the display of data resumes

FIGURE 14-15 Options for the TTITLE and BTITLE commands

Also in Figure 14-14, note that [text|variable] means that you can apply the alignment and FORMAT options either to text entered as a literal string or to SQL*Plus variables. Some of the valid SQL*Plus variables are shown in Figure 14-16.

VARIABLE	DESCRIPTION
SQL.LNO	Current line number
SQL.PNO	Current page number
SQL.RELEASE	Current Oracle release number
SQL.USER	Current user name

FIGURE 14-16 SQL*Plus variables

Let's look at each of these variables.

- The **SQL.LNO** variable can be included in a report to indicate a line number on a report. Many users will include this variable to determine the length of a report or to set how many lines can be printed on a page.
- The **SQL.PNO** variable is used to include a page number on a report.
- As a matter of documentation, some users include the software's release number in a report by using **SQL.RELEASE**. This variable is usually included if a report does not execute properly after an upgrade.
- The **SQL.USER** variable can be used to include the name of the user running a report on the report.

Next, let's add a title to the previous script to give the report a professional appearance. The title should include a heading that indicates the report's contents. An appropriate heading for the report is "Books in Inventory," and it should be centered over the report's contents. In addition, the page number should be displayed on a separate line and on the right side of the report. To properly add headers and footers, you also need to understand the LINESIZE and PAGESIZE settings. These two settings are discussed next.

NOTE

If you do not include a format model for the title, the date and page number will appear automatically.

LINESIZE

The TTITLE command can be used to create the report heading. However, how will the center or the right side of the report be determined? The placement of aligned text will be based on the environment variable **LINESIZE**. Therefore, if LINESIZE has been set to 100, the center of the report will be 50 characters from each margin. To ensure proper placement of the components of the heading in text output, LINESIZE needs to be calculated to find the exact width of the report based on the column widths.

The script shown in Figure 14-17 demonstrates how to add the LINESIZE and TTITLE definitions.

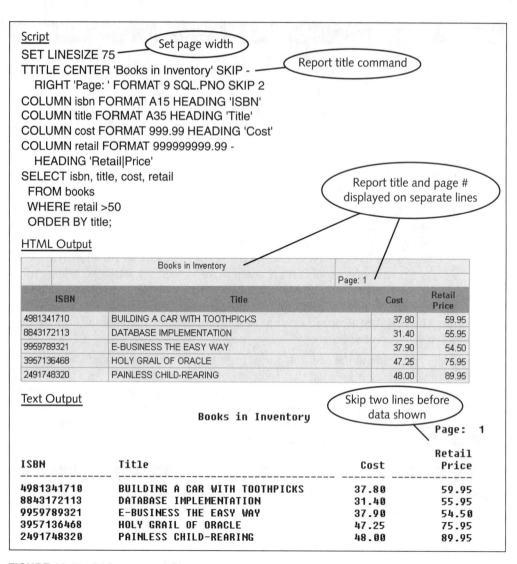

FIGURE 14-17 Adding a report title

Let's look at the individual elements in Figure 14-17:

- The first line added to the beginning of the script file defines a LINESIZE of 75 spaces to establish the right margin of the report.
- The second line adds a heading for the report. The first part of the TTITLE command has the words "Books in Inventory" as a centered heading. The command then advances to the next line of the report, using the **SKIP** option to begin the next portion of the heading on a separate line.

- The label "Page:" is then displayed, followed by the value stored in the SQL. PNO variable, to indicate the page number. A format model is applied to the page number to indicate the expected number of digits to be displayed. If the format model is not included, several blank spaces may appear between the "Page:" label and the page number.
- After the page number is displayed, the report advances two more lines before beginning the column headings. Using the SKIP option to display additional blank lines does not affect the HTML output.

Now that you've added the report's title, it is time to format the footer—the items displayed at the end of the report. Technically, a report's header and footer are displayed on every page. The policy at many organizations, however, is to have a message such as "End of Report" or "End of Job" displayed at the end of a report to make certain no pages of the report are missing. However, because the Books in Inventory report is only one page long, the message "End of Report" will be displayed as the footer, to indicate that there are no other pages to the report. In addition, the management at JustLee Books requires that the name of the user running a report also be displayed on the report. Therefore, the user name will be included as part of the footer.

PAGESIZE

Because the placement of the footer depends on the length of the page, the PAGESIZE environment variable needs to be set to indicate the length of a page. Although standard printed reports generally have 60 lines per page (leaving a small margin at the top and bottom of the page), a standard computer monitor displays approximately 32 lines, depending on the resolution settings of the monitor. Because this report is being displayed on a computer screen, the PAGESIZE variable will be set to 15 lines. This will allow both the report header and footer to be displayed on the screen at the same time.

The footer is created using the BTITLE command. Figure 14-18 shows how to add the BTITLE command so the words "End of Report" appear centered at the bottom of the page and the words "Run By:" appear at the right side of the bottom of the page. The name of the current user is then displayed. A format model allowing five spaces has been applied to the user name, so the name will appear flush with the right margin. However, if other users might run this report, extra spaces may be needed to accommodate longer names.

483

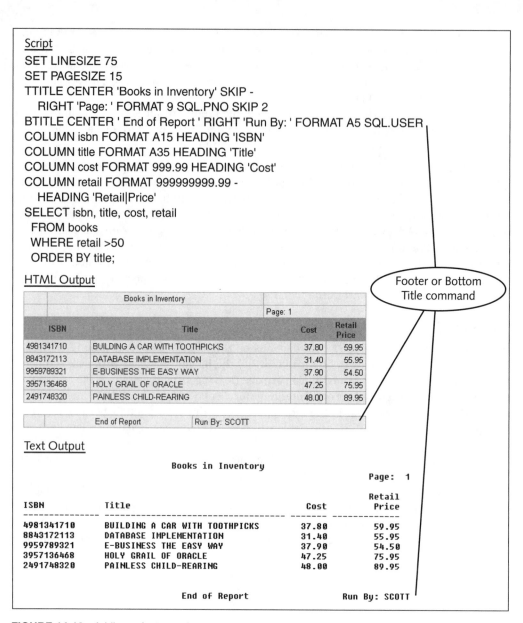

Script
```
SET LINESIZE 75
SET PAGESIZE 15
TTITLE CENTER 'Books in Inventory' SKIP -
   RIGHT 'Page: ' FORMAT 9 SQL.PNO SKIP 2
BTITLE CENTER ' End of Report ' RIGHT 'Run By: ' FORMAT A5 SQL.USER
COLUMN isbn FORMAT A15 HEADING 'ISBN'
COLUMN title FORMAT A35 HEADING 'Title'
COLUMN cost FORMAT 999.99 HEADING 'Cost'
COLUMN retail FORMAT 999999999.99 -
   HEADING 'Retail|Price'
SELECT isbn, title, cost, retail
  FROM books
  WHERE retail >50
  ORDER BY title;
```

HTML Output

	Books in Inventory		
		Page: 1	

ISBN	Title	Cost	Retail Price
4981341710	BUILDING A CAR WITH TOOTHPICKS	37.80	59.95
8843172113	DATABASE IMPLEMENTATION	31.40	55.95
9959789321	E-BUSINESS THE EASY WAY	37.90	54.50
3957136468	HOLY GRAIL OF ORACLE	47.25	75.95
2491748320	PAINLESS CHILD-REARING	48.00	89.95

	End of Report	Run By: SCOTT	

Footer or Bottom Title command

Text Output

```
                     Books in Inventory
                                                    Page:   1

                                                   Retail
ISBN             Title                       Cost   Price
---------------  ----------------------------------  -------  -------------
4981341710       BUILDING A CAR WITH TOOTHPICKS  37.80     59.95
8843172113       DATABASE IMPLEMENTATION        31.40     55.95
9959789321       E-BUSINESS THE EASY WAY        37.90     54.50
3957136468       HOLY GRAIL OF ORACLE           47.25     75.95
2491748320       PAINLESS CHILD-REARING         48.00     89.95

                     End of Report              Run By: SCOTT
```

FIGURE 14-18 Adding a footer to the report

Suppose that after the report has been displayed, you can't remember the exact structure of the TTITLE command used to generate the report header. You can execute the TTITLE command in SQL*Plus to display the current title settings, as shown in Figure 14-19.

Internet SQL*Plus interface

```
TTITLE
```

(Execute) (Load Script) (Save Script) (Cancel)

ttitle ON and is the following 76 characters: CENTER 'Books in Inventory' SKIP RIGHT 'Page: ' FORMAT 9 SQL.PNO SKIP 2

Client SQL*Plus interface

```
SQL> TTITLE
ttitle ON and is the following 76 characters:
CENTER 'Books in Inventory' SKIP          RIGHT 'Page: ' FORMAT 9 SQL.PNO SKIP 2
```

FIGURE 14-19 Checking current title settings

You can also check the current settings for the BTITLE and COLUMN commands. If referencing a column command, make certain to list the name of the column being requested (such as COLUMN isbn) to display the settings for that specific column.

BREAK COMMAND

The **BREAK** command is used to suppress duplicate data in a report. This is especially useful for reports containing groups of data. For example, suppose that you are creating a list of all books in inventory, sorted by category. This would result in the Computer, Business, and other Category column names being displayed multiple times. If you use the BREAK command, each Category column name will appear only once. The BREAK command also includes the options of skipping lines after each group and of having each group appear on a separate page. The syntax for the BREAK command is shown in Figure 14-20.

```
BREAK ON columnname|columnalias [ON ...] [skip n|page]
```

FIGURE 14-20 Syntax of the BREAK command

The effectiveness of the BREAK command is determined by how data is sorted in the SELECT statement. The contents of the specified column are printed once, and the printing for all subsequent rows is suppressed until the value in the column changes. For example, if the BREAK command has been applied to the Category column in a report containing a list of books sorted by category, then the first time the Computer Category is encountered, the word "COMPUTER" will be displayed. The category name will be suppressed until the next category name is encountered. However, if the data being displayed in the report is not in a sorted order, "COMPUTER" may be displayed several times throughout the report. Therefore, you should always include the ORDER BY clause when using the BREAK command. Duplicate values are suppressed only when they occur in sequence. As soon as the name of the category changes, the new value will be displayed.

If you want to include subgroupings, you can include an additional ON clause, followed by the name of the column to be used for the subgrouping.

Suppose that you want to produce a report that shows the amount due for orders in the ORDERS table. The orders should be sorted by customer number in the results. Rather than have customer numbers duplicated in the report, you can use the BREAK command. The script shown in Figure 14-21 can be used to create the report. (The WHERE clause is used to filter customers to avoid creating a long report for the demonstration.)

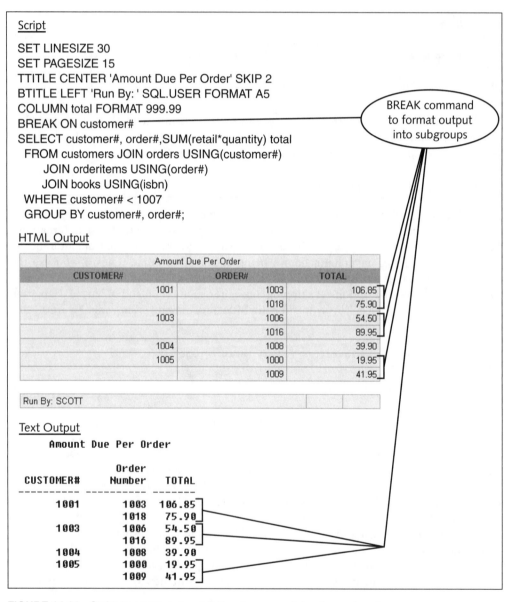

FIGURE 14-21   Script to create a report listing the amount due for each order

The BREAK command suppresses duplicate values in the Customer# column. Notice that an ORDER BY clause is not included in the SELECT statement. Because the SELECT statement has a GROUP BY clause, the data will automatically be sorted in the order of the Customer# and Order# columns.

To make the report easier to read, the SKIP keyword could have been included with the BREAK command to have a blank line occur between each group.

CLEAR COMMAND

The **CLEAR** command is used to clear the settings applied to the BREAK and COLUMN commands. The settings applied to these commands are valid even after a report has been finished. For example, if your next query contains a column or alias named "title," the 999.99 format would be applied, because this setting was created in the previous script file. Therefore, any time that you use the COLUMN or BREAK commands in a script, you should always clear those commands at the end of the script, so they will not adversely affect any subsequent reports. The syntax for the CLEAR command is shown in Figure 14-22.

```
CLEAR COLUMN|BREAK
```

FIGURE 14-22 Syntax of the CLEAR command

To instruct SQL*Plus to clear the COLUMN and BREAK settings defined in a script, the CLEAR command can be added at the end of the script, as shown in Figure 14-23.

```
SET LINESIZE 30
SET PAGESIZE 15
TTITLE CENTER 'Amount Due Per Order' SKIP 2
BTITLE LEFT 'Run By: ' SQL.USER FORMAT A5
COLUMN total FORMAT 999.99
BREAK ON customer#
SELECT customer#, order#,SUM(retail*quantity) total
  FROM customers JOIN orders USING(customer#)
     JOIN orderitems USING(order#)
     JOIN books USING(isbn)
  WHERE customer# < 1007
  GROUP BY customer#, order#;
CLEAR BREAK
CLEAR COLUMN
```

FIGURE 14-23 CLEAR command added to the end of a script

Notice that the CLEAR command is issued once to clear the settings defined by the BREAK command and again to clear the settings defined by any COLUMN commands. After the CLEAR command has been added, any settings that are applied when the report is generated will automatically be cleared after the report has been displayed.

Formatting Readable Output

COMPUTE COMMAND

In addition to determining the total amount due for each order, you can determine how much each customer owes by including the **COMPUTE** command. The COMPUTE command can be included with the AVG, SUM, COUNT, MAX, or MIN keywords to determine averages, totals, number of occurrences, the highest value, or the lowest value, respectively. Figure 14-24 shows the syntax for the COMPUTE command.

```
COMPUTE statisticalfunction OF columnname|REPORT ON groupname
```

FIGURE 14-24   Syntax of the COMPUTE command

For example, suppose that you want to determine the total amount due from each customer. The command COMPUTE SUM OF total ON customer# can be added to the script file, as shown in Figure 14-25, to instruct Oracle 10g to determine the total (SUM) amount due (OF total) for each customer (ON customer#). The LINESIZE and PAGESIZE variables have been changed to 30 and 25, respectively, to accommodate the extra data being added to the report. Because the report will now display the total amount due from each customer, the title of the report has also been changed.

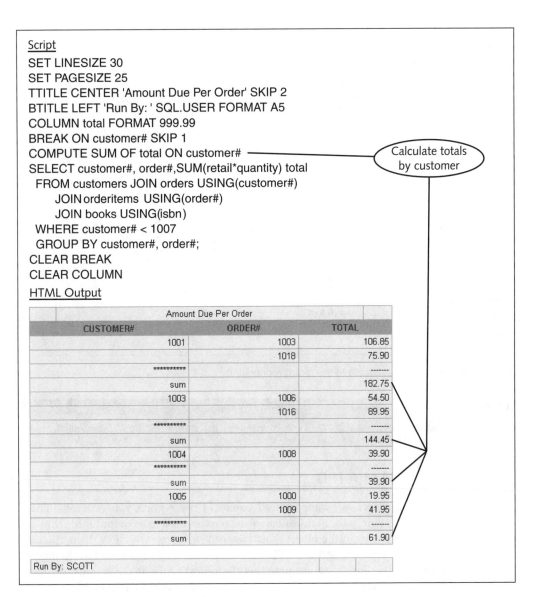

Script
```
SET LINESIZE 30
SET PAGESIZE 25
TTITLE CENTER 'Amount Due Per Order' SKIP 2
BTITLE LEFT 'Run By: ' SQL.USER FORMAT A5
COLUMN total FORMAT 999.99
BREAK ON customer# SKIP 1
COMPUTE SUM OF total ON customer#
SELECT customer#, order#,SUM(retail*quantity) total
  FROM customers JOIN orders USING(customer#)
     JOIN orderitems  USING(order#)
     JOIN books USING(isbn)
  WHERE customer# < 1007
  GROUP BY customer#, order#;
CLEAR BREAK
CLEAR COLUMN
```

Calculate totals by customer

HTML Output

Amount Due Per Order		
CUSTOMER#	ORDER#	TOTAL
1001	1003	106.85
	1018	75.90
**********		-------
sum		182.75
1003	1006	54.50
	1016	89.95
**********		-------
sum		144.45
1004	1008	39.90
**********		-------
sum		39.90
1005	1000	19.95
	1009	41.95
**********		-------
sum		61.90

Run By: SCOTT

FIGURE 14-25   Adding subtotals to the report

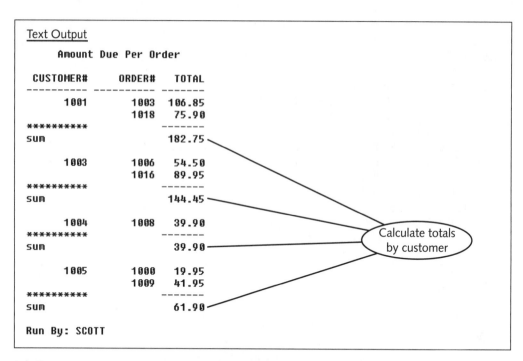

FIGURE 14-25 Adding subtotals to the report (continued)

SPOOL COMMAND

In most cases, management will want printed copies of the reports you generate. The results displayed on the computer monitor can also be redirected to a computer file that can then be opened and printed. In the client SQL*Plus interface, use the SPOOL command to save the results of a report to a text file. If the SPOOL command is added at the beginning of the script file, the entire script, including the report's format commands, will be saved to the file. The syntax for the SPOOL command is shown in Figure 14-26.

```
SPOOL drive:/filename
```

FIGURE 14-26 Syntax of the SPOOL command for client SQL*Plus

For example, if you want the results of a script **written** to a text file named reportresults.txt on your C drive for printing, you would add the command `SPOOL c:\reportresults.txt` to the beginning of the script. The command SPOOL OFF should be included at the end of your script to terminate the spooling process. Figure 14-27 shows an example script with the SPOOL command.

```
SPOOL c:\reportresults.txt              Starts spooling action
SET LINESIZE 30
SET PAGESIZE 25
TTITLE CENTER 'Amount Due Per Order' SKIP 2
BTITLE LEFT 'Run By: ' SQL.USER FORMAT A5
COLUMN total FORMAT 999.99
BREAK ON customer# SKIP 1
COMPUTE SUM OF total ON customer#
SELECT customer#, order#,SUM(retail*quantity) total
  FROM customers JOIN orders USING(customer#)
      JOIN orderitems USING(order#)
      JOIN books USING(isbn)
  WHERE customer# < 1007
  GROUP BY customer#, order#;
SPOOL OFF
CLEAR BREAK                             Stops spooling action
CLEAR COLUMN
```

FIGURE 14-27   Script example using the SPOOL command

You might want to save just the output to a file for printing, rather than the script and output. In this case, you can save the script to a file and include a SET ECHO OFF command at the top of the script. This setting suppresses the script commands in the output, so your spooled file will contain only the report data.

The SPOOL command is not available in the Internet SQL*Plus interface. However, there is a setting in the Preferences area under Interface Configuration that allows you to choose whether the output should be directed to the screen (default) or a file. If you select output to a file, when a script is executed you will be prompted as to where to save and what to name the output file.

HTML OUTPUT

At times you may want the output of a report to be in HTML form so that you can post it to a Web site. In the Internet SQL*Plus interface section, we have covered how to produce both text and HTML output; however, we have not addressed producing HTML in the client SQL*Plus tool. This can be accomplished by using the SET MARKUP HTML ON command in conjunction with the spooling option. Figure 14-28 displays an example script with these commands added.

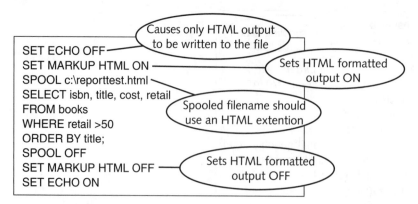

FIGURE 14-28 Script example for HTML output

You will need to save the script to a file and execute it by using the START command so that only HTML for the report output is saved to the file. The SET ECHO OFF command suppresses the formatting and query statements from being included in the output file. If you open the resulting file in a text editor, you will see HTML coding. To see the displayed Web page, open the file with a browser.

Chapter Summary

- Reports can be created by using a variety of formatting commands available through SQL*Plus. These are SQL*Plus commands, *not* SQL statements.

- Commands to format reports are usually entered into a script file and then executed. The SELECT statement appears after the SQL*Plus formatting commands in the script.

- A script file can be run in SQL*Plus by using the START command or an "at" sign (@) before the path and filename.

- The COLUMN command has the HEADING, FORMAT, and NULL options available to control the appearance of columns in the report.

- The HEADING option can be used to provide a column heading. A single vertical bar (|) is used to indicate where a line break should occur in the column heading.

- The SQL*Plus environment variable UNDERLINE indicates the symbol that should be used to separate column headings from the column data.

- The FORMAT option can be used to create a format model for displaying numeric data or to specify the width of nonnumeric columns.

- The NULL option indicates a text message that should be displayed if a NULL value exists in the specified column.

- The TTITLE command identifies the header, or title, that should appear at the top of the report. It can include format models, alignment options, and variables if needed.

- The BTITLE command identifies the footer, or title, that should appear at the bottom of the report. It follows the same syntax as the TTITLE command.

- Alignment used in the TTITLE and BTITLE commands are affected by the LINESIZE and PAGESIZE variables.

- The BREAK command is used to suppress duplicate data.

- At the end of each script file that includes a COLUMN or BREAK command, the CLEAR command should be provided to clear any setting changes made by the script.

- The COMPUTE command can be used to calculate totals in a report based on groupings.

- The SPOOL command instructs Oracle 10*g* to redirect output to a file. This command is available only in client SQL*Plus.

- The MARKUP HTML setting allows output in client SQL*Plus to be saved in HTML format.

Chapter 14 Syntax Summary

The following table presents a summary of the syntax that you have learned in this chapter. You can use the table as a study guide and reference.

SYNTAX GUIDE		
Element	Description	Example
iSQL*Plus Formatting Commands		

START *or* @	Executes a script file	`START titlereport.sql`
		or
		`@C:\titlereport.sql`

TTITLE [option [text|variable]...] [ON|OFF]	Defines a report's header	`TTITLE 'Top of|Report'`
BTITLE [option [*text*|*variable*]...] [ON|OFF]	Defines a report's footer	`BTITLE 'Bottom` `of|Report'`
BREAK ON *columnname*| *columnalias* [ON ...] [skip *n*|page]	Suppresses duplicate data in a specified column	`BREAK ON title`
COMPUTE *statisticalfunction* OF *columnname* | REPORT ON *groupname*	Performs calculations in a report, based on the AVG, SUM, MIN, and MAX statistical functions	`COMPUTE SUM OF` `total ON customer#`
SPOOL *filename*	Redirects the output to a text file	`SPOOL` `c:\reportresults.txt`
CLEAR	Clears settings applied with the COLUMN and BREAK commands	`CLEAR COLUMN`
COLUMN Command Options		
FORMAT *formatmodel*	Applies a specified format to the column data	`COLUMN title` `FORMAT A30`
HEADING *columnheading*	Specifies the heading to be used for the specified column	`COLUMN title HEADING` `'Book|Title'`
NULL *textmessage*	Identifies the text message to be displayed in the place of NULL values	`COLUMN shipdate NULL` `'Not Shipped'`
TTITLE and BTITLE Command Options		
CENTER	Centers the data to be displayed between a report's left and right margins	`TTITLE CENTER` `'Report Title'`

Element	Description	Example
TTITLE and BTITLE Command Options		
FORMAT	Applies a format model to the data to be displayed (uses the same format model elements as the COLUMN command)	`BTITLE 'Page:'` `SQL.PNO FORMAT 9`
LEFT	Aligns data to be displayed to a report's left margin	`TTITLE LEFT` `'Internal Use Only'`
RIGHT	Aligns data to be displayed to a report's right margin	`TTITLE RIGHT` `'Internal Use Only'`

SKIP *n*	Indicates the number of lines to skip before the display resumes	`TTITLE CENTER 'Report`
		`Title' SKIP 2 RIGHT`
		`'Internal Use Only'`

SQL*Plus Variables		

SQL.LNO	Retrieves the current line number	`BTITLE 'End of Page on`
		`Line: ' SQL.LNO`
		`Format 99`

SQL.PNO	Retrieves the current page number	`BTITLE 'Page: '` `SQL.PNO FORMAT 9`

SQL.RELEASE	Retrieves the current Oracle release number	`BTITLE 'Processed on`
		`Oracle release number:'`
		`SQL.RELEASE`

SQL.USER	Retrieves the current user name	`BTITLE 'Run By: '` `SQL.USER FORMAT AB`
SQL*Plus Environment Variables		
UNDERLINE *x*|OFF	Specifies the symbol used to separate a column heading from the contents of the column	SET UNDERLINE '='
LINESIZE *n*	Establishes the maximum number of characters that can appear on a single line of output	SET LINESIZE 35
PAGESIZE *n*	Establishes the maximum number of lines that can appear on one page of output	SET PAGESIZE 27
MARKUP HTML ON|OFF	Specifies if output from client SQL*Plus is to be saved in HTML format	SET MARKUP HTML ON

Formatting Readable Output

Review Questions

1. How can you instruct SQL*Plus to place a column heading on two lines?

2. Which command and option will force insignificant zeros to be shown in the decimal position(s) of a numeric column?

3. What is the effect on a report's headers and footers if a report is 30 characters wide, but LINESIZE is set to 50 characters?

4. How can you display text messages in place of NULL values in a report?

5. What happens if the BREAK command is used for data that is not retrieved by the SELECT statement in any type of sorted order?

6. What is the purpose of saving a script file in SQL*Plus, and how can the file be executed?

7. What is the purpose of the UNDERLINE variable?

8. How can a format be applied to a date column in a report?

9. What are some of the differences between the client and Internet SQL*Plus tools regarding SQL*Plus formatting features? What are the default output formats in each of these tools?

10. What is the purpose of the CLEAR command?

Multiple Choice

1. Which of the following commands can be used to clear column formats?
   a. COLUMN CLEAR
   b. CLEAR COLUMN
   c. CLEAR FORMAT
   d. all of the above
   e. None of the above—you must include a column name.

2. Which of the following commands creates a header for a report?
   a. HEADER
   b. TITLE
   c. REPORTTITLE
   d. TOPLINE
   e. none of the above

3. Which of the following format model elements cannot be applied to a column with a numeric datatype?
   a. $
   b. A*n*
   c. ,
   d. 9.99
   e. All of the above can be applied to a numeric column.

Chapter 14

4. Which of the following commands instructs SQL*Plus to advance to the next line before resuming the display of the report?

    a. SKIP

    b. NEXT

    c. ADVANCE

    d. JUMP

    e. LINEFEED

5. Which option is used with the COLUMN command to substitute a NULL value with a text message?

    a. ON NULL

    b. FORMAT

    c. HEADING

    d. NULL

6. Which of the following commands indicates that a numeric column should be formatted to a total width of six?

    a. `FORMAT A6`

    b. `FORMAT 9999.99`

    c. `FORMAT 9999999`

    d. `FORMAT 9,999`

7. Which of the following environment variables specifies the maximum number of characters per line in SQL*Plus?

    a. PAGESIZE

    b. LINENUMBER

    c. LINESIZE

    d. CPI

8. Which of the following variables can be used to display a page number in a report?

    a. SQL.PAGE

    b. SQL.PGNO

    c. SQL.PNO

    d. SQL.Page_Number

9. Which of the following variables can be used to display the user name in a report?

    a. SQL.USER

    b. SQL.USERID

    c. SQL.NAME

    d. SQL.USERNAME

10. Which of the following can be used to display the date 31–JAN–07 in a report using the format January 31, 2007?

    a. `FORMAT ADATE`

    b. `FORMAT Month DD, YYYY`

    c. `TO_CHAR`

    d. `FORMAT AAAAAA 99, 9999`

    Questions 11–20 refer to the following script to create a report:

    ```
    SET PAGESIZE 38
    SET LINESIZE 45
    TTITLE CENTER 'Listing of | Books by Publisher'
    BTITLE LEFT 'Page: ' SQL.PNO FORMAT
    BREAK ON "Publisher Name"
    COLUMN "Publisher Name" FORMAT A30
    COLUMN title
    COLUMN cost HEADING 'Cost Per | Book' FORMAT 9,999.99
    SELECT name, title, cost
    FROM books NATURAL JOIN publisher;
    CLEAR COLUMN
    ```

11. Which of the following commands is not valid?

    a. `BTITLE LEFT 'Page: ' SQL.PNO FORMAT`

    b. `COLUMN cost HEADING 'Cost Per | Book' FORMAT 9,999.99`

    c. `SET PAGESIZE 38`

    d. `CLEAR COLUMN`

12. When the following COLUMN command is used, which of the following statements will cause an error to occur?

    ```
    COLUMN "Publisher Name" FORMAT A30
    1 BREAK ON "Publisher Name"
    2 SELECT name, title, cost
    3 FROM books NATURAL JOIN publisher;
    ```

    a. line 1

    b. line 2

    c. line 3

    d. None of the above will cause an error.

13. The CLEAR COLUMN command given at the end of the script applies to which of the following?

    a. `SET PAGESIZE 38`

    b. `TTITLE CENTER 'Listing of | Books by Publisher'`

    c. `BREAK ON "Publisher Name"`

    d. `SELECT name, title, cost`

    e. none of the above

14. Which of the following clauses will cause an error to occur in a script file for a report?

    a.   `SELECT name, title, cost`

    b.   `SET PAGESIZE 38`

    c.   `COLUMN title`

    d.   `COLUMN cost HEADING 'Cost Per | Book' FORMAT 9,999.99`

15. Assuming that any errors existing in the script are corrected, how many columns will be displayed in the report?

    a.   1

    b.   2

    c.   3

    d.   4

16. What is the maximum number of characters that can appear on one line of output in the report?

    a.   38

    b.   45

    c.   30

    d.   9

17. The report will display data from how many tables?

    a.   1

    b.   2

    c.   3

    d.   4

18. Which of the following is a correct statement?

    a.   No header will be displayed because there is an error in the TTITLE command.

    b.   The BREAK command may not have the expected results because the SELECT statement does not include an ORDER BY clause.

    c.   The exact location of the report header will be determined by the PAGESIZE variable.

    d.   The exact location of the report footer will be determined by the LINESIZE variable.

19. Which of the following should be changed to correct the error in the BTITLE command?

    a.   SQL.PNO

    b.   A2

    c.   LEFT

    d.   None of the above—there is no error in the command.

20. Which of the following should be changed to correct the error in the TTITLE command?

    a.   The CENTER option should be removed.

    b.   FORMAT A18 should be added to the command.

    c.   The text should be enclosed in double quotation marks.

    d.   None of the above—there is no error in the command.

Hands-On Assignments

To perform these activities, refer to the tables in Appendix A.

1. Create a script that will retrieve the title of each book, the name of its publisher, and the first and last name of each author of the book. Save the script in a file named **BookandAuthor1.sql**. Add a title to be displayed with the query results. The title should be "Book and Author Report." Make any changes necessary to ensure that the title is centered across the top of the query results.

2. Add appropriate column headings for each column in the script file **BookandAuthor1.sql**. Save the modified script as **BookandAuthor2.sql**.

3. Modify the query in the **BookandAuthor2.sql** file so the results will be listed in ascending order by the title of the book. Save the modified script as **BookandAuthor3.sql**.

4. Change the report settings in the **BookandAuthor3.sql** file so the title of the book will not be duplicated for each author listed. Save the modified script as **BookandAuthor4.sql**.

5. Modify the **BookandAuthor4.sql** script to suppress the duplication of the publisher's name when more than one author writes a book. Save the new file as **BookandAuthor5.sql**.

6. Modify the **BookandAuthor5.sql** file and create a footer for your report that displays your user name at the bottom of each page. Save the new file as **BookandAuthor6.sql**.

7. Modify the **BookandAuthor6.sql** file to add a blank line to separate the different books listed in the report. Save the new file as **BookandAuthor7.sql**.

8. Modify the **BookandAuthor7.sql** file to clear any column formats used in the report. Save the modified file as **BookandAuthor8.sql**.

9. Modify the **BookandAuthor8.sql** file to clear any breaks used in the report. Save the modified file as **BookandAuthor9.sql**.

10. Modify the **BookandAuthor9.sql** file to save the output in HTML format to a file named report.html. Save the modified file as **BookandAuthor10.sql**.

Advanced Challenge

To perform this activity, refer to the tables in Appendix A.

The management of JustLee Books has requested a report that shows the profit generated by each order placed between April 1 and April 4, 2005. The report should include the order number for each order, the total amount of the order, and the total profit generated by each order. A report header should be displayed that is descriptive of the report's contents, and a report footer should indicate who ran the report. Apply any other elements or formats necessary to make the report look professional. The report will be used for two purposes: to be printed and to be posted to a Web site. Create one version of the script to save the output in text format for printing. Name the file **ProfitReport.txt**. Create a second version that will save the output in HTML format for Web posting, and name the file **ProfitReport.html**. Save the completed report definitions in script files named **ProfitReport_1.sql** and **ProfitReport_2.sql**, respectively.

Case Study: *City Jail*

The City Jail administration has identified a set of standard reports to be produced weekly and posted to the intranet Web site for review by internal division chiefs. Create a script to produce each of the following reports, using SQL*Plus formatting to improve the appearance of the output and/or to add totals.

1. Provide an alphabetical list of all criminals along with each of the crime charges filed. The report needs to include the criminal id, name, crime code, date charged, and fine amount.

2. List the total number of charges by crime classification and the charge status.

3. List the total amount of collections (fines and fees) and the total amount owed by crime classification and crime code.

CHAPTER **15**

EXPLORING SQL TOPICS IN APPLICATION DEVELOPMENT

LEARNING OBJECTIVES

After completing this chapter, you should be able to do the following:

- Explore how SQL is used in application development
- Learn about SQL tuning concepts and issues
- Review selected tuning examples
- Compare SQL in several databases

INTRODUCTION

The past 14 chapters have covered the fundamentals of SQL, and you are now prepared to venture into more advanced database development subjects. This chapter serves as an introduction to some topics you may want to pursue next. An introduction to these topics will assist you in using your SQL knowledge to understand some aspects of application development and SQL in non-Oracle databases. An overview of SQL in application development, SQL statement tuning, and SQL in non-Oracle databases is provided. Each of these topics can fill a book on their own, so keep in mind that this is intended only as an introduction to make you aware of these topics and to set the stage for further development.

First, most application environments require database interaction and, therefore, include SQL processing. How does SQL fit into application code? This chapter presents examples of embedding SQL within application code.

Second, tuning is an ongoing effort to make our applications process efficiently to minimize the resources demanded and to increase the response speed. An application typically has three areas to consider in optimizing performance: hardware/networking, database configuration, and application source code. SQL is typically a major component of application code tuning efforts because database interaction can represent a significant portion of the processing time. This chapter introduces SQL tuning topics within the context of an Oracle environment. However, these same concepts apply to tuning SQL in most databases.

Third, understanding the fundamentals of SQL is powerful because all relational databases include SQL capabilities. However, it is important to recognize the nuances of SQL when moving from one environment to another. This chapter provides some SQL statement comparisons between Oracle, MySQL, and Microsoft SQL Server.

NOTE

This chapter is not intended to provide step-by-step instructional exercises but to present general topics. However, you should review the prech15.sql script to identify the three tables used in the examples. If you want to experiment with some of the examples presented, execute the script file to create the tables.

SQL IN APPLICATION CODE

SQL skills are quite useful for almost any development environment in which your application needs to interact with the database. This section demonstrates two application examples to help you see how you might embed SQL in an application. First, many applications include controls that need to be populated from the database. For example, drop-down selection boxes can simplify data selection for the end user. Figure 15-1 displays a drop-down list of all the categories for the JustLee book inventory.

This drop-down list could be used on a screen that allows a user to select a Book Category when performing a search. Or, it could be used on a screen an employee uses to add a new book to inventory, because a category is assigned to every book added to the BOOKS table. Because categories can change over time (new categories can be added, and discontinued categories deleted), the drop-down list should be populated from the database. This ensures that the most current category list is available and it eliminates the need to change code as categories change.

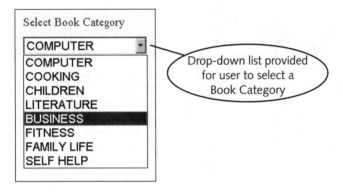

FIGURE 15-1 Drop-down list on application screen

Figure 15-2 displays the code for a Web page developed using ASP.net (Microsoft Active Server Pages—the coding language is Visual Basic). The routine executes the query to retrieve the data and attaches the data to the drop-down list object.

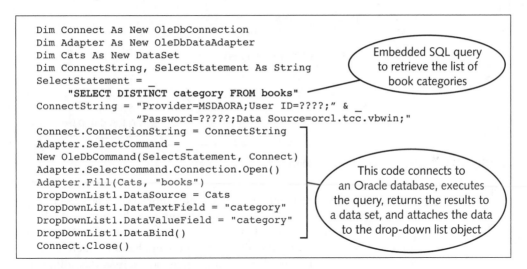

```
Dim Connect As New OleDbConnection
Dim Adapter As New OleDbDataAdapter
Dim Cats As New DataSet
Dim ConnectString, SelectStatement As String
SelectStatement = _
        "SELECT DISTINCT category FROM books"
ConnectString = "Provider=MSDAORA;User ID=????;" & _
                "Password=?????;Data Source=orcl.tcc.vbwin;"
Connect.ConnectionString = ConnectString
Adapter.SelectCommand = _
New OleDbCommand(SelectStatement, Connect)
Adapter.SelectCommand.Connection.Open()
Adapter.Fill(Cats, "books")
DropDownList1.DataSource = Cats
DropDownList1.DataTextField = "category"
DropDownList1.DataValueField = "category"
DropDownList1.DataBind()
Connect.Close()
```

Embedded SQL query to retrieve the list of book categories

This code connects to an Oracle database, executes the query, returns the results to a data set, and attaches the data to the drop-down list object

FIGURE 15-2 ASP.net code with embedded SQL

Applications need to not only query data, but also to support DML activity. Figure 15-3 displays a data entry screen in which the user would enter an e-mail address and name. The code supporting this screen will need to retrieve the data from the text boxes on the screen and perform an INSERT to add the data to the database.

Figure 15-4 shows the code used to support the when-button-pressed event for the application screen. The screen was produced using the Oracle Developer Forms tool. The code uses the PL/SQL language, which is Oracle's built-in procedural language. The values clause of the INSERT statement references the names of each of the text boxes.

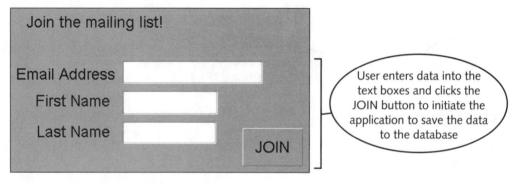

FIGURE 15-3 User entry screen

```
DECLARE
        alert_show NUMBER;
BEGIN
   IF :MAIL.email IS NULL
        OR :MAIL.fname IS NULL
        OR :MAIL.lname IS NULL THEN
           alert_show := Show_Alert('email_alert');
        ELSE
      INSERT INTO maillist (email, firstname, lastname)
        VALUES (:MAIL.email,:MAIL.fname,:MAIL.lname);
      COMMIT;
      alert_show := Show_Alert('done_alert');
      CLEAR_FORM;
   END IF;
END;
```

Checks if user entered a value for each text box

Embedded SQL to add the user-entered data into the database

FIGURE 15-4 PL/SQL code with embedded SQL

You have numerous choices of development environments; however, almost every environment supports the use of SQL. As you can see in these examples, being very familiar with SQL syntax is a great advantage in embedding SQL into application code. Notice that the SQL statement must now interact with the objects from the application, such as drop-down lists or text boxes. Imagine that you construct an application screen that allows a user to make selections to filter data. The embedded SQL would include a WHERE clause with conditions based on user input. Constructing these statements could be quite challenging without SQL knowledge.

As we build and test applications, we must be concerned with how efficiently the code executes. The next section introduces the topic of SQL statement tuning.

TUNING CONCEPTS AND ISSUES

To begin tuning code, we must first become familiar with methods that help identify issues affecting coding efficiency. First, we need to be able to identify which statements are causing lengthy execution times. In addition, not only do we need to know the execution time, but we also need to understand how the database server is processing the statement to determine what potential improvements could be applied. Then, after we attempt a modification, we need to be able to determine if it succeeded in improving performance.

In this part of the chapter, we first explore some methods of identifying SQL statements that are resource intensive. Second, we examine how statement processing and options are managed in Oracle. Third, we investigate database features such as the explain plan, to obtain processing information helpful in reviewing performance.

Identifying Problem Areas in Coding

To begin tuning efforts, we need to first identify the areas in source code that are likely candidates for tuning. Several basic methods are available to assist in this identification:

- The application-testing phase should simulate actual operations and should include end users. Feedback from end users helps pinpoint problem areas, especially regarding slow response. Set up a testing procedure that makes it easy for the testing group to document both where in the application they encountered trouble and what specific actions were problematic. This information can lead you directly to the coding that requires review.
- Oracle 10g includes a number of new automatic tuning capabilities. One example is the SQL Tuning Advisor, which assists in identifying problem SQL statements and makes recommendations regarding items such as collecting object statistics or creating an index. The SQL Tuning Advisor is available within the Oracle Enterprise Manager interface.
- The V_$SQLAREA view provides execution details, such as disk and memory reads for all statements processed since the database startup and as memory allows. Use the DESCRIBE command on the view to list all the available columns of data. Here are examples of statistics that are provided in this view:
  - The number of executions, which indicates how many times a statement has been processed, identifying the more heavily used statements
  - The number of disk reads, which reflects the total number of physical reads that were accomplished for a given statement
  - The disk reads divided by the number of executions, which determines the number of reads per execution of a given statement

  When you are deciding which statements need tuning, focus on the number of reads per execution. A high number of reads indicates that you might be able to improve performance by making modifications to the statement. Another important statistic in the V_$SQLAREA view is the number of buffer gets, which is the number of memory reads performed for the statement. The Buffer_Gets column can be queried to show the number per execution. A high number of buffer reads can indicate that an index is needed or that a join could be improved.

The SQL TRACE facility can be turned on for a session, and statistics regarding SQL statement execution during that session can be stored for review. The statistics are saved in an operating system file that must be converted to a readable format using the TKPROF executable file. The converted file can then be viewed using simple word-processing software such as WordPad.

After you've identified problem SQL statements, you need to determine possible tuning actions. Questions such as, "Could an index improve the performance of this statement?" should be explored. Before examining some specific tuning examples, you must be able to understand and identify how Oracle processes a SQL statement.

Processing and the Optimizer

Before tackling performance tuning, we need to establish an understanding of the SQL processing architecture to be able to determine how statements can be modified to execute more efficiently. SQL processing contains the components pictured in Figure 15-5: parser, optimizer, row source generator, and execution engine.

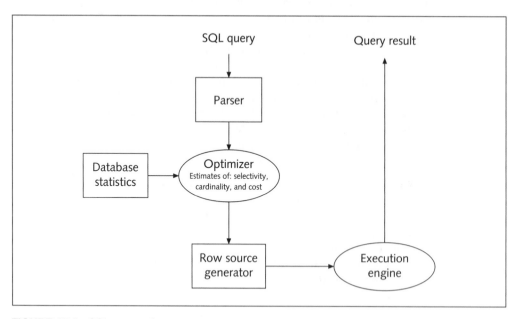

FIGURE 15-5 SQL processing components

The parser checks for correct statement syntax and that all objects referenced exist. The optimizer determines the most efficient way to process the statement creating an execution plan that is followed. For example, the optimizer makes decisions as to whether an index will be used or in what order tables will be joined. These are examples of factors that will determine how efficient your statements process. The row source generator sends the execution plan and the row source for each step in the plan to the execution engine. A row source returns a set of rows for the applicable step. The execution engine processes each row source and completes the execution plan to produce the final results.

In versions prior to Oracle 10g, the Oracle database server contains two statement optimizers, rule-based and cost-based, which follow different methodologies to determine how a statement is processed. The rule-based optimizer is the older method, which uses a list of rules to determine processing. For example, the rule-based engine typically uses an index if one is available, even if it might not be beneficial.

Oracle 10g now includes only the cost-based optimizer (CBO). The cost-based optimizer is a newer methodology that uses database object statistics, such as distribution of data to determine how best to process statements. If particular rows are being retrieved from a small table, the cost-based optimizer may decide not to use an index, as it would not increase the performance of the query. To determine the execution plan, the optimizer estimates the following items based on database statistics:

- *Selectivity*: the fraction of rows from the row set to be used
- *Cardinality*: the number of rows in the row set
- *Cost*: the units of resources used, including disk I/O, CPU usage, and memory usage

For the cost-based optimizer to make the best decisions, current database object statistics are needed. By default, Oracle 10g gathers statistics on all database objects that have a daily scheduled job. The schedule for updating statistics can be modified as needed depending on the volatility of the database. In addition, the analysis of objects used to gather statistics might need to be undertaken manually by issuing an ANALYZE command or by using the DBMS_STATS built-in package when new objects are added to the database. In versions prior to Oracle 10g, gathering automatic statistics is not set up by default, and it is critical to manually initiate database object analysis to take advantage of the cost-based optimizer.

If the cost-based optimizer cannot find any statistics for an object, dynamic sampling takes effect and performs random sampling during statement execution. This could dramatically slow the performance of statements.

The optimizer is one of the many database settings, and the parameter is named OPTIMIZER_MODE. The following OPTIMIZER_MODE settings are available in Oracle 10g:

- `ALL_ROWS`: Optimizes with a goal of achieving the best throughput (the use of minimum resources). Use this setting to complete the entire statement. This is the default value.
- `FIRST_ROWS_n`: Optimizes with a goal of best response time to return the first n number of rows; n can equal 1, 10, 100, or 1000.
- `FIRST_ROWS`: Uses a mix of cost and heuristics to find the best plan for fast delivery of the first few rows.

The cost-based optimizer uses a goal of best throughput by default. Best throughput means it chooses the path of execution that uses the least amount of resources needed to process all the rows in the statement. However, the CBO can run with a goal of optimizing the response time. With this goal, it generates an execution path that uses the least amount of resources to process the first row accessed by the statement.

Note that applications that center around large or batch requests such as with Oracle Reports typically optimize for best throughput. Interactive or operational applications, such as those created with Oracle Forms, should be optimized for best response time, because users are waiting to view feedback. If many rows are returned, the end user is still interested in getting the first rows of feedback fast to begin analyzing the results.

So how do we determine how a SQL statement is being processed? The next section introduces some basic tools to assist in examining statement processing.

The Explain Plan

One of the most important items to review regarding the performance of a statement is the execution or explain plan for that statement. Recall that the optimizer develops an execution plan that outlines the specific steps that are taken to process a statement. Oracle provides several methods to review the explain plan of a SQL statement. This section introduces several common methods: the Tuning Advisor, the EXPLAIN PLAN FOR command, and the AUTOTRACE facility.

You can access the Tuning Advisor via the Enterprise Manager console. After you have selected a particular statement to review, you are offered a choice to view the explain plan. The explain plan appears as shown in Figure 15-6.

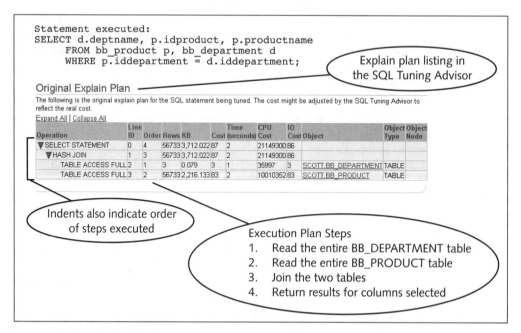

FIGURE 15-6　The SQL Tuning Advisor displays the explain plan

Let's review the contents of the explain plan. First, the access path for each table is determined. The access path in this example is a full table scan. Another common access path is the use of an index. If the query contained a WHERE clause limiting the rows returned from the BB_PRODUCT table and an index is available for the filtering column, the table might have been accessed via the index (an example of this is demonstrated later in this section).

Second, the order in which the tables are accessed in a multiple table query is determined. In this example, the BB_DEPARTMENT table is read first. It is most efficient to drive a join by the smallest table or by the table with the most selective filter.

Third, the join method for a multiple table query is determined. Three common join methods are nested join loops, sort merge, and hash joins. In the example, a hash join is indicated. This type of join method is appropriate when a large table is being joined to a small table via an equi-join. A hash table based on the small table is placed in memory and is used to more quickly perform the join with the larger table. Nested join loops are typically used if small subsets of data are being joined and if the join condition is based on a column that can be accessed via an index. In this join method, one table is considered the outer table and the other the inner table. The inner table is accessed for each row in the outer table. The sort merge operation sorts both tables by the join key and then merges the tables. The sort merge operation can be used if the data is already sorted or if the join type is a non-equi-join.

The Oracle AUTOTRACE tool enables the display of both the execution plan and of the execution statistics. The AUTOTRACE tool is started by issuing the SET AUTOTRACE ON command in SQL*Plus. A number of options can be used when starting this tool, as listed in Figure 15-7.

511

NOTE

The terms "execution plan" and "explain plan" are used interchangeably in this text.

SQL*PLUS COMMAND	DESCRIPTION
SET AUTOTRACE ON	Displays explain plan, statistics, and result set
SET AUTOTRACE ON EXPLAIN	Displays explain plan and result set
SET AUTOTRACE ON STATISTICS	Displays statistics and result set
SET AUTOTRACE TRACEONLY	Displays explain plan and statistics

FIGURE 15-7 AUTOTRACE tool options

Figure 15-8 shows the same SQL statement used earlier with output from the AUTOTRACE facility.

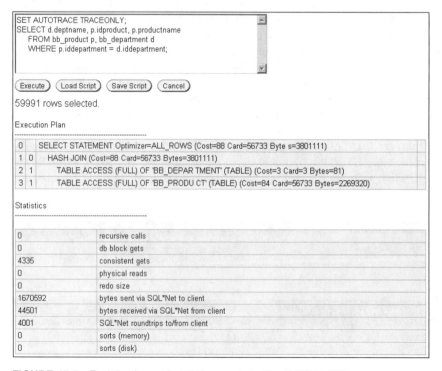

FIGURE 15-8   Explain plan and statistics produced by AUTOTRACE

Figure 15-9 lists a description for each of the statistics displayed by the AUTOTRACE tool. The explain plan shown matches the explain plan displayed in the Tuning Advisor. Notice the SQL*Net statistics, which are helpful in identifying network traffic. In addition, the consistent reads are the same as the buffer gets we covered earlier, and the physical reads are the same as the disk reads.

STATISTIC	DESCRIPTION
Recursive calls	Number of recursive calls generated at both the user and system level. Oracle maintains tables used for internal processing. When Oracle needs to make a change to these tables, it internally generates a SQL statement, which, in turn, generates a recursive call.
Db block gets	Number of times a CURRENT block was requested.
Consistent gets	Number of times a consistent read was requested for a block.
Physical reads	Total number of data blocks read from disk. This number equals the value of "physical reads direct" plus all reads into buffer cache.
Redo size	Total amount of redo generated in bytes.
Bytes sent via SQL*Net to client	Total number of bytes sent to the client from the foreground to client processes.
Bytes received via SQL*Net from client	Total number of bytes received from the client over Oracle Net.
SQL*Net roundtrips to/from client	Total number of Oracle Net messages sent to and received from the client.
Sorts (memory)	Number of sort operations that were performed completely in memory and did not require any disk writes.
Sorts (disk)	Number of sort operations that required at least one disk write.
Rows processed	Number of rows processed during the operation.

FIGURE 15-9 Statistic definitions

The EXPLAIN PLAN FOR command produces an explain plan along with a list of conditions that are used to perform the query and any appropriate warnings. Figure 15-10 shows the output of the EXPLAIN PLAN FOR command. Notice the warning at the bottom concerning the use of dynamic sampling. The tables used in the query were just created and no statistics existed for the tables. Therefore, sampling had to be performed.

Let's modify the query and examine the impact on the explain plan. Figure 15-11 displays the explain plan for the query with an additional condition in the WHERE clause. Also, the tables are analyzed to provide statistics for the optimizer. Notice that because fewer rows will now be processed by the query, the optimizer has decided to use table indexes to access the data and merge the data rows with a sort merge operation. In addition, the dynamic sampling warning is no longer displayed.

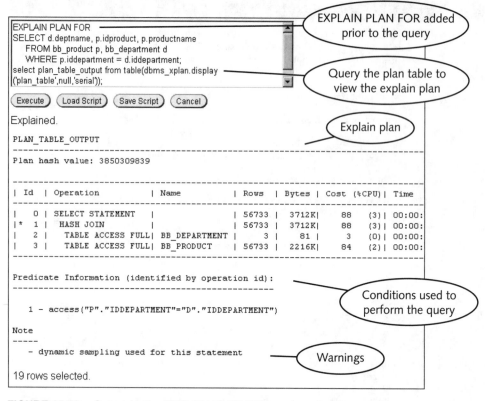

FIGURE 15-10   Output for the EXPLAIN PLAN FOR command

The explain plan cost estimates can be used to determine if a statement change has resulted in a more efficient performance. However, another useful measure is the execution time. The next section introduces the TIMING feature that enables the display of statement execution time.

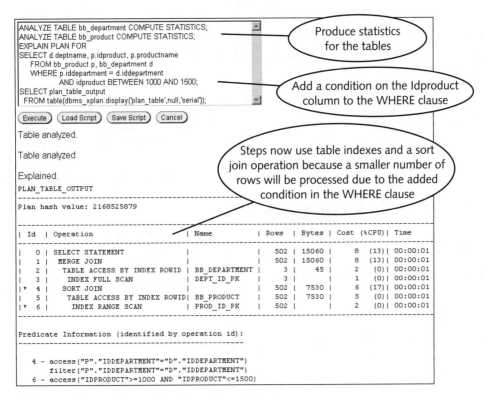

```
ANALYZE TABLE bb_department COMPUTE STATISTICS;
ANALYZE TABLE bb_product COMPUTE STATISTICS;
EXPLAIN PLAN FOR
SELECT d.deptname, p.idproduct, p.productname
    FROM bb_product p, bb_department d
    WHERE p.iddepartment = d.iddepartment
            AND idproduct BETWEEN 1000 AND 1500;
SELECT plan_table_output
  FROM table(dbms_xplan.display('plan_table',null,'serial'));
```

(Execute)   (Load Script)   (Save Script)   (Cancel)

Produce statistics for the tables

Add a condition on the Idproduct column to the WHERE clause

Table analyzed.

Table analyzed.

Steps now use table indexes and a sort join operation because a smaller number of rows will be processed due to the added condition in the WHERE clause

Explained.

PLAN_TABLE_OUTPUT

Plan hash value: 2168525879

| Id | Operation | Name | Rows | Bytes | Cost (%CPU)| Time |

0	SELECT STATEMENT		502	15060	8 (13)	00:00:01
1	MERGE JOIN		502	15060	8 (13)	00:00:01
2	TABLE ACCESS BY INDEX ROWID	BB_DEPARTMENT	3	45	2 (0)	00:00:01
3	INDEX FULL SCAN	DEPT_ID_PK	3		1 (0)	00:00:01
* 4	SORT JOIN		502	7530	6 (17)	00:00:01
5	TABLE ACCESS BY INDEX ROWID	BB_PRODUCT	502	7530	5 (0)	00:00:01
* 6	INDEX RANGE SCAN	PROD_ID_PK	502		2 (0)	00:00:01

Predicate Information (identified by operation id):
----
   4 - access("P"."IDDEPARTMENT"="D"."IDDEPARTMENT")
       filter("P"."IDDEPARTMENT"="D"."IDDEPARTMENT")
   6 - access("IDPRODUCT">=1000 AND "IDPRODUCT"<=1500)
```

**FIGURE 15-11**   Examining changes in the explain plan

## Timing Feature

Before we jump into some SQL statement tuning examples, it is worth looking at another basic tuning tool that can be used in SQL*Plus. The timing feature allows a developer to measure execution time of a statement. Issue the SET TIMING ON command in SQL*Plus to display the timing information following each statement. The timing information is displayed in terms of hours, minutes, and seconds. Figure 15-12 displays three executions of the same query with the timing feature on.
```

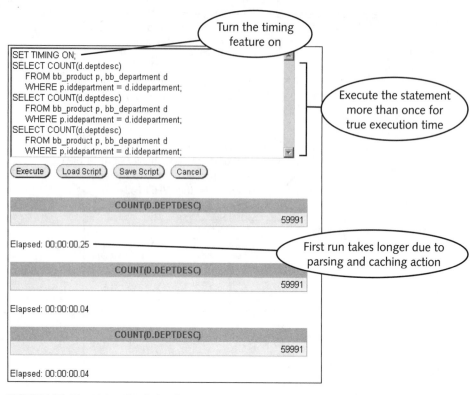

FIGURE 15-12   Using the timing feature

In this query, the first execution shows a time of .25 seconds. The second and third queries display an execution time of .04 seconds. Note that the execution time is dependent on the computer system being used. Why the difference? The first run of a query is cached into memory of the Oracle server, and, therefore, successive runs of the same query can skip the parsing, creating the execution plan, loading into the SQL area, and storing the result set in memory. Before modifying statements in a tuning effort, make sure to use the timing from a second run of the statement as your base time to improve. Otherwise, you are led to believe you improved performance when, in fact, your modification might not have done this at all.

SELECTED SQL TUNING GUIDELINES AND EXAMPLES

In this section, we look at some specific tuning examples concentrating on the explain plan and identifying how statement modifications affect the plan and potentially affect the performance. It is important to try out statements on a test database that resembles the production database in design and size. The database tables used in the following examples are rather small, and most statements are fairly simple. This is done purposely to clearly demonstrate the explain plan and to show the specific effects on them.

Avoiding Unnecessary Column Selection

Including columns that are not actually needed in a select list can have quite a detrimental effect on performance. Let's look at the explain plan for a query on the BB_PRODUCT table. The AUTOTRACE and TIMING commands are issued prior to the example statements to display the explain plan and execution time. Figure 15-13 displays a query on the BB_PRODUCT table including three columns.

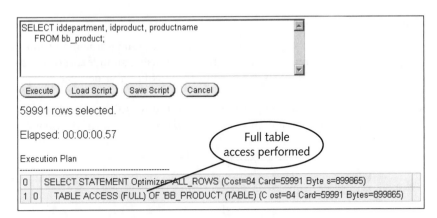

FIGURE 15-13   Explain plan and execution time for query

Notice that a full table scan is performed on the BB_PRODUCT table, and the query execution time is .57 seconds. However, what if only the Idproduct column was actually needed from the query? Could this affect performance? Let's run the same statement, this time selecting only the Idproduct column, as shown in Figure 15-14.

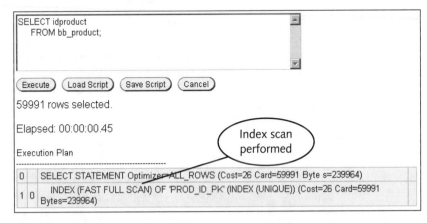

FIGURE 15-14   Explain plan and execution time for modified query

Now, the only step that needs to be processed is an index scan. A table scan is not necessary because the index can return the IDPRODUCT value. Keep in mind that an index contains the value of the column(s) indexed and the ROWID for the associated row. The ROWID is a physical address for a table row and provides the fastest method for retrieving rows. The execution time was reduced to .45 seconds. This might not seem like a big difference, but you can imagine that with larger tables and more complex queries, the increase in performance can be significant.

Index Suppression

Because indexes are one of the central topics within performance tuning, it is important to recognize when SQL statements suppress the optimizer from using an index. Index suppression can occur when a WHERE clause uses a function on a column or compares different datatype values.

Figure 15-15 displays a query that includes a condition in the WHERE clause. No functions are used in the WHERE condition, and an index scan is used to perform the query.

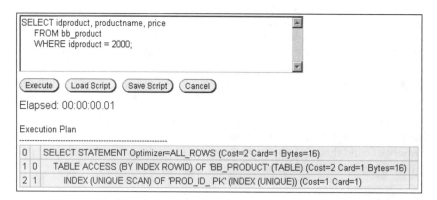

FIGURE 15-15 No function in the WHERE clause—index scan used

Let's change the WHERE clause to add a function so that the query will return only the products with an id starting with 200. Figure 15-16 displays the query with a modified WHERE clause, using the SUBSTR function on the Idproduct column. Notice that an index scan is no longer used and the execution time is increased.

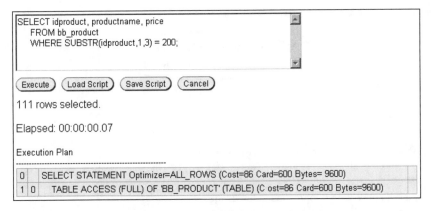

```
SELECT idproduct, productname, price
    FROM bb_product
    WHERE SUBSTR(idproduct,1,3) = 200;
```

Execute   Load Script   Save Script   Cancel

111 rows selected.

Elapsed: 00:00:00.07

Execution Plan

| 0 | | SELECT STATEMENT Optimizer=ALL_ROWS (Cost=86 Card=600 Bytes= 9600) |
| 1 | 0 | TABLE ACCESS (FULL) OF 'BB_PRODUCT' (TABLE) (C ost=86 Card=600 Bytes=9600) |

FIGURE 15-16   Function used in the WHERE clause—index scan suppressed

How can this query be accomplished without using the SUBSTR function? In this scenario, the BETWEEN operator could be used to perform the same operation in the query. Figure 15-17 displays the query using the BETWEEN operator in the WHERE clause. Notice that an index scan is used and execution time is improved.

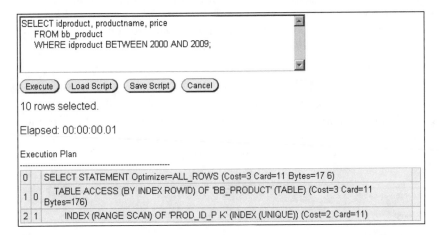

```
SELECT idproduct, productname, price
    FROM bb_product
    WHERE idproduct BETWEEN 2000 AND 2009;
```

Execute   Load Script   Save Script   Cancel

10 rows selected.

Elapsed: 00:00:00.01

Execution Plan

0		SELECT STATEMENT Optimizer=ALL_ROWS (Cost=3 Card=11 Bytes=17 6)
1	0	TABLE ACCESS (BY INDEX ROWID) OF 'BB_PRODUCT' (TABLE) (Cost=3 Card=11 Bytes=176)
2	1	INDEX (RANGE SCAN) OF 'PROD_ID_P K' (INDEX (UNIQUE)) (Cost=2 Card=11)

FIGURE 15-17   Between operator used in the WHERE clause—index scan used

Comparison of different datatypes can also cause index suppression in statement execution. If the Character column is compared to a numeric value, the index usage is suppressed. In this case, the column value is internally converted to a numeric value and, therefore, the optimizer considers this activity the same as when we use a function on the column value. Figure 15-18 demonstrates this index suppression. The Idprod2 column is an exact copy of the Idproduct column, but it is a Character column.

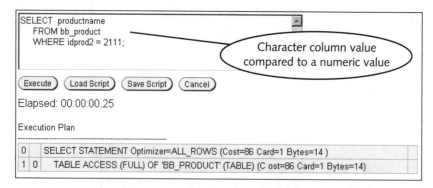

FIGURE 15-18   Implicit conversion suppresses index usage

Concatenated Indexes

At times, we create concatenated indexes or indexes that involve more than one column. The optimizer uses these indexes only when the leading column indexed is included in the criteria of the SQL statement. For example, a concatenated index named BB_SHOPNAME_IDX that indexes the Lastname and Firstname columns of the BB_SHOPPER table would be used only if the query uses the Lastname column in the criteria and if the index is created with the Lastname column first.

Let's review a few queries to see how the index is used or suppressed. First, the concatenated index is created using the following statement:

```
CREATE INDEX bb_shopname_idx
  ON bb_shopper (lastname, firstname);
```

Figure 15-19 displays a query on the BB_SHOPPER table which uses the Lastname column in the criteria. Notice that an index scan is accomplished as the first column of the index is used in the WHERE clause condition (the index would also be used if both the Lastname and Firstname columns were included in the criteria).

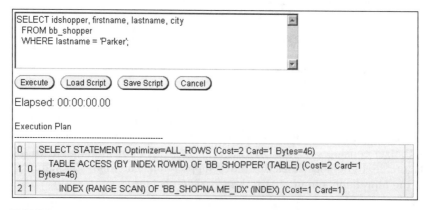

FIGURE 15-19   The concatenated index used as the criteria includes the leading column

Now, we will modify the WHERE clause so that only the Firstname is used in the criteria. Figure 15-20 displays the modified query and shows that the index is no longer used. Because the leading column of the concatenated index, Lastname, is not included in the query criteria, it cannot be used.

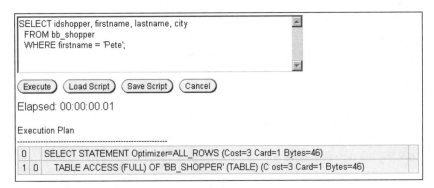

FIGURE 15-20 Concatenated index not used

Subqueries

Another area worth exploring is the type of subqueries used. Correlated subqueries are typically considered more efficient; however, this is not always the case. If the selective predicate is in the subquery, an IN operator is typically most efficient. If the selective predicate is in the parent query, the EXISTS operator (correlated subquery) is typically more efficient. When a subquery is noncorrelated, the inner query executes first and returns a result set that is treated like an IN list. Then, the outer query is executed and compared to the result set. In a correlated subquery, the outer query executes first and then the inner query executes for each record returned from the outer query. Also, using the EXISTS operator terminates the inner query of a subquery when a match is found, whereas the IN operator continues until fetching all the rows in the inner query. Figure 15-21 displays two queries and identifies the appropriate use of noncorrelated and correlated subqueries.

Optimizer Hints

Based on knowledge of the data and review of the execution plan, we might determine the need to alter the execution plan to improve performance. We can alter the execution plan of SELECT, INSERT, UPDATE, and DELETE statements by using hints within the statement. Hints are included in statements as comments including a plus sign.

Numerous hints are available and are grouped in the following categories:

- The optimization approach for a SQL statement
- The goal of the cost-based optimizer for a SQL statement
- The access path for a table accessed by the statement
- The join order for a join statement
- A join operation in a join statement

```
SELECT productname, price
    FROM bb_product
    WHERE productname IN
        (SELECT productname
            FROM bb_prod_list
            WHERE price = 10);
```

Selective predicate in the subquery; use the IN operator

```
SELECT productname, price
    FROM bb_product p
    WHERE price = 10
    AND EXISTS
        (SELECT 'x'
            FROM bb_prod_list
            WHERE productname = p.productname);
```

Selective predicate in the parent query; use the EXISTS operator

FIGURE 15-21   Selecting the type of subquery

One hint you can see in the access path category is FULL, which forces a full table scan, regardless of whether an index exists that could be used. The following statement shows the FULL hint included in a SELECT statement:

```
SELECT /*+FULL(b)*/ idshopper, firstname, lastname
 FROM bb_shopper b
 WHERE lastname = 'Parker';
```

As seen in the example, hints are embedded into statements using comments. Notice that the opening comment symbol, forward slash asterisk (/*), must be immediately followed by a plus sign (+). Also notice that the table name must be indicated in the hint. Or, if a table alias is included in the statement, the alias must be referenced in the hint. All hints must be included in the statement immediately following the keyword of SELECT, INSERT, UPDATE, or DELETE. If a hint is incorrectly added to a statement, it typically does not produce an error, but it is ignored. Therefore, to determine not only if an improvement resulted, but also to confirm whether the hint is being used, as always, it is important to review the execution plan after adding a hint.

This section provides only a cursory introduction to hints; the main idea is to recognize that you can control the optimizer by using hints. Use hints to modify execution plans of statements and compare the execution efficiency of each plan. This is an excellent way to become more familiar with the optimizer and to determine which processing modifications improve efficiency. Read the *Oracle9i Database Performance Tuning Guide and Reference* on the OTN Web site for in-depth coverage of this topic.

TIP

Multiple hints can be used in a single statement. This is accomplished by listing all the hints, each separated by a space, within a single comment area.

As we embed SQL in applications, we must also consider the target database to be used and the potential for the application to move or be used with different databases. The next section introduces the differences in SQL among several relational databases.

USING SQL WITH DIFFERENT DATABASES

ANSI standards have provided a common ground on which all database vendors build to include SQL capabilities in their products. Having common SQL standards makes application code portable and allows the application to be used with different databases with few or no changes. This capability could be critical to developing an application product to market to organizations that may have different databases. In addition, your current department or company might later switch database vendors due to cost or capabilities sought.

On the other hand, it is very challenging to develop SQL code that is 100% portable. The ANSI standards do not cover every area of SQL features, and many vendors add SQL extensions to enhance capabilities and differentiate their product from others. Therefore, every developer needs to be aware of the nuances of SQL among various database products. The most significant differences are found in the SQL functions area. The listing below provides a sampling of specific SQL differences between Oracle, MySQL, and Microsoft SQL Server.

> **NOTE**
>
> The database versions used in all the following SQL comparisons are Oracle 10*g*, MySQL 4.1, and SQL Server 2000.

Unduplicate Rows

ORACLE	MYSQL	MICROSOFT SQL SERVER
DISTINCT or **UNIQUE**	**DISTINCT**	**DISTINCT**
----------------------------------	------------------------------------	------------------------------------
SELECT UNIQUE category FROM books;	SELECT DISTINCT category FROM books;	SELECT DISTINCT category FROM books;

Locate Value Within String

ORACLE	MYSQL	MICROSOFT SQL SERVER
INSTR INSTR(value, expression)	**LOCATE** LOCATE(expression, value, start)	**CHARINDEX** CHARINDEX(expression, value, start)
----------------------------------	------------------------------------	------------------------------------
SELECT INSTR(title, 'SQL') FROM books;	SELECT LOCATE('SQL',title,1) FROM books;	SELECT CHARINDEX('SQL',title,1) FROM books;

Current Date

ORACLE	MYSQL	MICROSOFT SQL SERVER
SYSDATE	NOW()	**GETDATE()**
SELECT SYSDATE FROM DUAL;	SELECT NOW() FROM DUAL;	SELECT GETDATE();

Default Date Format

ORACLE	MYSQL	MICROSOFT SQL SERVER
26-SEP-06	2006-09-26	09-26-2006
SELECT * FROM orders WHERE orderdate= '26-SEP-06';	SELECT * FROM orders WHERE orderdate= '2006-09-26';	SELECT * FROM orders WHERE orderdate= '09-26-2006';

NULL Value Replacement for Text Data

ORACLE	MYSQL	MICROSOFT SQL SERVER
NVL2 NVL2(value, if NULL, if not NULL)	**IFNULL** IFNULL(value, if NULL)	**ISNULL** ISNULL(value, if NULL)
SELECT NVL2(referred, to_ char(referred),'Not referred') FROM customers;	SELECT IFNULL(referred, 'Not referred') FROM customers;	SELECT ISNULL(referred, 'Not referred') FROM customers;

Add Time to Dates

ORACLE	MYSQL	MICROSOFT SQL SERVER
ADD_MONTHS ADD_MONTHS(date, # months) ---------------------------------- SELECT ADD_ MONTHS(SYSDATE, 12) FROM DUAL;	**ADDDATE** ADDDATE(date, interval) ---------------------------------- SELECT ADDDATE(NOW(), Interval 1 Year) FROM DUAL;	**DATEADD** DATEADD(interval, #, date) ---------------------------------- SELECT DATEADD(YYYY, 1, GETDATE());

NOTE

Example statement adds one year to the current date.

Extracting Values from a String

ORACLE	MYSQL	MICROSOFT SQL SERVER
SUBSTR or SUBSTRING SUBSTR(value, start, length) ---------------------------------- SELECT SUBSTR(zip, 1, 3) FROM customers;	**SUBSTR or SUBSTRING** SUBSTR(value, start, length) ---------------------------------- SELECT SUBSTR(zip, 1, 3) FROM customers;	**SUBSTRING** SUBSTRING(value, start, length) ---------------------------------- SELECT SUBSTRING(zip, 1, 3) FROM customers;

NOTE

SQL Server can use only SUBSTRING, not the abbreviated name SUBSTR.

Concatenation

ORACLE	MYSQL	MICROSOFT SQL SERVER
|| ---------------------------------- SELECT lastname|| ', '|| firstname FROM customers;	**CONCAT** CONCAT(value, value, ...) ---------------------------------- SELECT CONCAT(lastname, ', ', firstname) FROM customers;	+ ---------------------------------- SELECT lastname + ', ' + firstname FROM customers;

NOTE

Oracle also has a CONCAT function, but it is limited to two arguments. MySQL in ANSI mode treats the \|\| symbol as concatenation rather than an OR.

Data Structures

In addition to functions, a number of differences exist among databases with regard to data structures. Differences in the column datatypes and constraints must also be recognized. For example, Oracle has a datatype named NUMBER; however, MySQL and SQL Server do not use this name for any of their numeric datatypes. Another example is the Oracle datatype of DATE, which is equivalent to DATETIME in MySQL and in SQL Server. In terms of constraints, MySQL currently has no check constraint capabilities that are provided by the other databases.

NOTE

Check constraint capabilities are planned for future versions of MySQL.

Chapter Summary

This chapter was provided to bridge the SQL topics covered in this text to the complex topic of using SQL within the world of database development. First, some examples of embedding SQL within application coding were presented to highlight how SQL can be used within many development environments to handle database interaction tasks. Second, a discussion on performance tuning underscored the need to create code that executes efficiently. Some of the methods used to investigate performance issues were also presented. Third, even though SQL is based on the ANSI standard, variations exist among the available relational databases with regard to SQL command syntax. This chapter highlighted some of the basic SQL differences for three widely used relational databases.

Because this chapter was mainly conceptual, end of chapter questions are not included.

TABLES FOR THE JUSTLEE BOOKS DATABASE

The tables created by the execution of the **Bookscript.SQL** file include CUSTOMERS, BOOKS, ORDERS, ORDERITEMS, AUTHOR, BOOKAUTHOR, PUBLISHER, and PROMOTION. The structure and contents for each table are provided in this appendix.

CUSTOMERS Table

Data about JustLee Books' customers is stored in the CUSTOMERS table. The structure of the CUSTOMERS table is shown in Figure A-1. The table's contents are shown in Figure A-2.

```
Name                                       Null?     Type
---------------------------------------    --------  -------------------------
CUSTOMER#                                  NOT NULL  NUMBER(4)
LASTNAME                                             VARCHAR2(10)
FIRSTNAME                                            VARCHAR2(10)
ADDRESS                                              VARCHAR2(20)
CITY                                                 VARCHAR2(12)
STATE                                               VARCHAR2(2)
ZIP                                                 VARCHAR2(5)
REFERRED                                            NUMBER(4)
```

FIGURE A-1   Structure of the CUSTOMERS table

```
CUSTOMER# LASTNAME   FIRSTNAME  ADDRESS                 CITY          ST ZIP    REFERRED
--------- ---------- ---------- ----------------------- ------------- -- -----  ----------
     1001 MORALES    BONITA     P.O. BOX 651            EASTPOINT     FL 32328
     1002 THOMPSON   RYAN       P.O. BOX 9835           SANTA MONICA  CA 90404
     1003 SMITH      LEILA      P.O. BOX 66             TALLAHASSEE   FL 32306
     1004 PIERSON    THOMAS     69821 SOUTH AVENUE      BOISE         ID 83707
     1005 GIRARD     CINDY      P.O. BOX 851            SEATTLE       WA 98115
     1006 CRUZ       MESHIA     82 DIRT ROAD            ALBANY        NY 12211
     1007 GIANA      TAMMY      9153 MAIN STREET        AUSTIN        TX 78710      1003
     1008 JONES      KENNETH    P.O. BOX 137            CHEYENNE      WY 82003
     1009 PEREZ      JORGE      P.O. BOX 8564           BURBANK       CA 91510      1003
     1010 LUCAS      JAKE       114 EAST SAVANNAH       ATLANTA       GA 30314
     1011 MCGOVERN   REESE      P.O. BOX 18             CHICAGO       IL 60606
     1012 MCKENZIE   WILLIAM    P.O. BOX 971            BOSTON        MA 02110
     1013 NGUYEN     NICHOLAS   357 WHITE EAGLE AVE.    CLERMONT      FL 34711      1006
     1014 LEE        JASMINE    P.O. BOX 2947           CODY          WY 82414
     1015 SCHELL     STEVE      P.O. BOX 677            MIAMI         FL 33111
     1016 DAUM       MICHELL    9851231 LONG ROAD       BURBANK       CA 91508      1010
     1017 NELSON     BECCA      P.O. BOX 563            KALMAZOO      MI 49006
     1018 MONTIASA   GREG       1008 GRAND AVENUE       MACON         GA 31206
     1019 SMITH      JENNIFER   P.O. BOX 1151           MORRISTOWN    NJ 07962      1003
     1020 FALAH      KENNETH    P.O. BOX 335            TRENTON       NJ 08607
```

FIGURE A-2   Data contained in the CUSTOMERS table

BOOKS Table

Data about the books sold by JustLee Books is stored in the BOOKS table. The structure of the BOOKS table is shown in Figure A-3. The table's contents are shown in Figure A-4.

```
Name                                              Null?     Type
-----------------------------------------------   --------  --------------------
ISBN                                              NOT NULL  VARCHAR2(10)
TITLE                                                       VARCHAR2(30)
PUBDATE                                                     DATE
PUBID                                                       NUMBER(2)
COST                                                        NUMBER(5,2)
RETAIL                                                      NUMBER(5,2)
CATEGORY                                                    VARCHAR2(12)
```

FIGURE A-3   Structure of the BOOKS table

```
ISBN         TITLE                              PUBDATE     PUBID      COST    RETAIL CATEGORY
----------   -------------------------------    ---------   --------  -------- ------- -----------
1059831198   BODYBUILD IN 10 MINUTES A DAY      21-JAN-01       4      18.75    30.95 FITNESS
0401140733   REVENGE OF MICKEY                  14-DEC-01       1      14.2        22 FAMILY LIFE
4981341710   BUILDING A CAR WITH TOOTHPICKS     18-MAR-02       2      37.8     59.95 CHILDREN
8843172113   DATABASE IMPLEMENTATION            04-JUN-99       3      31.4     55.95 COMPUTER
3437212490   COOKING WITH MUSHROOMS             28-FEB-00       4      12.5     19.95 COOKING
3957136468   HOLY GRAIL OF ORACLE               31-DEC-01       3      47.25    75.95 COMPUTER
1915762492   HANDCRANKED COMPUTERS              21-JAN-01       3      21.8        25 COMPUTER
9959789321   E-BUSINESS THE EASY WAY            01-MAR-02       2      37.9     54.5 COMPUTER
2491748320   PAINLESS CHILD-REARING             17-JUL-00       5        48     89.95 FAMILY LIFE
0299282519   THE WOK WAY TO COOK                11-SEP-00       4        19     28.75 COOKING
8117949391   BIG BEAR AND LITTLE DOVE           08-NOV-01       5      5.32      8.95 CHILDREN
0132149871   HOW TO GET FASTER PIZZA            11-NOV-02       4      17.85    29.95 SELF HELP
9247381001   HOW TO MANAGE THE MANAGER          09-MAY-99       1      15.4     31.95 BUSINESS
2147428890   SHORTEST POEMS                     01-MAY-01       5      21.85    39.95 LITERATURE
```

FIGURE A-4   Data contained in the BOOKS table

ORDERS Table

Data about recent orders received by JustLee Books is stored in the ORDERS table. The structure of the ORDERS table is shown in Figure A-5. The table's contents are shown in Figure A-6.

```
Name                                              Null?     Type
-----------------------------------------------   --------  --------------------
ORDER#                                            NOT NULL  NUMBER(4)
CUSTOMER#                                                   NUMBER(4)
ORDERDATE                                                   DATE
SHIPDATE                                                    DATE
SHIPSTREET                                                  VARCHAR2(18)
SHIPCITY                                                    VARCHAR2(15)
SHIPSTATE                                                   VARCHAR2(2)
SHIPZIP                                                     VARCHAR2(5)
```

FIGURE A-5   Structure of the ORDERS table

```
    ORDER#  CUSTOMER# ORDERDATE  SHIPDATE    SHIPSTREET            SHIPCITY         SHIPSTATE SHIPZIP
---------- ---------- ---------- ---------- -------------------- ---------------- --------- -------
      1000       1005 31-MAR-05  02-APR-05  1201 ORANGE AVE      SEATTLE          WA        98114
      1001       1010 31-MAR-05  01-APR-05  114 EAST SAVANNAH    ATLANTA          GA        30314
      1002       1011 31-MAR-05  01-APR-05  58 TILA CIRCLE       CHICAGO          IL        60605
      1003       1001 01-APR-05  01-APR-05  958 MAGNOLIA LANE    EASTPOINT        FL        32328
      1004       1020 01-APR-05  05-APR-05  561 ROUNDABOUT WAY   TRENTON          NJ        08601
      1005       1018 01-APR-05  02-APR-05  1008 GRAND AVENUE    MACON            GA        31206
      1006       1003 01-APR-05  02-APR-05  558A CAPITOL HWY.    TALLAHASSEE      FL        32307
      1007       1007 02-APR-05  04-APR-05  9153 MAIN STREET     AUSTIN           TX        78710
      1008       1004 02-APR-05  03-APR-05  69821 SOUTH AVENUE   BOISE            ID        83707
      1009       1005 03-APR-05  05-APR-05  9 LIGHTENING RD.     SEATTLE          WA        98110
      1010       1019 03-APR-05  04-APR-05  384 WRONG WAY HOME   MORRISTOWN       NJ        07960
      1011       1010 03-APR-05  05-APR-05  102 WEST LAFAYETTE   ATLANTA          GA        30311
      1012       1017 03-APR-05             1295 WINDY AVENUE    KALMAZOO         MI        49002
      1013       1014 03-APR-05  04-APR-05  7618 MOUNTAIN RD.    CODY             WY        82414
      1014       1007 04-APR-05  05-APR-05  9153 MAIN STREET     AUSTIN           TX        78710
      1015       1020 04-APR-05             557 GLITTER ST.      TRENTON          NJ        08606
      1016       1003 04-APR-05             9901 SEMINOLE WAY    TALLAHASSEE      FL        32307
      1017       1015 04-APR-05  05-APR-05  887 HOT ASPHALT ST   MIAMI            FL        33112
      1018       1001 05-APR-05             95812 HIGHWAY 98     EASTPOINT        FL        32328
      1019       1018 05-APR-05             1008 GRAND AVENUE    MACON            GA        31206
      1020       1008 05-APR-05             195 JAMISON LANE     CHEYENNE         WY        82003
```

FIGURE A-6   Data contained in the ORDERS table

ORDERITEMS Table

Because an order can consist of more than one book, the ORDERITEMS table is used to store information about the individual books purchased on each order. It bridges the ORDERS and BOOKS tables. The structure of the ORDERITEMS table is shown in Figure A-7. The table's contents are shown in Figure A-8.

```
Name                                              Null?    Type
------------------------------------------------- -------- -------------------------------
ORDER#                                            NOT NULL NUMBER(4)
ITEM#                                             NOT NULL NUMBER(2)
ISBN                                                       VARCHAR2(10)
QUANTITY                                                   NUMBER(3)
```

FIGURE A-7   Structure of the ORDERITEMS table

```
    ORDER#     ITEM# ISBN        QUANTITY
---------- ---------- --------- ----------
      1000         1 3437212490          1
      1001         1 9247381001          1
      1001         2 2491748320          1
      1002         1 8843172113          2
      1003         1 8843172113          1
      1003         2 1059831198          1
      1003         3 3437212490          1
      1004         1 2491748320          2
      1005         1 2147428890          1
      1006         1 9959789321          1
      1007         1 395713648           3
      1007         2 9959789321          1
      1007         3 8117949391          1
      1007         4 8843172113          1
      1008         1 3437212490          2
      1009         1 3437212490          1
      1009         2 0401140733          1
      1010         1 8843172113          1
      1011         1 2491748320          1
      1012         1 8117949391          1
      1012         2 1915762492          2
      1012         3 2491748320          1
      1012         4 0401140733          1
      1013         1 8843172113          1
      1014         1 0401140733          2
      1015         1 3437212490          1
      1016         1 2491748320          1
      1017         1 8117949391          2
      1018         1 3437212490          1
      1018         2 8843172113          1
      1019         1 0401140733          1
      1020         1 3437212490          1
```

FIGURE A-8   Data contained in the ORDERITEMS table

AUTHOR Table

The names of book authors are stored in the AUTHOR table. The structure of the AUTHOR table is shown in Figure A-9. The table's contents are shown in Figure A-10.

```
Name                                               Null?    Type
------------------------------------------------   -------- -----------------------------------
AUTHORID                                           NOT NULL VARCHAR2(4)
LNAME                                                       VARCHAR2(10)
FNAME                                                       VARCHAR2(10)
```

FIGURE A-9   Structure of the AUTHOR table

```
AUTH LNAME       FNAME
---- ----------  ----------
S100 SMITH       SAM
J100 JONES       JANICE
A100 AUSTIN      JAMES
M100 MARTINEZ    SHEILA
K100 KZOCHSKY    TAMARA
P100 PORTER      LISA
A105 ADAMS       JUAN
B100 BAKER       JACK
P105 PETERSON    TINA
W100 WHITE       WILLIAM
W105 WHITE       LISA
R100 ROBINSON    ROBERT
F100 FIELDS      OSCAR
W110 WILKINSON   ANTHONY
```

FIGURE A-10   Data contained in the AUTHOR table

BOOKAUTHOR Table

Because a book could have more than one author, and an author could have written more than one book, the BOOKAUTHOR table bridges the BOOKS and AUTHOR tables. The structure of the BOOKAUTHOR table is shown in Figure A-11. The table's contents are shown in Figure A-12.

```
Name                                            Null?     Type
----------------------------------------------- --------  -----------------------------------
ISBN                                            NOT NULL  VARCHAR2(10)
AUTHORID                                        NOT NULL  VARCHAR2(4)
```

FIGURE A-11   Structure of the BOOKAUTHOR table

```
ISBN       AUTH
---------- ----
1059831198 S100
1059831198 P100
0401140733 J100
4981341710 K100
8843172113 P105
8843172113 A100
8843172113 A105
3437212490 B100
3957136468 A100
1915762492 W100
1915762492 W105
9959789321 J100
2491748320 R100
2491748320 F100
2491748320 B100
0299282519 S100
8117949391 R100
0132149871 S100
9247381001 W100
2147428890 W105
```

FIGURE A-12   Data contained in the BOOKAUTHOR table

PUBLISHER Table

All data about book publishers is stored in the PUBLISHER table. The structure of the PUBLISHER table is shown in Figure A-13. The table's contents are shown in Figure A-14.

```
Name                                               Null?     Type
-------------------------------------------------- --------  -------------------------
PUBID                                              NOT NULL  NUMBER(2)
NAME                                                         VARCHAR2(23)
CONTACT                                                      VARCHAR2(15)
PHONE                                                        VARCHAR2(12)
```

FIGURE A-13   Structure of the PUBLISHER table

```
PUBID NAME                    CONTACT          PHONE
----- ----------------------- ---------------- ------------
    1 PRINTING IS US          TOMMIE SEYMOUR   000-714-8321
    2 PUBLISH OUR WAY         JANE TOMLIN      010-410-0010
    3 AMERICAN PUBLISHING     DAVID DAVIDSON   800-555-1211
    4 READING MATERIALS INC.  RENEE SMITH      800-555-9743
    5 REED-N-RITE             SEBASTIAN JONES  800-555-8284
```

FIGURE A-14   Data contained in the PUBLISHER table

PROMOTION Table

The gifts that are provided to customers during JustLee Books' annual sales promotion are listed in the PROMOTION table, along with the range of retail prices required to obtain each gift. The structure of the PROMOTION table is shown in Figure A-15. The table's contents are shown in Figure A-16.

```
Name                                               Null?     Type
-------------------------------------------------- --------  -------------------------
GIFT                                                         VARCHAR2(15)
MINRETAIL                                                    NUMBER(5,2)
MAXRETAIL                                                    NUMBER(5,2)
```

FIGURE A-15   Structure of the PROMOTION table

```
GIFT              MINRETAIL  MAXRETAIL
----------------  ---------- ----------
BOOKMARKER                0         12
BOOK LABELS          12.01         25
BOOK COVER           25.01         56
FREE SHIPPING        56.01     999.99
```

FIGURE A-16   Data contained in the PUBLISHER table

SQL*PLUS AND *i*SQL*PLUS USER'S GUIDE

INTRODUCTION

SQL*Plus is a basic software tool commonly available with Oracle. This software allows a user to connect

to the database and execute SQL commands and view results. In versions that have been released after

Oracle9*i*, a user can employ either the client based SQL*Plus software or the browser-based SQL*Plus

software (*i*SQL*Plus). This appendix introduces the operation of both the client-based and Internet-based

versions of SQL*Plus software and identifies some of the execution differences that you may encounter

when using the two programs.

CLIENT SQL*PLUS

Because the client SQL*Plus software is command line-oriented, it can seem a bit awkward for those accustomed to graphical user interfaces. This tool is almost always available for an Oracle installation, and so it's worth becoming familiar with it.

Connecting to Oracle

Within the programs listing on your operation system, you should find a selection for SQL Plus under your Oracle entries. When you start the software, you are prompted for login information with the screen displayed in Figure B-1.

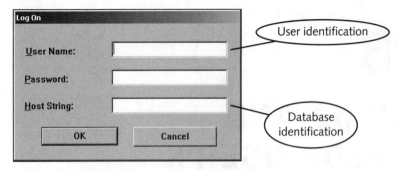

FIGURE B-1   Client SQL*Plus login screen

The login involves three items of information, as depicted in the figure. Once you log in, you connect, and the SQL*Plus screen appears as shown in Figure B-2.

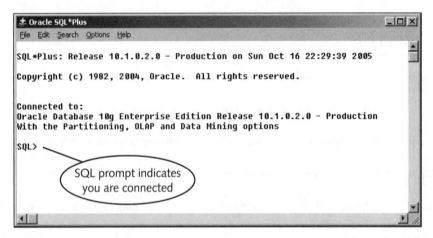

FIGURE B-2   Client SQL*Plus work screen

You can enter SQL statements at the SQL prompt. As you enter each line and press the Enter key, the line number is displayed. When you enter a semicolon and press the Enter key, the statement is executed. An example of an executed statement is shown in Figure B-3.

Unfortunately, if you make an error in your statement, you cannot just press the Backspace key to delete previous lines. There are several approaches to entering statements for modifications. One approach is to enter your statements into a text editor. The default text editor for most Windows operating systems is Notepad. If you wish to use another program (or if no editor is defined), enter DEFINE_EDITOR = NOTEPAD (or the appropriate program name) at the SQL> prompt, and then press the Enter key.

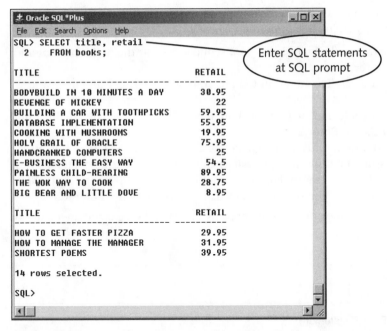

FIGURE B-3 Executing a statement in Client SQL*Plus

NOTE

If you attempt to change the default editor and receive a permissions error message from the operating system, it is possible that you are not allowed to make changes to the system setting. If this happens, you should notify your instructor.

To enter your commands in the editor, use the following steps:

1. To access the editor, type **edit** at the SQL> prompt. Because you did not provide a filename, the contents of the buffer (the SQL statement you last executed) appears in the editor, which is using the filename **afiedt.buf**. This allows you to use any features—such as cut and paste, insert, and delete—that are available in the editor to make changes to your SQL statement.
2. After you close the editor, the SQL statement that was in the editor when you exited is displayed on the SQL*Plus screen.
3. To execute the statement, type a **slash (/)** at the SQL> prompt and press the **Enter** key. The statement will then be executed.

Some users take this a step further and enter all their statements in a text file and then copy and paste from the text file to SQL*Plus.

If you prefer not to use an editor, you can use the SQL*Plus editing commands. To view what is currently in the SQL*Plus buffer, do the following:

1. Type the letter **L**, the word **LIST**, or a semicolon (;) at the SQL> prompt and press the Enter key.
2. To display the last line stored in the buffer, type **LIST LAST** and press the **Enter** key.
3. If you need to delete a line from the buffer, type **DEL** followed by the line number and press the **Enter** key.
4. You can add lines to the stored SQL statement by typing **INPUT** (or the letter I) and pressing the **Enter** key. You will then be able to add the remaining lines of text. To add text to the end of the current line in the buffer, enter **APPEND** (or the letter **A**) and the text to be added, and then press the **Enter** key.

You can experiment with the editing commands by entering the SQL statement shown in Figure B–4.

```
SELECT title, cost
FROM books;
```

FIGURE B-4   Simple SELECT statement

Suppose that after pressing the **Enter** key, you realize that you wanted the books just from Publisher 4. To revise the query you just entered, do the following:

1. Type **LIST** to redisplay the SQL statement you just entered. Notice the asterisk (*) next to line 2. This indicates that the current line is line 2.
2. Since you want to add another line to the SQL statement, type **INPUT** and then press the **Enter** key. After pressing the Enter key, you will be able to enter additional lines to the statement immediately after the current line. Line 2 was the current line when you entered INPUT, so a 3 appeared after you pressed the **Enter** key.
3. Type **WHERE pubid=4;** Press the Enter key.

After looking over the results, you realize you forgot to include the retail price for each book. To include Retail in the display, you need to alter the first line of the SQL statement you entered. Here's how to make that change:

1. If you type **LIST** to review the current statement in the buffer, you should notice that the asterisk is now indicating that line 3 is the current line. If you append the additional column name at this time, it would be appended to line 3.
2. To set line 1 as the current line, type a **1** at the SQL> prompt and press the **Enter** key. Line 1 is redisplayed with an asterisk, indicating that this is now the current line.
3. At the SQL> prompt, type **A , retail** and then press the **Enter** key. This adds a comma and the word retail to the end of line 1. After pressing the Enter key, line 1 is redisplayed with the additional text displayed.

4. Type **L** (or **LIST**) and then press the **Enter** key to see the revised SQL statement.

5. To execute the revised SQL statement, enter a **slash** (**/**) (or type **RUN**), and then press the **Enter** key.

Suppose that after examining the new output, you realize that the ISBN of each book, not its title, is needed. You can use the **CHANGE** command to perform a simple search-and-replace operation to change Title to ISBN in the first line of the SQL statement. The format for the **CHANGE** command is **C\old\new** (the last backslash is optional). Make that change by following these steps:

1. To make the correction to the first line, you need to make line 1 the current line. Type **1** at the SQL> prompt and then press the **Enter** key.

2. Once the first line is displayed, enter **C\title\ISBN** and press the **Enter** key.

3. The revision for line 1 is displayed. If it is correct, run the statement and examine the new output.

You can type EXIT; at the SQL prompt and press the **Return** key to close the SQL*Plus session.

*i*SQL*PLUS

This section covers connecting to iSQL*Plus, loading scripts, as well as features and preferences.

Connecting to *i*SQL*Plus

Connect to your Internet service via a network or ISP. Enter your Oracle HTTP server's URL in the browser address area. This should be in the format:

http://*your_ machine_name.domain*:7777/isqlplus.

The 7777 represents the port number set to listen for connections upon installation of the *i*SQL*Plus server. The Login screen appears as shown in Figure B-5.

Enter your user name and password. The Connection Identifier can be left blank if you are connecting to the default database. Otherwise, you will need the correct database connection string as included in the server's tnsnames.ora file.

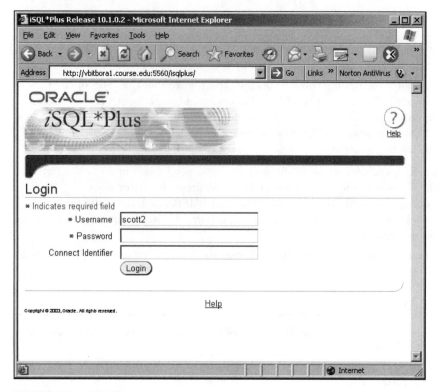

FIGURE B-5 iSQL*Plus login screen

Upon clicking Login, the screen shown in Figure B-6 will be displayed. Notice that you enter SQL statements in the main area, called the workspace area. You can move about freely within the workspace to easily edit statements.

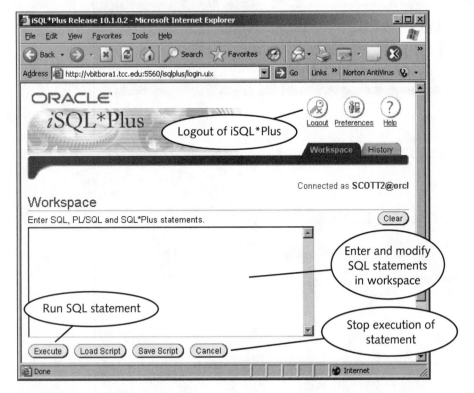

FIGURE B-6 iSQL*Plus work screen

Type a statement such as a SELECT directly into the workspace area, as shown in Figure B-7. Editing here is easier than in the client SQL*Plus tool. If you make a mistake, you can edit the code right in the workspace area using the familiar mechanisms of highlighting, cutting, and pasting, as well as simply clicking where the cursor is needed to add or change text. Click the **Execute** button to execute the statement. Note that the results or error message appear below the workspace area. The statement remains available so that you can easily make modifications and execute the statement again.

If you want to maintain a log file of your session, including statements executed and the results, you can cut and paste items into a text editor such as Notepad or to a word processing tool, such as Word. To put your results in a Word table, highlight the HTML table output and then copy it into Word.

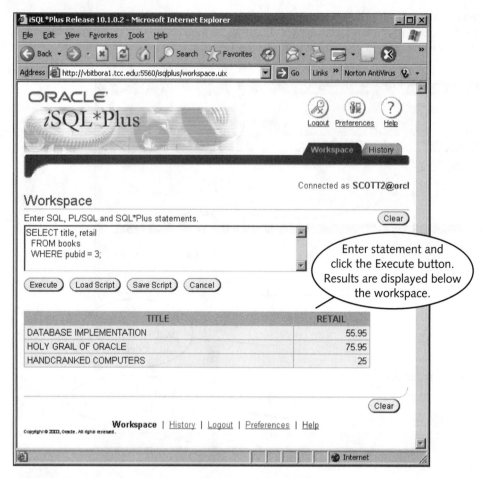

FIGURE B-7 Executing a statement in iSQL*Plus

If you want to save the statement in a script file to rerun at a later time, click the **Save Script** button. This prompts you for a filename and a location in which to save the file. You may want to enter a file extension of .sql to identify the file as an SQL script file.

Loading a Script

If you have a script file containing SQL statements, you can easily bring that file into iSQL*Plus and execute it using the **Load Script** button. Upon clicking the Load Script button, a screen as shown in Figure B-8 is displayed.

You can type a filename and path to load a file, or use the **Browse** button to locate the file. Once the file is identified, click the **Load** button and the contents of the script are brought into the workspace area. You can click the **Execute** button to run or modify the code at this point.

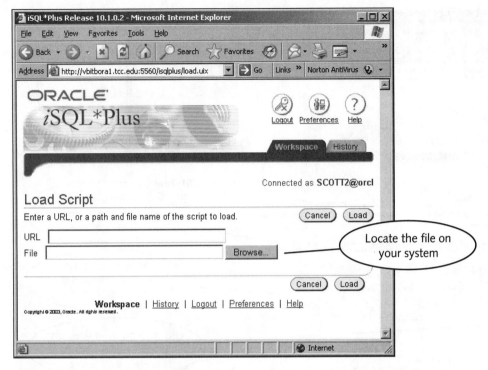

FIGURE B-8 Locate file for loading

History Feature

Clicking the **History** tab displays of a set of the most recent SQL statements you have executed in this session (the default setting saves the most recent ten statements). Figure B-9 shows an example. Notice you can select one or more of the statements and click **Load** to bring those statements back into the workspace area.

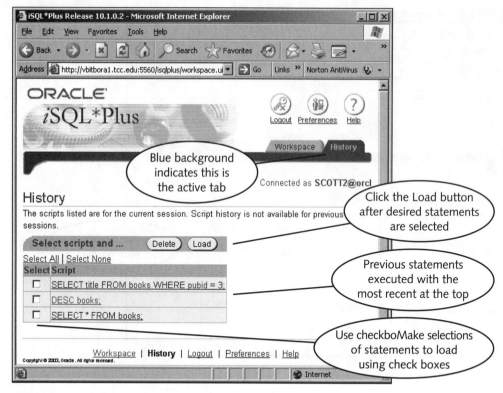

FIGURE B-9 iSQL*Plus History screen

Preferences

Clicking the **Preferences** button displays the screen shown in Figure B-10. This allows the user to set many of the environment settings, such as the size of the workspace area, the location for output, and the number of statements to be stored in history. Notice that one of the options is to reset your password.

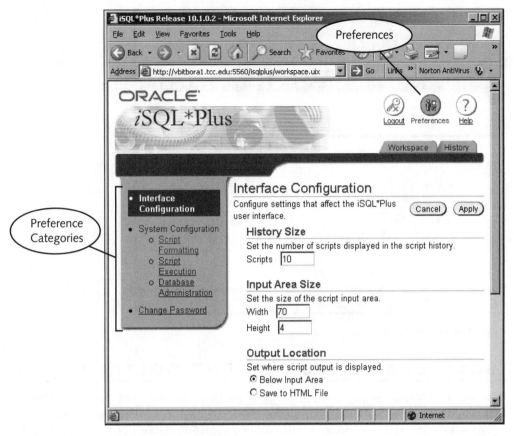

FIGURE B-10   iSQL*Plus preferences screen

Help

The Help screen, as shown in Figure B-11, contains a wealth of information pertaining to the iSQL*Plus environment and command syntax. Notice the various tabs located at the top right. The Help screen appears in a new browser window and does not affect the current session window.

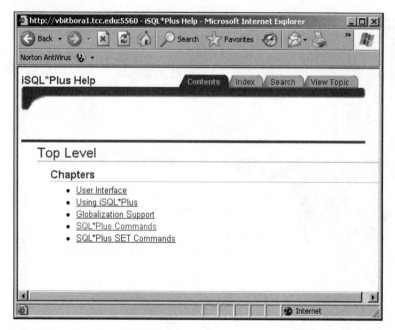

FIGURE B-11   iSQL*Plus help screen

*i*SQL*PLUS VERSUS CLIENT SQL*PLUS

Two specific differences can cause frustration for those familiar with using the client SQL*Plus tool who are switching to using the *i*SQL*Plus tool. This section addresses the differences in using the @ command and output formatting options of line breaks and indenting.

The @ or START Command

The client SQL*Plus tool allows you to execute files located on the client machine by using the @ or START commands. However, using the @ or START command in *i*SQL*Plus specifies a script by the URL from a Web server. This is fine if you have located script files on the Web server, but much of the time, you will want to run a file located on your computer. To do this in *i*SQL*Plus, use the load script feature discussed earlier in this guide.

Text Output Option

Options can be used in SQL and PL/SQL statements to format output with line breaks or indenting. These options work fine in client SQL*Plus because the output is in text format. However, the default output of *i*SQL*Plus is in HTML format, which does not recognize the line break or indenting commands. This appendix presents an example of using the CHR(10) line break option in an SQL command.

Notice that the SELECT statement executed in client SQL*Plus, shown in Figure B-12, processes the line breaks in the statement.

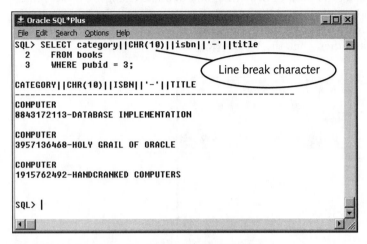

FIGURE B-12   Using line breaks in Client SQL*Plus

However, if you execute the same statement in *i*SQL*Plus, as shown in Figure B-13, notice that the line breaks are ignored.

FIGURE B-13   Default iSQL*Plus output ignores line break command

This can be resolved by modifying the preferences in the *i*SQL*Plus session to choose a text output format rather than the default HTML format. Click the **Preferences** button at the top right and then select the **Script Formatting** link to make changes to your preference settings. Move down the alphabetical list of settings until you find the Preformatted Output area, as shown in Figure B-14. Within this area, set the Preformat option to **On**. Click **Apply** to save the change and then click the **Workspace** tab.

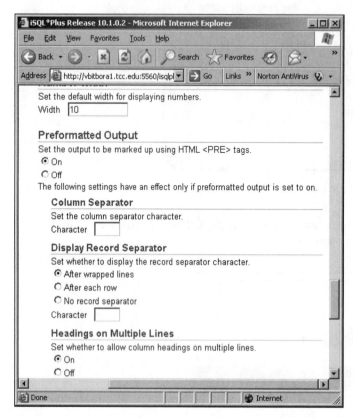

FIGURE B-14 Set Preformatted Output to ON

Now, if you execute the same SQL statement again, as shown in Figure B-15, notice that the output is in text format and the line breaks are processed. If you are cutting and pasting your results to a text editor such as Notepad, this setting is preferable because it is already in text format.

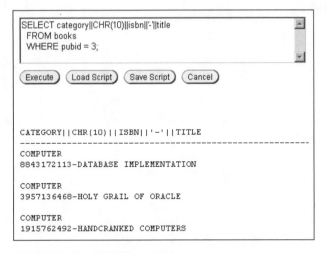

```
SELECT category||CHR(10)||isbn||'-'||title
  FROM books
  WHERE pubid = 3;
```

(Execute) (Load Script) (Save Script) (Cancel)

```
CATEGORY||CHR(10)||ISBN||'-'||TITLE
-----------------------------------------------------------
COMPUTER
8843172113-DATABASE IMPLEMENTATION

COMPUTER
3957136468-HOLY GRAIL OF ORACLE

COMPUTER
1915762492-HANDCRANKED COMPUTERS
```

FIGURE B-15 iSQL*Plus with text output

APPENDIX **C**

SQL SYNTAX GUIDE

SQL COMMANDS		
Command	Syntax	Example
ALTER ROLE	ALTER ROLE *rolename* IDENTIFIED BY *password*;	**ALTER ROLE dataentry** **IDENTIFIED BY apassword;**

ALTER SEQUENCE	ALTER SEQUENCE *sequencename*
	  [INCREMENT BY *value*]
	  [{MAXVALUE *value* |
	NOMAXVALUE}]
	  [{MINVALUE *value* |
	NOMINVALUE}]
	  [{CYCLE | NOCYCLE}]
	  [{ORDER | NOORDER}]
	  [{CACHE *value* | NOCACHE}];
		ALTER SEQUENCE
		orders_ordernumber
		INCREMENT BY 10;

ALTER TABLE... ADD	ALTER TABLE *tablename* ADD (*columnname* datatype);	**ALTER TABLE acctmanager** **ADD (ext NUMBER(4));**
ALTER TABLE... DROP COLUMN	ALTER TABLE *tablename* DROP COLUMN *columnname*;	**ALTER TABLE acctmanager** **DROP COLUMN ext;**
ALTER TABLE... MODIFY	ALTER TABLE *tablename* MODIFY (*columnname* datatype);	**ALTER TABLE acctmanager** **MODIFY (amnameVARCHAR2(25));**
ALTER TABLE... SET UNUSED *or* SET UNUSED COLUMN	ALTER TABLE *tablename* SET UNUSED (*columnname*);	**ALTER TABLE secustomerorders** **SET UNUSED (cost);**
ALTER USER	ALTER USER *username* [IDENTIFIED BY *newpassword*] [PASSWORD EXPIRE];	**ALTER USER rthomas** **IDENTIFIED BY** **monstertruck42;**
ALTER USER... DEFAULT ROLE	ALTER USER *username* DEFAULT ROLE *rolename*;	**ALTER USER rthomasDEFAULT ROLE** **dataentry;**
COMMIT	COMMIT;	**COMMIT;**
CREATE INDEX (B-tree)	CREATE INDEX *indexname* ON *tablename(columnname, ...)*;	**CREATE INDEX** **customers_lastname_idx** **ON** **customers(lastname);**
CREATE INDEX (bitmap)	CREATE BITMAP INDEX *indexname* ON *tablename* *(columnname, ...)*	**CREATE BITMAP INDEX** **customers_region_idx** **ON** **customers (region);**

SQL COMMANDS		
Command	Syntax	Example
CREATE INDEX (Functionbased)	CREATE INDEX indexname ON tablename; (expression);	CREATE INDEX books_profit_idx ON books(retail-cost);
CREATE INDEX (Index Organized Table)	CREATE TABLE tablename(columnname datatype,columnname datatype)ORGANIZATION INDEX;	CREATE TABLE books(isbn NUMBER(10), title VARCHAR2(30)) ORGANIZATION INDEX;
CREATE MATERIALIZED VIEW	Replicates data by physically storing the results of the view query	CREATE MATERIALIZED VIEW custbal_mv REFRESH COMPLETE START WITH SYSDATE NEXT SYSDATE+7 AS SELECT customer#, city, state, order#, SUM(quantity*retail) Amtdue FROM customers JOIN orders USING(customer#) JOIN orderitems USING(order#) JOIN books USING(isbn) GROUP BY customer#, city, state, order#;
CREATE OR REPLACE VIEW	CREATE OR REPLACE [FORCE|NOFORCE] VIEW viewname (columnname, ...)] AS subquery [WITH CHECK OPTION [CONSTRAINT constraintname]] [WITH READ ONLY];	CREATE OR REPLACE VIEW inventory AS SELECT isbn, title, retail price FROM books;
CREATE ROLE	CREATE ROLE rolename;	CREATE ROLE dataentry;
CREATE SEQUENCE	CREATE SEQUENCE sequencename [INCREMENT BY value] [START WITH value] [{MAXVALUE value | NOMAXVALUE}] [{MINVALUE value | NOMINVALUE}] [{CYCLE | NOCYCLE}] [{ORDER | NOORDER}] [{CACHE value | NOCACHE}];	CREATE SEQUENCE orders_ordernumber INCREMENT BY 1 START WITH 1021 NOCYCLE NOCACHE;
CREATE SYNONYM	CREATE [PUBLIC] SYNONYM synonymname FOR objectname;	CREATE PUBLIC SYNONYM orderentry FOR orders;

SQL COMMANDS

Command	Syntax	Example
CREATE TABLE	CREATE TABLE *tablename* (*columnname* datatype [DEFAULT] [, *columnname* ...];	CREATE TABLE acctmanager (amid VARCHAR2(4), amname VARCHAR2(20), amedate DATE DEFAULT SYSDATE, region CHAR(2));

CREATE TABLE...AS	CREATE TABLE *tablename*
	AS (*subquery*);
		CREATE TABLE
		secustomerorders
		  AS (SELECT customer#, state,
		     ISBN, category, quantity,
		     cost, retail
		  FROM customers NATURAL
		     JOIN orders NATURAL
		     JOIN orderitems
		     NATURAL JOIN books
		  WHERE state IN ('FL',
		     'GA', 'AL'));

CREATE USER	CREATE USER *username* IDENTIFIED BY *password*:	CREATE USER *rthomas* IDENTIFIED BY *little25car*;

CREATE VIEW	CREATE [FORCE|NOFORCE]
	VIEW *viewname*
	(*columnname*, ...)]
	 AS *subquery*
	[WITH CHECK OPTION [CONSTRAINT
	constraintname]]
	[WITH READ ONLY];
		CREATE VIEW inventory
		  AS SELECT isbn, title,
		     retail price
		     FROM books
		WITH READ ONLY;

DELETE	DELETE FROM *tablename* [WHERE *expression*];	DELETE FROM acctmanager WHERE amid = 'D500';
DROP INDEX	DROP INDEX *indexname*:	DROP INDEX books_profit_idx;
DROP ROLE	DROP ROLE *rolename*:	DROP ROLE dataentry;
DROP SEQUENCE	DROP SEQUENCE *sequencename*;	DROP SEQUENCE orders_ordernumber;
DROP SYNONYM	DROP [PUBLIC] SYNONYM *synonymname*:	DROP PUBLIC SYNONYM orderentry;
DROP TABLE	DROP TABLE *tablename*;	DROP TABLE setotals;
DROP TABLE ... PURGE	Removes an entire table permanently from the Oracle10g database with the PURGE option	DROP TABLE cust_mkt_092005 PURGE;
DROP UNUSED COLUMNS	ALTER TABLE *tablename* DROP UNUSED COLUMNS;	ALTER TABLE secustomerorders DROP UNUSED COLUMNS;
DROP USER	DROP USER *username*;	DROP USER rthomas;

SQL COMMANDS

Command	Syntax	Example
DROP VIEW	`DROP VIEW viewname;`	`DROP VIEW inventory;`
FLASHBACK TABLE . . . TO BEFORE DROP	Recovers a dropped table if option not used when table dropped	`PURGE FLASHBACK TABLE cust_mkt_092005 TO BEFORE DROP;`
GRANT...ON	`GRANT {objectprivilege|ALL} [(columnname), objectprivilege (columnname)] ON objectname TO {username|rolename|PUBLIC} [WITH GRANT OPTION];`	`GRANT select, insert ON customers TO rthomas WITH GRANT OPTION;`
GRANT...TO (system privilege)	`GRANT systemprivilege [, systemprivilege, ...] TO username|rolename [,username|rolename, ....] [WITH ADMIN OPTION];`	`GRANT CREATE SESSION TO rthomas;`
GRANT...TO (role)	`GRANT rolename [, rolename] TO username [,username];`	`GRANT dataentry TO rthomas;`
INSERT	`INSERT INTO tablename [(columnname,...)] VALUES (datavalues, ...) [WHERE expression];` *or* `INSERT INTO tablename; subquery;`	`INSERT INTO acctmanager VALUES ('T500', 'NICK TAYLOR', '05-SEP-01', 'NE');` *or* `INSERT INTO acctmanager2 SELECT amid, amname, amedate, region FROM acctmanager WHERE amedate< '01-OCT-02';`
LOCK TABLE	`LOCK TABLE tablename IN modetype MODE;`	`LOCK TABLE customers IN SHARE MODE;` *or* `LOCK TABLE customers IN EXCLUSIVE MODE;`

SQL COMMANDS

Command	Syntax	Example

Merge	Conditionally performs a series of DML actions	`MERGE INTO books_1 a`
		`USING books_2 b`
		`ON (a.isbn = b.isbn)`
		`WHEN MATCHED THEN`
		`UPDATE SET a.retail = b.retail,`
		`a.category = b.category`
		`WHEN NOT MATCHED THEN`
		`INSERT (isbn, title, pubdate,`
		`retail, category)`
		`VALUES (b.isbn, b.title,`
		`b.pubdate, b.retail,`
		`b.category);`

PURGE TABLE	Permanently deletes a table in the recycle bin	`PURGE TABLE` `"BIN$IDMdosJceWxgg041==$0";`
RENAME...TO	`RENAME oldtablename TO` `newtablename;`	`RENAME secustomersspent TO` `setotals;`
REVOKE...FROM	`REVOKE rolename` `FROM username|rolename;`	`REVOKE dataentry` `FROM rthomas;`

REVOKE...ON	`REVOKE objectprivilege`
	`[,...objectprivilege]`
	`ON objectname`
	`FROM username|rolename;`
		`REVOKE INSERT`
		`ON customers`
		`FROM rthomas;`

ROLLBACK	`ROLLBACK;`	`ROLLBACK;`

SELECT	`SELECT columnname [, ...]`
	`FROM tablename`
	`[WHERE condition]`
	`[ORDER BY columnname]`
	`[GROUP BY columnname`
	`[, columnname, ...]]`
	`[HAVING groupfunction`
	`comparisonoperator value];`
		`SELECT category, AVG(cost)`
		`FROM books`
		`WHERE retail >25`
		`GROUP BY category`
		`HAVING AVG(cost)>21;`

SELECT...FOR UPDATE	`SELECT statement`
	`FOR UPDATE;`
		`SELECT cost`
		`FROM books`
		`WHERE category = 'COMPUTER'`
		`FOR UPDATE;`

SET ROLE	`SET ROLE rolename;`	`SET ROLE DBA;`

SQL COMMANDS

Command	Syntax	Example
TRUNCATE TABLE	`TRUNCATE TABLE tablename;`	`TRUNCATE TABLE setotals;`

UPDATE	`UPDATE tablename`
	`SET columnname = datavalue`
	`[WHERE expression];`
		`UPDATE acctmanager`
		`SET amedate=`
		`   TO_DATE('JANUARY 12,`
		`   2002','MONTH DD, YYYY')`
		`WHERE amid = 'J500';`

SQL KEYWORDS AND SYMBOLS

Keyword/Symbol	Description	Example
‖ concatenation)	Combines the display of contents from multiple columns into a single column	`SELECT city || state` `FROM customers;`
*	Returns all data in a table when used in a SELECT clause	`SELECT *` `FROM books;`
%	A "wildcard" character used with the LIKE operator to perform pattern searches, it represents zero or more characters	`WHERE lastname LIKE 'T%'`
_	A "wildcard" character used with the LIKE operator to perform pattern searches; it represents exactly one character in the specified position	`WHERE lastname LIKE 'MONT_` `ASA'`

* multiplication
/ division
+ addition
- subtraction
	Solves arithmetic operations (Oracle 10g first solves * and /, then solves + and -)	`SELECT title, retail-cost`
		`profit`
		`FROM books;`

,	Separates column names in a list when retrieving multiple columns from a table	`SELECT title, pubdate` `FROM books;`
' ' (string literal)	Indicates the exact set of characters, including spaces, to be displayed	`SELECT city || ' ' ||state` `FROM customers;`
" "	Preserves spaces, symbols, or case in a column heading alias	`SELECT title AS "Book Name"` `FROM books;`

AS	Indicates a column alias to change the heading of a column in output—optional	`SELECT title AS titles,`
		`pubdate`
		`FROM books;`
		or
		`SELECT title titles, pubdate`
		`FROM books;`

SQL KEYWORDS AND SYMBOLS

Keyword/Symbol	Description	Example

CHR(10)	Inserts a line break	`SELECT customer# ||CHR(10)`
		`||city || ' ' ||state`
		`FROM customers;`

CROSS JOIN	Matches each record in one table with each record in another table; also known as a *Cartesian product or Cartesian join*
	
	Syntax:
	`SELECT columnname[,...]`
	`FROM tablename1 CROSS JOIN`
	`tablename2;`
		`SELECT title, name`
		`FROM books CROSS JOIN`
		`publisher;`

DISTINCT	Eliminates duplicate lists	`SELECT DISTINCT state` `FROM customers;`

JOIN...ON	The JOIN keyword is used in the FROM clause; the ON clause identifies the column to be used to join the tables
	
	Syntax:
	`SELECT columnname [,...]`
	`FROM tablename1 JOIN tablename2`
	`ON tablename1.columnname`
	`<comparison operator>`
	`tablename2.columnname;`
		`SELECT customers.customer#,`
		`order#`
		`FROM customers JOIN orders`
		`ON customers.customer# =`
		`orders.customer#;`

JOIN...USING	The JOIN keyword is used in the FROM clause and, combined with the USING clause, it identifies the common column to be used to join the tables. It is normally used if the tables have more than one commonly named column, and only one is being used for the join.
	
	Syntax:
	`SELECT columnname [,...]`
	`FROM tablename1 JOIN tablename2`
	`USING (columnname);`
		`SELECT customer#, order#`
		`FROM customers JOIN orders`
		`USING (customer#);`

SQL KEYWORDS AND SYMBOLS

Keyword/Symbol	Description	Example
NATURAL JOIN	These keywords are used in the FROM clause to join tables containing a common column with the same name and definition. Syntax: SELECT *columnname*[,...] FROM *tablename1* NATURAL JOIN *tablename2*;	SELECT customer#, order# FROM customers NATURAL JOIN orders;
OUTER [RIGHT|LEFT| FULL] JOIN	This indicates that at least one of the tables does not have a matching row in the other table. Syntax: SELECT *columnname* [,...] FROM *tablename1* [RIGHT|LEFT|FULL] OUTER JOIN *tablename2* ON *tablename1.columnname* = *tablename2.columnname*;	SELECT customer#, order# FROM customers RIGHT JOIN orders ON customers.customer# = orders.customer#;
UNIQUE	Eliminates duplicate lists	SELECT UNIQUE state FROM customers;

SQL SINGLE-ROW FUNCTIONS

Function	Description	Syntax
ABS	Returns absolute value of a numeric value	ABS(n) n = numeric value
ADD_MONTHS	Adds months to a date to signal a target date in the future	ADD_MONTHS(d, m) d = date (beginning) for the calculation m = months—the number of months to add to the date
CONCAT	Concatenates two data items	CONCAT$(c1, c2)$ $c1$ = first data item to be concatenated $c2$ = second data item to be included in the concatenation

SQL SINGLE-ROW FUNCTIONS

Function	Description	Syntax
DECODE	Takes a given value and compares it to values in a list. If a match is found, the specified result is returned. If no match is found, a default result is returned. If no default result is defined, a NULL is returned as the result.	DECODE(V, L1, R1, L2, R2,..., D) V = value sought L1 = the first value in the list R1 = result to be returned if L1 and V match D = default result to return if no match is found
INITCAP	Converts words to mixed case, with initial capital letters	INITCAP(c) c = character string or field to be converted to mixed case
INSTR	Identifies position of search string	INSTR(c, s, p, o) c = character string to be searched s = search string p = search starting position o = occurrence of search string to identify
LENGTH	Returns the numbers of characters in a string	LENGTH(c) c = character string to be analyzed
LOWER	Converts characters to lower-case letters	LOWER(c) c = character string or field to be converted to lower case
LPAD/RPAD	Pads, or fills in, the area to the Left (or Right) of a character string, using a specific character— or even a blank space	LPAD(c, l, s) c = character string to be padded l = length of character string after being padded s = symbol or character to be used as padding
MOD	Returns remainder of division operation	MOD(n, d) n = numerator d = denominator
MONTHS_BETWEEN	Determines the number of months between two dates	MONTHS_BETWEEN(d1, d2) d1 and d2 = dates in question d2 is subtracted from d1

SQL SINGLE-ROW FUNCTIONS		
Function	Description	Syntax
NEXT_DAY	Determines the next day—a specific day of the week after a given date	NEXT_DAY(*d, DAY*) *d* = date (starting) *DAY* = the day of the week to be identified
NVL	Solves problems arising from arithmetic operations having fields that may contain NULL values. When a NULL value is used in a calculation, the result is a NULL value. The NVL function is used to substitute a value for the existing NULL.	NVL(*x, y*) *y* = the value to be substituted if *x* is NULL
NVL2	Provides options based on whether a NULL value exists	NVL2(*x, y, z*) *y* = what should be substituted if *x* is not NULL *z* = what should be substituted if *x* is NULL
REGEXP_LIKE	Searches for pattern values within a character string	REGEXP_LIKE(*c, p*) *c* = character string *p* = pattern operators
REGEXP_SUBSTR	Extracts a part of a string that matches a pattern of values	REGEXP_SUBSTR(*c, p*) *c* = character string *p* = pattern operators
REPLACE	Used to perform a search-and-replace operation on displayed results	REPLACE(*c, s, r*) *c* = the data or column to be searched *s* = the string of characters to be found *r* = the string of characters to be substituted for *s*
ROUND	Rounds date fields by month or year	ROUND(*d, fmt*) *d* = Date value *fmt* = Date format (YEAR or MONTH)
ROUND	Rounds numeric fields	ROUND(*n, p*) *n* = numeric data, or a field, to be rounded *p* = position of the digit to which the data should be rounded
RTRIM/LTRIM	Trims, or removes, a specific string of characters from the Right (or Left) of a set of data	LTRIM(*c, s*) *c* = characters to be affected *s* = string to be removed from the left of the data

SQL SINGLE-ROW FUNCTIONS		
Function	Description	Syntax
SOUNDEX	Converts alphabetic characters to their phonetic representation, using an alphanumeric algorithm	SOUNDEX *(c)* *c* = characters to be phonetically represented
SUBSTR	Returns a substring, or portion of a string, in output	SUBSTR *(c, p, l)* *c* = character string *p* = position (beginning) for the extraction *l* = length of output string
TO_CHAR	Converts dates and numbers to a formatted character string	TO_CHAR *(n, 'f')* *n* = number or date to be formatted *f* = format model to be used
TO_DATE	Converts a date in a specified format to the default date format	TO_DATE *(d, f)* *d* = date entered by the user *f* = format of the entered date
TRANSLATE	Converts single characters to a substitution value	TRANSLATE *(c, s, r)* *c* = character string to be searched *r* = search character *s* = substitution character
TRUNC	Truncates, or cuts, numbers to a specific position	TRUNC *(n, p)* *n* = numeric data, or a field, to be truncated *p* = position of the digit to which the data should be truncated
UPPER	Converts characters to upper-case letters	UPPER *(c)* *c* = character string or field to be converted to upper case

SQL GROUP (MULTIPLE-ROW) FUNCTIONS		
Function	Description	Syntax
AVG	Returns the average value of the selected numeric field. Ignores NULL values.	`AVG([DISTINCT|ALL] n)`
COUNT	Returns the number of rows that contain a value in the identified column. Rows containing NULL values in the column will not be included in the results. To count all rows, including those with NULL values, use an * rather than a column name.	`COUNT(*|[|DISTINCT|ALL] c)`
MAX	Returns the highest (maximum) value from the selected field. Ignores NULL values.	`MAX([DISTINCT|ALL] c)`
MIN	Returns the lowest (minimum) value from the selected field. Ignores NULL values.	`MIN([DISTINCT|ALL] c)`
STDDEV	Returns the standard deviation of the selected numeric field. Ignores NULL values.	`STDDEV([DISTINCT|ALL] n)`
SUM	Returns the sum, or total value, of the selected numeric field. Ignores NULL values.	`SUM([DISTINCT|ALL] n)`
VARIANCE	Returns the variance of the numeric field selected. Ignores NULL values.	`VARIANCE([DISTINCT|ALL] n)`
GROUP BY Extensions		
GROUPING SETS	Enables multiple GROUP BY clauses to be performed with a single query	`SELECT name, category, AVG(retail) FROM publisher JOIN books USING (pubid) GROUP BY GROUPING SETS (name, category, (name,category),());`
CUBE	Performs aggregations for all possible combinations of columns included	`SELECT name, category, AVG(retail) FROM publisher JOIN books USING (pubid) GROUP BY CUBE(name, category) ORDER BY name, category;`
ROLLUP	Performs increasing levels of cumulative subtotals based on column list provided	`SELECT name, category, AVG(retail) FROM publisher JOIN books USING (pubid) GROUP BY ROLLUP(name, category) ORDER BY name, category;`

SQL COMPARISON OPERATORS

Operator	Description	Example
=	Equality operator—requires an exact match of the record data and the search value	`WHERE cost = 55.95`
>	"Greater than" operator—requires a record to be greater than the search value	`WHERE cost > 55.95`
<	"Less than" operator—requires a record to be less than the search value	`WHERE cost < 55.95`

<>, !=, or ^=	"Not equal to" operator—requires a record not to match the search value	`WHERE cost <> 55.95`
		or
		`WHERE cost != 55.95`
		or
		`WHERE cost ^= 55.95`

<=	"Less than or equal to" operator—requires a record to be less than or an exact match with the search value	`WHERE cost <= 55.95`
>=	"Greater than or equal to" operator—requires record to be greater than or an exact match with the search value	`WHERE cost >= 55.95`
[NOT] BETWEEN x AND y	Searches for records in a specified range of values	`WHERE cost BETWEEN 40 AND 65`
[NOT] IN(x,y,...)	Searches for records that match one of the items in the list	`WHERE cost IN(22, 55.95, 13.50)`
[NOT] LIKE	Searches for records that match a search pattern—used with wildcard characters	`WHERE lastnameLIKE '_A%'`
IS [NOT] NULL	Searches for records with a NULL value in the indicated column	`WHERE referred IS NULL`

>ALL	More than the highest value returned by the subquery	`WHERE cost >ALL`
		   `(SELECT MAX(cost)`
		   `FROM books`
		   `GROUP BY category);`

<ALL	Less than the lowest value returned by the subquery	`WHERE cost <ALL`
		   `(SELECT MAX(cost)`
		   `FROM books`
		   `GROUP BY category);`

SQL COMPARISON OPERATORS		
Operator	Description	Example

<ANY	Less than the highest value returned by the subquery	`WHERE cost <ANY`
		`    (SELECT MAX(cost)`
		`    FROM books`
		`    GROUP BY category);`

>ANY	More than the lowest value returned by the subquery	`WHERE cost >ANY`
		`    (SELECT MAX(cost)`
		`    FROM books`
		`    GROUP BY category);`

=ANY	Equal to any value returned by the subquery (same as IN)	`WHERE cost >ALL`
		`    (SELECT MAX(cost)`
		`    FROM books`
		`    GROUP BY category);`

[NOT] EXISTS	Row must match a value in the subquery	`WHERE isbn EXISTS`
		`    (SELECT DISTINCT isbn`
		`    FROM orderitems);`

AND	Combines two conditions together—a record must match both conditions	`WHERE cost > 20` `AND retail < 50`
OR	Requires a record to only match one of the search conditions	`WHERE cost > 20` `OR retail < 50`
OTHER SQL OPERATORS		
Operator	Description	Example

INTERSECT	Lists only the results returned by both queries	`SELECT customer# FROM`
		`customers`
		`INTERSECT`
		`SELECT customer# FROM orders;`

MINUS	Lists only the results returned by the first query and not returned by the second query	`SELECT customer# FROM`
		`customers`
		`MINUS`
		`SELECT customer# FROM orders;`

Outer Join Operator (+)	Used to indicate the table containing the deficient rows. The operator is placed next to the table that should have NULL rows added to create a match.	`SELECT lastname, firstname,`
		`order#`
		`FROM customers c, orders o`
		`WHERE c.customer# =`
		`o.customer#(+);`

UNION	Used to combine the distinct results returned by multiple SELECT statements	`SELECT customer# FROM`
		`customers`
		`UNION`
		`SELECT customer# FROM orders;`

OTHER SQL OPERATORS		
Operator	Description	Example

UNION ALL	Used to combine all the results returned by multiple SELECT statements	`SELECT customer# FROM`
		`customers`
		`UNION ALL`
		`SELECT customer# FROM orders;`

&	Identifies a substitution variable. Allows the user to be prompted to enter a specific value for the substitution variable.	`UPDATE customers`
		`SET region = '&Region'`
		`WHERE state = '&State';`

SQL CONSTRAINTS		
Constraint	During Table Creation	After Table Creation

CHECK
 	
Ensures that a specified condition is true before the data value is added to the table. For example, an order's ship date cannot be "less than" its order date.
	`CREATE TABLE newtable`
	`(firstcol NUMBER,`
	`secondcol VARCHAR2(20),`
	`thirdcol NUMBER CHECK BETWEEN`
	`20 AND 30);`
	or
	`CREATE TABLE newtable`
	`(firstcol NUMBER,`
	`secondcol VARCHAR2(20),`
	`thirdcol NUMBER,`
	`CHECK (thirdcol BETWEEN 20`
	`AND 80));`
		`ALTER TABLE newtable`
		`ADD CHECK`
		`(thirdcol BETWEEN`
		`20 AND 80);`

FOREIGN KEY
 	
In a one-to-many relationship, the constraint is added to the "many" table. The constraint ensures that if a value is entered to the specified column, it must exist in the table being referred to, or the row is not added.
	`CREATE TABLE newtable`
	`(firstcol NUMBER,`
	`secondcol VARCHAR2(20)`
	`REFERENCES`
	`anothertable(col1));`
	or
	`CREATE TABLE newtable`
	`(firstcol NUMBER,`
	`secondcol VARCHAR2(20),`
	`FOREIGN KEY (secondcol)`
	`REFERENCES`
	`anothertable(col1);`
		`ALTER TABLE newtable`
		`ADD FOREIGN KEY`
		`(secondcol)`
		`REFERENCES`
		`anothertable(col1);`

NOT NULL
 	
Requires that the specified column cannot contain a NULL value. It can only be created with the column-level approach to table creation.
	`CREATE TABLE newtable`
	`(firstcol NUMBER,`
	`secondcol VARCHAR2(20),`
	`thirdcol NUMBER NOT NULL);`
		`ALTER TABLE newtable`
		`MODIFY (thirdcol NOT`
		`NULL);`

SQL CONSTRAINTS

Constraint	During Table Creation	After Table Creation

PRIMARY KEY
 	
Determines which column(s) uniquely identifies each record. The primary key cannot be NULL, and the data value(s) must be unique.
	`CREATE TABLE newtable`
	`(firstcol NUMBER PRIMARY KEY,`
	`secondcol VARCHAR2(20));`
	or
	`CREATE TABLE newtable`
	`(firstcol NUMBER,`
	`secondcol VARCHAR2(20),`
	`PRIMARY KEY (firstcol));`
		`ALTER TABLE newtable`
		`ADD PRIMARY KEY`
		`(firstcol);`

UNIQUE
 	
Ensures that all data values stored in the specified column must be unique. The UNIQUE constraint differs from the PRIMARY KEY constraint in that it allows NULL values.
	`CREATE TABLE newtable`
	`(firstcol NUMBER,`
	`secondcol VARCHAR2(20)`
	`UNIQUE);`
	or
	`CREATE TABLE newtable`
	`(firstcol NUMBER,`
	`secondcol VARCHAR2(20),`
	`UNIQUE (secondcol));`
		`ALTER TABLE newtable`
		`ADD UNIQUE (secondcol);`

SQL*PLUS COMMANDS

Command	Description	Example

/	Executes a SQL statement— must be on a separate line	`SELECT zip`
		`FROM customers`
		`/`

;	View contents in the buffer	`SQL>;`
CONN *or* CONNECT	Used to connect to the Oracle 10g database from within SQL*Plus CONNECT *username/* *password@connectstring*	`CONNECT rthomas/` `little25car`

DESC
or
DESCRIBE
	Displays the structure of table	`DESCRIBE books`
		or
		`DESC books`

L
or
LIST
	Displays the contents of the buffer	`SQL> LIST`
		or
		`SQL>L`

R
or
RUN
	Executes a SQL statement stored in the buffer	`SQL>run (Enter)`
		or
		`SQL>r (Enter)`

SQL*PLUS EDITING COMMANDS

Command	Description	Example
A *or* APPEND	Adds the entered text to the end of the active line	`A WHERE retail < 42.89`

C \old\new\
 or
 CHANGE \old\new\
	Finds an indicated string of characters and replaces it with a new string of characters	`C \42.89\52.35\`
		 or
		 `CHANGE \42.89\52.35\`

DEL *n*
 or
 DELETE *n*
	Deletes the indicated line number from the SQL statement	`DEL 3`
		 or
		 `DELETE 3`

I
 or
 INPUT
	Allows new rows to be entered after the current active line	`I OR cost > 15`
		 or
		 `INPUT OR cost > 15`

n	Sets the current active line in the SQL*Plus buffer	`2`

SQL*PLUS ENVIRONMENT VARIABLES

Variable	Description	Example
UNDERLINE *x*│OFF	Specifies the symbol used to separate a column heading from the contents of the column	`SET UNDERLINE '='`
LINESIZE *n*	Establishes the maximum number of characters that can appear on a single line of output	`SET LINESIZE 35`
PAGESIZE *n*	Establishes the maximum number of lines that can appear on one page of output	`SET PAGESIZE 27`

SQL*PLUS REPORT COMMANDS

Command/Syntax	Description	Example
START or @	Executes a script file	START titlereport.sql or @titlereport.sql
TTITLE [option [text|variable] ...] ON|OFF]	Defines a report's header	TTITLE 'Top of|Report'
BTITLE [option [text|variable] ...] ON|OFF]	Defines a report's footer	BTITLE 'Bottom of|Report'
BREAK ON columnname| columnalias [ON ...] [skip n|page]	Suppresses duplicate data in a specified column	BREAK ON title
COMPUTE statistical function OF columnname| REPORT ON groupname	Performs calculations in a report based on the AVG, SUM, COUNT, MIN, and MAX statistical functions	COMPUTE SUM OF total ON customer#
MARKUP HTML	Enables output from client SQL*Plus to be saved as HTML	SET MARKUP HTML ON
SPOOL filename	Redirects the output to a text file	SPOOL c:\reportresults.txt
CLEAR	Clears settings applied with the COLUMN and BREAK commands	CLEAR COLUMN

COLUMN Command Options

Command/Syntax	Description	Example
FORMAT formatmodel	Applies a specified format to the column data	COLUMN title FORMAT A30
HEADING columnheading	Specifies the heading to be used for the specified column	COLUMN title HEADING 'Book|Title'
NULL textmessage	Identifies the text message to be displayed in the place of NULL values	COLUMN shipdate NULL 'Not Shipped'

SQL*PLUS REPORT COMMANDS		
Command/Syntax	Description	Example
TTITLE and BTITLE Command Options		
CENTER	Centers the display between a report's left and right margins	`TTITLE CENTER 'Report Title'`
FORMAT	Applies a format model to the data to be displayed. It uses the same format model elements as the COLUMN command.	`BTITLE 'Page: ' SQL.PNO FORMAT 9`
LEFT	Aligns data to be displayed to a report's left margin	`TTITLE LEFT 'Internal Use Only'`
RIGHT	Aligns data to be displayed to a report's right margin	`TTITLE RIGHT 'Internal Use Only'`
SKIP *n*	Indicates the number of lines to skip before the display resumes	`TTITLE CENTER 'Report Title'` `SKIP 2 RIGHT 'Internal Use Only'`

INTRODUCTION TO THE TOAD THIRD-PARTY TOOL

INTRODUCTION

This text demonstrated issuing SQL commands in Oracle via the SQL*Plus software. The SQL*Plus software tool is packaged with Oracle, and, therefore, is most always available in an Oracle environment. However, we must recognize that many third-party products exist for database utility, and that these programs can make DBA and developer tasks more manageable. Because many companies invest in these complementary products, database developers should know about them. This appendix identifies one of these products named TOAD (Tool for Oracle Application Developers), available from Quest Software. Anyone interested in pursuing a database development career should investigate database utility products of this nature.

TOAD

Figure D-1 shows the initial screen of the TOAD software. The main menu is a good indication of the variety of features available in this tool.

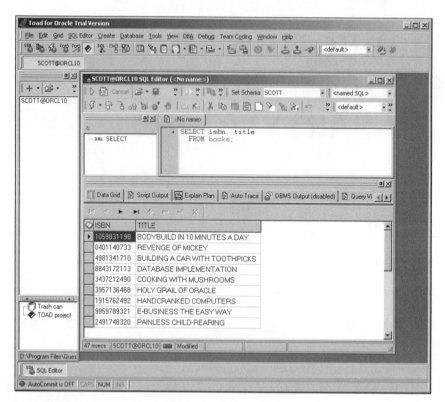

FIGURE D-1 Initial TOAD software screen

This appendix points out only a couple of TOAD features so you can begin to visualize the potential of the program. On the initial screen, you can see the SQL editor offers color-coded syntax checking, a feature that is typically available with programming editors. In addition, you can view the output or analyze the statement by viewing the explain plan. As you can see, this is a highly graphical interface to the database.

By viewing the schema browser, you can scan through all the database objects, as shown in Figure D-2.

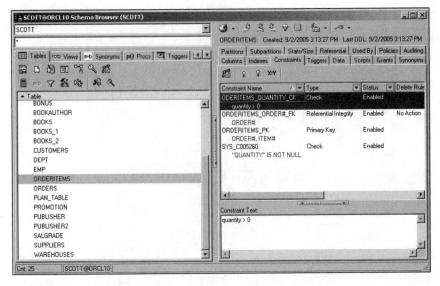

FIGURE D-2   TOAD schema browser

Notice that the left side of the schema browser lets the user move through various types of objects, and the right side provides numerous tabs of information regarding each object. In this case, the user has selected the ORDERITEMS table object on the left and the Constraints tab on the right, so all constraint details for this table are displayed.

The graphical interface can be quite helpful in developing or reviewing the database design. An ERD representation of the database can be viewed within TOAD, as displayed in Figure D-3.

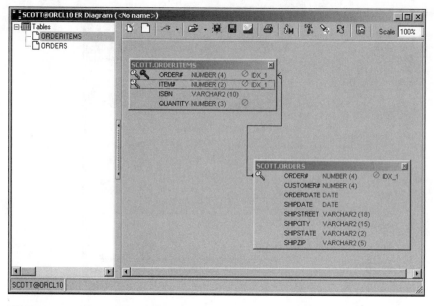

FIGURE D-3 TOAD ERD diagram

Most companies that sell database utility software provide free trial versions of their products for testing. Search and explore the options available.

ORACLE 10*g* PRACTICE EXAMS (A–E)

Practice Exam A

The questions in this practice exam test your knowledge of the concepts presented in Chapters 2, 8 and 9 (basic questions) of this textbook. Solutions to the exam questions are available from your instructor. When necessary, reference the table structures that follow. Full column names are listed after the tables.

EMP TABLE	
Column Name	Definition
Empno	NUMBER(4), PK
Ename	VARCHAR2(10)
Job	VARCHAR2(9)
Mgr	NUMBER(4)
Hiredate	DATE
Sal	NUMBER(7,2)
Comm	NUMBER(7,2)
Deptno	NUMBER(2), FK

DEPT TABLE	
Column Name	Definition
Deptno	NUMBER(2)
Dname	VARCHAR2(14)
Loc	VARCHAR2(13)

- Empno — Employee
- Ename — Employee Name
- Job — Job Title
- Mgr — Manager of Employee
- Hiredate — Hire Date
- Sal — Salary (monthly)
- Comm — Commission (for sales)
- Deptno — Department Number

- Dname — Department Name
- Loc — Location of Operation

1. Which two of the following queries can be used to determine the name of the department in which employee Blake works?

   a. `SELECT dname FROM dept WHERE ename = ('BLAKE');`

   b. `SELECT d.dname FROM dept d NATURAL JOIN emp e`
      `WHERE e.ename = 'BLAKE';`

   c. `SELECT dname FROM dept d, emp e`
      `WHERE e.ename = ('BLAKE')`
      `AND d.deptno = e.deptno;`

   d. `SELECT dname FROM dept JOIN emp USING (dname)`
      `WHERE ename = 'BLAKE';`

   e. `SELECT dname FROM dept JOIN emp`
      `ON dept.deptno = emp.deptno`
      `WHERE ename = 'BLAKE';`

2. Which of the following queries will display the annual salary for each employee in the EMP table if the Sal column contains each employee's monthly salary?

   a. `SELECT sal * 12 'Annual Salary' FROM emp;`

   b. `SELECT salary*12 annual FROM emp;`

   c. `SELECT annual sal*12 FROM emp;`

   d. `SELECT sal*12 FROM emp;`

3. Which of the following queries will display all data stored in the EMP table?

   a. `SELECT * FROM emp;`

   b. `SELECT % FROM emp;`

   c. `SELECT ^ FROM emp;`

   d. `SELECT _ FROM emp;`

4. Which statement reflects what will occur when the following query is executed?

   ```
   SELECT ename
   FROM emp e, emp m
   WHERE e.mgr = m.empno;
   ```

   a. The query will result in a self-join that will display the name of each employee's manager.

   b. An ambiguity error will be displayed and the statement will not be executed.

   c. The query will execute a full outer join and the names of employees who have not been assigned a manager will be displayed.

   d. The query will execute a right outer join and the names of employees who are not considered managers will be displayed.

5. Which of the following keywords can be used to create a non-equality join? Choose all that apply.

   a. NATURAL JOIN

   b. JOIN...USING

   c. OUTER JOIN

   d. JOIN...ON

   e. None—a non-equality join cannot be created using any of the JOIN keywords.

6. Which of the following queries will return only the department numbers contained in the DEPT table that are not also listed in the EMP table?

   a. `SELECT deptno FROM dept NATURAL JOIN emp`
      `WHERE deptno NOT IN emp;`
   b. `SELECT deptno FROM dept MINUS deptno FROM emp;`
   c. `SELECT deptno FROM dept MINUS`
      `SELECT deptno FROM emp;`
   d. `SELECT deptno FROM dept JOIN emp`
      `ON dept.deptno <> emp.deptno;`
   e. `SELECT deptno FROM emp MINUS`
      `SELECT deptno FROM dept;`

7. Which of the following queries will display the employee number of the employee named King?

   a. `SELECT empno FROM emp WHERE ename = '%KING';`
   b. `SELECT empno FROM emp WHERE ename = '_ING';`
   c. `SELECT empno FROM emp WHERE ename LIKE KING;`
   d. `SELECT empno FROM emp WHERE ename = KING;`
   e. none of the above

8. Which of the following queries will display all employees in the Sales Department who were hired in 1981?

   a. `SELECT * FROM emp`
      `WHERE dname = 'SALES' AND hiredate LIKE '%81';`
   b. `SELECT * FROM emp NATURAL JOIN dept`
      `WHERE dname = 'SALES' AND hiredate LIKE '%81';`
   c. `SELECT * FROM emp`
      `WHERE dname = 'SALES' OR hiredate LIKE '%81';`
   d. `SELECT * FROM emp NATURAL JOIN dept`
      `WHERE dname = 'SALES' AND hiredate LIKE '%1981';`

9. Which of the following queries will display the name and job title of each employee stored in the EMP table? (Choose all that apply.)

   a. `SELECT ename, job AS "Job Title" FROM emp;`
   b. `SELECT ename, job "Job Title" FROM emp;`
   c. `SELECT ename, job FROM emp;`
   d. `SELECT ename, job 'JOB TITLE' FROM emp;`
   e. `SELECT ename, job 'Job Title' FROM emp;`

10. Which sentence most accurately describes the results of the following SELECT statement?

    `SELECT DISTINCT job, ename FROM emp;`
    a. Each row returned in the results will be unique.
    b. Each job title will only be displayed once in the results.
    c. Each job title will be displayed once, along with the names of each employee assigned to that job.
    d. The results will be sorted in order of employee names.

11. Which of the following clauses is used to project certain columns from a table?
    a. SELECT
    b. FROM
    c. WHERE
    d. ORDER BY

12. Which of the following queries will display the names of all employees who earn an annual salary of at least $10,000?
    a. `SELECT ename FROM emp WHERE sal*12 > 10,000;`
    b. `SELECT ename FROM emp WHERE sal*12 > '10,000';`
    c. `SELECT ename FROM emp WHERE sal*12 => 10000;`
    d. `SELECT ename FROM emp WHERE sal *12 >= 10000.00;`
    e. The correct statement is not given.

13. Which of the following queries will display each employee's number in a sorted order, by employee name? (Choose all that apply.)
    a. `SELECT empno, ename FROM emp ORDER BY empno;`
    b. `SELECT empno, ename FROM emp ORDER BY ename;`
    c. `SELECT empno, ename FROM emp ORDER BY 1;`
    d. `SELECT empno, ename FROM emp ORDER BY 2;`
    e. `SELECT empno, ename ORDER BY ename;`

14. Which of the following queries will display the name of each employee who earns a salary of at least $1200 per month but less than $2000 per month?
    a. `SELECT ename FROM emp`
       `WHERE sal BETWEEN (1200, 2000);`
    b. `SELECT ename FROM emp`
       `WHERE sal BETWEEN 1200 AND 2000;`
    c. `SELECT ename FROM emp`
       `WHERE sal >=1200 AND <2000;`
    d. `SELECT ename FROM emp`
       `WHERE sal >=1200 AND sal<2000;`
    e. `SELECT ename FROM emp`
       `WHERE sal >1200 AND sal<2000;`

15. Which of the following clauses is used to restrict the rows returned by a query?
    a. SELECT
    b. FROM
    c. WHERE
    d. ORDER BY

16. Which of the following operators is used to perform pattern searches?
    a. IN
    b. BETWEEN
    c. IS NULL
    d. LIKE

17. Which of the following queries will not include any employees in Department 30 in its results? (Choose all that apply.)

    a. `SELECT * FROM emp WHERE deptno !=30;`

    b. `SELECT * FROM emp WHERE deptno <>30;`

    c. `SELECT * FROM emp WHERE deptno ^30;`

    d. `SELECT * FROM emp WHERE deptno =30;`

18. Which of the following queries will display all employees who do not earn a commission?

    a. `SELECT ename FROM emp WHERE comm = NULL;`

    b. `SELECT ename FROM emp WHERE comm IS NULL;`

    c. `SELECT ename FROM emp WHERE comm LIKE NULL;`

    d. `SELECT ename FROM emp WHERE comm LIKE 'NULL';`

19. Which of the following queries will return the names of all employees who work in the Sales Department or Accounting Department and earn at least $2000 per month? (Choose all that apply.)

    a.
```
SELECT ename FROM emp NATURAL JOIN dept
WHERE dname IN ('SALES', 'ACCOUNTING')
AND sal >= 2000;
```

    b.
```
SELECT ename FROM emp JOIN dept
ON emp.deptno = dept.deptno
WHERE sal >= 2000 AND dname = 'SALES'
OR dname = 'ACCOUNTING';
```

    c.
```
SELECT ename FROM emp JOIN dept USING (deptno)
WHERE sal >= 2000 AND (dname = 'SALES' OR dname =
'ACCOUNTING');
```

    d.
```
SELECT ename FROM emp NATURAL JOIN dept
WHERE sal >= 2000 AND (dname = 'SALES' OR dname =
'ACCOUNTING');
```

20. Which of the following clauses is used to present a query's results in sorted order?

    a. SELECT

    b. FROM

    c. WHERE

    d. ORDER BY

Practice Exam B

The questions in this practice exam test your knowledge of the concepts presented in Chapters 10-12 (functions and subqueries) of this textbook. Solutions to the exam questions are available from your instructor. When necessary, reference the table structures that follow. Full column names are listed after the tables.

EMP TABLE	
Column Name	Definition
Empno	NUMBER(4), PK
Ename	VARCHAR2(10)
Job	VARCHAR2(9)
Mgr	NUMBER(4)
Hiredate	DATE
Sal	NUMBER(7,2)
Comm	NUMBER(7,2)
Deptno	NUMBER(2), FK

DEPT TABLE	
Column Name	Definition
Deptno	NUMBER(2)
Dname	VARCHAR2(14)
Loc	VARCHAR2(13)

- Empno — Employee
- Ename — Employee Name
- Job — Job Title
- Mgr — Manager of Employee
- Hiredate — Hire Date
- Sal — Salary (monthly)
- Comm — Commission (for sales)
- Deptno — Department Number
- Dname — Department Name
- Loc — Location of Operation

1. Which of the following queries will display the name of each department in low-ercase letters?
   a. `SELECT LOW(dname) FROM dept;`
   b. `SELECT LOWER(dname) FROM dept;`
   c. `SELECT LOWERCASE(dname) FROM dept;`
   d. `SELECT NOTUPPER(dname) FROM dept;`

2. Which of the following queries will display the gross salary of each employee if the Sal column contains the salary of each employee and the Comm column contains the commission earned by the sales representatives? (Choose all that apply.)

    a. `SELECT ename, sal + NVL(comm, 0) AS "Gross Salary"`
        `FROM emp;`

    b. `SELECT ename, NVL2(sal+comm, sal, comm) "Gross" FROM`
        `emp;`

    c. `SELECT ename, NVL(comm,0) + sal FROM emp;`

    d. `SELECT ename, NVL(sal+comm, sal) FROM emp;`

3. Which of the following queries will return the total salary earned by all individuals working in Department 10? (Choose all that apply.)

    a. `SELECT SUM(sal) FROM emp WHERE deptno = 10;`

    b. `SELECT TOTAL(sal) FROM emp WHERE deptno = 10;`

    c. `SELECT SUM(sal) FROM emp`
        `WHERE deptno = 10 GROUP BY deptno;`

    d. `SELECT SUM(sal) FROM emp HAVING deptno = 10;`

    e. `SELECT SUM(sal) FROM emp HAVING deptno = 10 GROUP BY`
        `deptno;`

4. Which of the following queries will display the name of all employees who work in the same department as an employee named King?

    a. `SELECT ename FROM emp WHERE ename = 'KING';`

    b. `SELECT ename FROM emp WHERE deptno =`
        `(SELECT deptno FROM emp WHERE ename = "KING";`

    c. `SELECT ename FROM emp WHERE ename =`
        `(SELECT deptno FROM emp WHERE ename = 'KING');`

    d. `SELECT ename FROM emp WHERE deptno =`
        `(SELECT deptno FROM emp WHERE ename = 'KING');`

5. Which of the following queries will display the name of the department whose employees earn an average monthly salary of at least $1500?

    a. `SELECT dname, AVERAGE(sal)`
        `FROM dept NATURAL JOIN emp`
        `WHERE AVERAGE(sal) > 1500;`

    b. `SELECT dname, AVERAGE(sal)`
        `FROM dept NATURAL JOIN emp`
        `HAVING AVERAGE(sal) > 1500;`

    c. `SELECT dname, AVG(sal)`
        `FROM dept NATURAL JOIN emp`
        `WHERE AVG(sal) > 1500;`

    d. `SELECT dname, AVG(sal)`
        `FROM dept NATURAL JOIN emp`
        `GROUP BY dname`
        `HAVING AVG(sal) > 1500;`

6. Which of the following queries will display the fourth number in each employee's employee number?
   a. `SELECT ename, SUBSTR(empno, 4, 1) FROM emp;`
   b. `SELECT ename, LENGTH(empno, 4) FROM emp;`
   c. `SELECT ename, TRUNC(empno, 4) FROM emp;`
   d. `SELECT ename, SOUNDEX(empno, 4, 1) FROM emp;`

7. Which of the following queries will return the total monthly salary of all employees in the company?
   a. `SELECT SUM(sal) FROM emp GROUP BY deptno;`
   b. `SELECT SUM(sal) FROM emp;`
   c. `SELECT SUM(DISTINCT sal) FROM emp;`
   d. `SELECT TOTAL(sal) FROM emp WHERE sal IS NOT NULL;`

8. Which of the following queries will return the names of only those employees who are clerks?
   a. `SELECT UPPER(ename) FROM emp`
      `WHERE LOWER(job) = 'CLERK';`
   b. `SELECT LOWER(ename) FROM emp`
      `WHERE LOWER(job) = 'CLERK';`
   c. `SELECT UPPER(ename) FROM emp`
      `WHERE LOWER(job) = 'clerk';`
   d. `SELECT LOWER(ename) FROM emp`
      `WHERE UPPER(job) = 'clerk';`

9. Which of the following statements is accurate?
   a. Group functions are used to calculate multiple values per row, but single-row functions are used to calculate only one value per row.
   b. A query containing a group function must also contain a GROUP BY clause.
   c. Group functions return one value per group of rows processed, but single-row functions return one value for each row processed.
   d. A HAVING clause cannot be used in a query that contains a single-row function.

10. Which of the following describes a scenario that would call for the use of a subquery?
    a. You need to know all the employees who have a salary higher than employee Blake's salary.
    b. You need to know the name of all the clerks who earn more than $1000 per month.
    c. You need a list of all employees who work in Department 30.
    d. You need to find the average salary for all the employees in each department.

11. Which of the following queries will calculate the difference between today's date and the date on which an employee was hired?

    a. `SELECT ename, MONTH_BETWEEN(SYSDATE, hiredate)`

        `FROM emp;`

    b. `SELECT ename, SYSDATE-hiredate`

        `FROM emp;`

    c. `SELECT ename, DIFF(SYSDATE, hiredate)`

        `FROM emp;`

    d. `SELECT ename, TO_DATE(SYSDATE, hiredate)`

        `FROM emp;`

12. Which of the clauses that come after the following query will cause the query to return an error message?

```
SELECT ename
FROM emp
WHERE sal >
        (SELECT AVG(sal)
         FROM emp
         GROUP BY deptno);
```

    a. `SELECT ename`

    b. `WHERE sal >`

    c. `SELECT AVG(sal)`

    d. `GROUP BY deptno`

13. Which of the following queries will return the number of employees who have the same job title?

    a. `SELECT COUNT(*), job FROM emp GROUP BY job;`

    b. `SELECT COUNT(job) FROM emp;`

    c. `SELECT COUNT(DISTINCT job) FROM emp;`

    d. `SELECT SUM(job) FROM emp;`

14. Which of the following queries will display the lowest salary earned by an employee?

    a. `SELECT MIN(ename) FROM emp;`

    b. `SELECT LOW(ename) FROM emp;`

    c. `SELECT LOWER(sal) FROM emp;`

    d. `SELECT MIN(sal) FROM emp;`

    e. `SELECT MIN(sal) FROM emp GROUP BY job;`

15. Which of the following queries will display the names of all employees who are in the same department as an employee named Smith, but who earn a higher salary than Smith?

a. ```
SELECT ename FROM emp
WHERE deptno = 'SMITH' AND sal > 'SMITH';
```

b. ```
SELECT ename FROM emp
WHERE (deptno, sal) >
    (SELECT deptno, sal FROM emp
     WHERE ename = 'SMITH');
```

c. ```
SELECT ename FROM emp WHERE deptno =
(SELECT deptno FROM emp WHERE ename = 'SMITH')
AND sal > (SELECT sal FROM emp WHERE ename = 'SMITH');
```

d. ```
SELECT ename FROM emp
WHERE (deptno, sal) >ANY
    (SELECT deptno, sal FROM emp
     WHERE ename = 'SMITH');
```

16. Which of the following is a valid multiple-row operator?

a. ANY

b. OR

c. =

d. >

17. Which of the following queries will display the name of all employees who work in the city of Boston? (Choose all that apply.)

a. ```
SELECT ename FROM emp NATURAL JOIN dept
WHERE loc = 'BOSTON';
```

b. ```
SELECT ename FROM emp WHERE loc = 'BOSTON';
```

c. ```
SELECT ename FROM dept WHERE loc = 'BOSTON';
```

d. ```
SELECT ename FROM emp WHERE deptno =
(SELECT deptno FROM dept WHERE loc = 'BOSTON');
```

e. ```
SELECT ename FROM emp WHERE deptno = 'BOSTON';
```

18. Which of the following operators is equivalent to using the IN operator in a multiple-row subquery?

a. =ANY

b. =ALL

c. >ANY

d. <ANY

19. Assuming the Comm column can contain NULL values, which of the following queries will display how many employees in the company earn a commission? (Choose all that apply.)

a. ```
SELECT COUNT(comm) FROM emp;
```

b. ```
SELECT COUNT(comm) FROM emp
WHERE comm IS NULL;
```

c. ```
SELECT COUNT(*) FROM emp
WHERE comm IS NOT NULL;
```

d. ```
SELECT COUNT(*) FROM emp
WHERE comm IS NULL;
```
```

20. Which of the following is a single-row function?
    a. AVERAGE
    b. VARIANCE
    c. SUM
    d. ADD_MONTHS

Practice Exam C

The questions in this practice exam test your knowledge of the concepts presented in Chapters 3-5 (Table creation and DML) of this textbook. Solutions to the exam questions are available from your instructor. When necessary, reference the table structures that follow. Full column names are listed after the tables.

EMP TABLE	
Column Name	Definition
Empno	NUMBER(4), PK
Ename	VARCHAR2(10)
Job	VARCHAR2(9)
Mgr	NUMBER(4)
Hiredate	DATE
Sal	NUMBER(7,2)
Comm	NUMBER(7,2)
Deptno	NUMBER(2), FK

DEPT TABLE	
Column Name	Definition
Deptno	NUMBER(2)
Dname	VARCHAR2(14)
Loc	VARCHAR2(13)

- Empno — Employee
- Ename — Employee Name
- Job — Job Title
- Mgr — Manager of Employee
- Hiredate — Hire Date
- Sal — Salary (monthly)
- Comm — Commission (for sales)
- Deptno — Department Number

- Dname — Department Name
- Loc — Location of Operation

1. Which of the following SQL statements will create a new table containing data for only the employees from Department 30?

   a. ```
      CREATE TABLE ee30
      AS (SELECT * FROM emp WHERE deptno = 30);
      ```
   b. ```
      CREATE TABLE ee30,
      AS (SELECT * FROM emp WHERE deptno = 30);
      ```
   c. ```
      CREATE TABLE (SELECT * FROM emp WHERE deptno = 30);
      ```
   d. ```
      CREATE TABLE 30department
      AS (SELECT * FROM emp WHERE deptno = 30);
      ```

2. Which of the following SQL statements will remove all rows from the DEPT table and release the storage space occupied by the rows?

   a. `DROP TABLE dept;`
   b. `DELETE FROM dept;`
   c. `TRUNCATE TABLE dept;`
   d. `DELETE *.* FROM dept;`

3. Which of the following SQL statements will add a numeric column named SSN to the EMP table?

   a. `ALTER TABLE emp MODIFY (add SSN NUMBER(9));`
   b. `ALTER TABLE emp ADD (SSN NUMBER(9);`
   c. `ALTER TABLE emp MODIFY (SSN NUMBER(9));`
   d. `ALTER TABLE emp ADD (SSN NUMBER(9));`

4. Which of the following SQL statements will change the name of the DEPT table to DEPARTMENT?

   a. `ALTER TABLE dept RENAME AS department;`
   b. `RENAME TO department FROM dept;`
   c. `RENAME dept TO department;`
   d. `RENAME dept AS department;`

5. Which of the following statements is correct? (Choose all that apply.)

   a. A column that has been marked as unused cannot be reclaimed or unmarked at a later time.
   b. When a column is dropped, the contents of the column can be restored by using the ROLLBACK command.
   c. When a column is dropped, the contents of the column cannot be restored by using the ROLLBACK command.
   d. A column that has been marked as unused can be reclaimed or unmarked at a later time.

6. Which of the following SQL statements will add a new department to the DEPT table?

   a. ```
      UPDATE dept
      SET deptno = 65, dname = 'HR', loc = 'SEATTLE';
      ```
   b. `INSERT VALUES (65, HR, SEATTLE) INTO dept;`
   c. `INSERT INTO dept VALUES (65, HR, SEATTLE);`
   d. None of these SQL statements will add a new department to the table.
      ```

7. Which of the following statements is correct? (Choose all that apply.)
   a. To ensure that an employee is assigned to a department that already exists in the DEPT table, a FOREIGN KEY constraint must exist on the DEPT table.
   b. To ensure that an employee is assigned to a department that already exists in the DEPT table, a FOREIGN KEY constraint must exist on the EMP table.
   c. To ensure that an employee is assigned to a department that already exists in the DEPT table, a NOT NULL constraint must exist on the DEPT table.
   d. To ensure that an employee is assigned to a department that already exists in the DEPT table, a UNIQUE constraint must exist on the EMP table.

8. Assuming that the PRIMARY KEY constraint for the EMP table is called EMP_EMPNO_PK, which SQL statement will remove the constraint?
   a. `DROP CONSTRAINT emp_empno_pk;`
   b. `ALTER TABLE emp DROP emp_empno_pk;`
   c. `ALTER TABLE emp DROP CONSTRAINT emp_empno_pk;`
   d. `ALTER TABLE emp DROP PRIMARY KEY;`

9. Which of the following SQL statements will add a NOT NULL constraint to the Sal column of the EMP table?
   a. `ALTER TABLE emp ADD NOT NULL(sal);`
   b. `ALTER TABLE emp MODIFY (sal NOT NULL);`
   c. `ALTER TABLE emp MODIFY NOT NULL(sal);`
   d. `ALTER TABLE emp ADD (sal NOT NULL);`

10. Which of the following statements is correct? (Choose all that apply.)
    a. A NOT NULL constraint can only be created using the column-level approach.
    b. A constraint that is composed of multiple columns must be created using the column-level approach.
    c. If a PRIMARY KEY constraint consists of more than one column, the constraint can be added to each column individually, using the column-level approach.
    d. A PRIMARY KEY constraint that consists of more than one column must be created using the table-level approach.
    e. To change the condition used by a CHECK constraint, the change must be made using the MODIFY clause of the ALTER TABLE command.

11. Which of the following letters is used to denote a NOT NULL constraint type in the USER_CONSTRAINTS view?
    a. FK
    b. NN
    c. R
    d. C
    e. U

12. Which of the following letters is used to denote a FOREIGN KEY constraint type in the USER_CONSTRAINTS view?
    a. FK
    b. NN
    c. R
    d. C
    e. U

13. Which of the following SQL*Plus commands is used to view the structure of a table?
    a. DESCRIBE
    b. LIST
    c. VIEW
    d. DISPLAY
    e. STRUCTURE

14. Which of the following SQL statements will add a new employee, Gary Lito, to the EMP table?
    a. `INSERT INTO emp VALUES (1462, 'GARY LITO');`
    b. `INSERT INTO emp (empno, ename)`
       `VALUES (1462, 'GARY LITO', NULL, NULL, NULL, NULL,`
       `NULL, NULL);`
    c. `INSERT INTO emp (empno, ename) VALUES (1462, 'GARY`
       `LITO');`
    d. `UPDATE emp SET empno = 1462 WHERE ename = 'GARY LITO';`

15. Which of the following symbols is used to indicate a substitution variable?
    a. _
    b. &
    c. %
    d. *

16. Which of the following is not a valid table name?
    a. #DeptEE
    b. EE#
    c. Dept_EE
    d. Dept30

17. Which of the following SQL statements will remove all data within the DEPT table as well as permanently delete the entire structure of the DEPT table?
    a. `DROP TABLE dept;`
    b. `DELETE TABLE dept;`
    c. `TRUNCATE TABLE dept;`
    d. `DELETE *.* FROM dept; [END CODE]`

18. If you do not specify a name for a constraint when it is created, Oracle 10g will automatically use which naming convention to internally assign a name to the constraint?
    a. n_pk
    b. SYSC_n
    c. SYS_Cn
    d. C_SYSn

19. Which of the following types of constraints is used to ensure referential integrity?
   a. NOT NULL
   b. PRIMARY KEY
   c. FOREIGN KEY
   d. CHECK
   e. UNIQUE

20. Executing which of the following commands will release any table locks previously held by the user? (Choose all that apply.)
   a. `COMMIT;`
   b. `EXIT`
   c. `ALTER TABLE emp ADD UNIQUE(ename);`
   d. `UPDATE emp SET sal = 3000 WHERE ename = 'SMITH';`

Practice Exam D

The questions in this practice exam test your knowledge of the concepts presented in Chapters 6, 7 and 13 of this textbook. Solutions to the exam questions are available from your instructor. When necessary, reference the table structures that follow. Full column names are listed after the tables.

EMP TABLE	
Column Name	Definition
Empno	NUMBER(4), PK
Ename	VARCHAR2(10)
Job	VARCHAR2(9)
Mgr	NUMBER(4)
Hiredate	DATE
Sal	NUMBER(7,2)
Comm	NUMBER(7,2)
Deptno	NUMBER(2), FK

DEPT TABLE	
Column Name	Definition
Deptno	NUMBER(2)
Dname	VARCHAR2(14)
Loc	VARCHAR2(13)

- Empno — Employee
- Ename — Employee Name
- Job — Job Title
- Mgr — Manager of Employee
- Hiredate — Hire Date
- Sal — Salary (monthly)
- Comm — Commission (for sales)
- Deptno — Department Number
- Dname — Department Name
- Loc — Location of Operation

1. Which of the following SQL statements will create a new user with ACCTSUPER as the user name and SUPERPWORD as the password?

   a. `CREATE USER acctsuper PASSWORD superpword;`

   b. `CREATE USER acctsuper PASS superpword;`

   c. `CREATE USER acctsuper IDENTIFIED BY superpword;`

   d. `CREATE acctsuper WITH PASSWORD superpword;`

2. Which of the following SQL statements will generate the next value in the EMP_EMPNO sequence?

   a. `SELECT emp_empno.nextvalue FROM dual;`

   b. `SELECT emp_empno.currentvalue FROM dual;`

   c. `SELECT emp_empno.nextval FROM dual;`

   d. `SELECT emp_empno.currentval FROM dual;`

3. Which of the following SQL statements will modify the existing view EMP_SAL_VU so the data it displays cannot be updated by a user?

   a. `CREATE OR REPLACE VIEW emp_sal_vu`
      `   AS SELECT  empno, ename, sal, comm FROM emp`
      `   WITH READ ONLY;`

   b. `REPLACE VIEW emp_sal_vu WITH READ ONLY;`

   c. `ALTER VIEW emp_sal_vu READ ONLY;`

   d. `CREATE OR REPLACE emp_sal_vu`
      `   AS SELECT  empno, ename, sal, comm FROM emp`
      `   WITH CHECK OPTION;`

4. Oracle10g will automatically create an index when which of the following occurs? (Choose all that apply.)

   a. A sequence is created.

   b. A PRIMARY KEY constraint is created.

   c. The CREATE INDEX command is successfully executed.

   d. A PUBLIC synonym is created.

5. Which of the following can be included in a simple view?

   a. grouped data

   b. joined tables

   c. SUM function

   d. column alias

6. Which of the following SQL commands grants a system privilege to the user named ACCTSUPER? (Choose all that apply.)
   a. `GRANT INSERT ON emp TO acctsuper;`
   b. `GRANT CREATE TABLE TO acctsuper;`
   c. `GRANT SELECT ON emp TO acctsuper;`
   d. `GRANT UPDATE ANY TABLE TO acctsuper;`
   e. `GRANT CREATE SESSION TO acctsuper;`

7. Which of the following SQL commands will create a view that will prevent a user from performing any operation that will make a row currently being displayed by the view inaccessible to the view in the future?
   a. ```
      CREATE VIEW eejobs30
        AS SELECT empno, ename, job
             FROM emp WHERE deptno = 30
      WITH CHECK OPTION;
      ```
   b. ```
      CREATE OR REPLACE VIEW eejobs30
        AS SELECT empno, ename, job
             FROM emp WHERE deptno = 30
      WITH READ ONLY;
      ```
   c. ```
      CREATE VIEW eejobs30
        AS SELECT empno, ename, job
             FROM emp WHERE deptno = 30
      WITH READ ONLY;
      ```
   d. ```
      CREATE OR REPLACE VIEW eejobs30
        AS SELECT empno, ename, job
             FROM emp WHERE deptno = 30;
      ```

8. Which of the following statements about simple and complex views is not valid?
   a. DML operations are not permitted on non-key-preserved tables.
   b. A row cannot be added to a table through a view if it violates an underlying constraint.
   c. NULL values cannot be added to a table through a view.
   d. DML operations are allowed on simple views that contain the pseudocolumn ROWNUM.

9. Which of the following SQL statements will require a user to create a new password the next time the user accesses his or her account?
   a. `CREATE USER acctsuper IDENTIFIED BY NULL;`
   b. `ALTER USER acctsuper PASSWORD EXPIRE;`
   c. `ALTER USER acctsuper EXPIRE PASSWORD;`
   d. `CREATE USER acctsuper IDENTIFIED BY PASSWORD EXPIRE;`

10. Which of the following SQL statements creates an inline view when executed?

   a. `CREATE FORCE VIEW inline_grosspay`
      `AS SELECT empno, ename, sal + NVL(comm,0)`
         `FROM emp;`

   b. `CREATE VIEW inline_grosspay AS`
      `inline SELECT empno, ename, sal + NVL(comm,0)`
         `FROM emp;`

   c. `CREATE VIEW inline_grosspay`
      `AS SELECT empno, ename, sal + NVL(comm,0)`
         `FROM emp;`

   d. `SELECT empno, ename, dname`
      `FROM (SELECT * FROM emp NATURAL JOIN dept);`

11. Which of the following SQL statements will delete the PUBLIC synonym name EMPLOYEE?

   a. `DELETE SYNONYM employee;`
   b. `DROP SYNONYM employee;`
   c. `DROP PUBLIC SYNONYM employee;`
   d. `DELETE PUBLIC SYNONYM employee;`
   e. `DROP PUBLIC employee;`

12. Which of the following statements about indexes is correct?

   a. Row retrievals are always slower when an index is used.
   b. DML operations are always faster when an index exists on the primary key of a table.
   c. An index always slows down DML operations.
   d. A function-based index will automatically result in slower query executions.

13. Which of the following terms applies to a collection or group of privileges?

   a. schema
   b. role
   c. data dictionary
   d. permission
   e. group account

14. Which of the following commands will grant the CONNECT role to two users, Smith and Blake?

   a. `GRANT CONNECT ON database TO SMITH BLAKE;`
   b. `GRANT CONNECT TO SMITH BLAKE;`
   c. `GRANT CONNECT TO 'SMITH', 'BLAKE';`
   d. `GRANT CONNECT TO SMITH, BLAKE;`

15. Which of the following commands will create a view even if the underlying table(s) does not exist?

    a. `CREATE FORCE VIEW inline_grosspay`
       `AS SELECT empno, ename, sal + NVL(comm,0)`
           `FROM emp;`

    b. `CREATE VIEW inline_grosspay AS`
       `inline SELECT empno, ename, sal + NVL(comm,0)`
           `FROM emp;`

    c. `CREATE VIEW inline_grosspay`
       `AS SELECT empno, ename, sal + NVL(comm,0)`
           `FROM emp;`

    d. `SELECT empno, ename, dname`
       `FROM (SELECT * FROM emp NATURAL JOIN dept);`

16. Which of the following terms is used to describe a collection of objects that belong to a particular user?

    a. user account
    b. role
    c. data dictionary
    d. schema

17. Which of the following commands will allow a user to change his or her password? (Choose all that apply.)

    a. ALTER USER...IDENTIFIED BY
    b. PASSWORD
    c. CREATE USER...IDENTIFIED BY
    d. MODIFY USER
    e. ALTER USER...PASSWORD

18. Which of the following will prevent the user SMITH from viewing the data stored in the EMP table but will still allow him access to the DEPT table?

    a. `REVOKE select ON emp FROM smith;`
    b. `LOCK TABLE emp FROM smith;`
    c. `REVOKE SELECT ANY TABLE FROM smith;`
    d. `ALTER USER smith RESTRICTED ACCESS on emp;`

19. Which of the following commands can be used to modify a view?

    a. ALTER VIEW
    b. MODIFY VIEW
    c. ALTER VIEW...MODIFY
    d. ALTER TABLE...MODIFY VIEW
    e. Views cannot be modified.

20. Which of the following clauses cannot be used with the ALTER SEQUENCE command?

    a. INCREMENT BY
    b. START WITH
    c. CACHE
    d. NOMAXVALUE

Practice Exam E

The questions in this practice exam test your knowledge of the concepts presented in Chapter 14 of this textbook. Solutions to the exam questions are available from your instructor.

1. Which of the following indicates numeric data in a column format model?
   a. #
   b. 9
   c. _
   d. &
   e. %

2. Which of the following is not an accepted parameter for a procedure?
   a. INPUT
   b. OUT
   c. INOUT
   d. IN

3. If a variable is declared with a NUMBER datatype in a PL/SQL block but is not initialized when it is created, what value will be assigned to the variable?
   a. 0
   b. 1
   c. NULL
   d. 999

4. Which of the following variables can be used to include the user's account name on a report?
   a. SQL.PNO
   b. SQL.LINENO
   c. SQL.USERID
   d. SQL.USER

5. Which of the following commands will suppress duplicate data within a column of a report?
   a. SUPPRESS
   b. AVOIDDUP
   c. BREAK
   d. GROUP BY

6. Which of the following commands identifies the title that should appear at the top of a report?
   a. BTITLE
   b. TITLE
   c. TTITLE
   d. HEADING

7. Which of the following COLUMN command options is used to identify the text message in place of a NULL value?
   a. TEXT
   b. NULL
   c. MESSAGE
   d. NLMSG

8. Which of the following commands is used to reset any formatting applied by the COLUMN command in a report?
   a. CLS
   b. CLEAR
   c. COLUMN
   d. STOP
9. Which of the following symbols can be used to execute a script file in Oracle 10g?

   a. %
   b. $
   c. @
   d. !

APPENDIX F

ORACLE RESOURCES

The following list of resources may assist students receiving Oracle 10g database training or considering careers that will require interaction with various Oracle products.

Oracle Academic Initiative (OAI)

The Oracle Academic Initiative (OAI) is a result of an effort by Oracle Corporation to provide curricula and other resources to the higher-education community. Individuals enrolled at an institution participating in the OAI receive benefits such as discount vouchers for certification exams and a free subscription to *Oracle Magazine*. Visit the Oracle Web site at *http://oai.oracle.com* for more information.

Oracle Certification Program (OCP)

The Oracle Certification Program (OCP) provides certification for both database administrators and application developers. Certification is based on successful completion of a series of exams. Current information about the OCP can be found at the Web site for this textbook or at *www.oracle.com/education/certification*.

Oracle Technology Network (OTN)

The Oracle Technology Network (OTN) provides several services to registered members. For example, members can download trial versions of various Oracle software products and access discussion groups for help with technical issues. In addition, the site also has an extensive documentation area, including reference manuals for SQL, PL/SQL, installation procedures, etc. Some of the documentation is available in PDF format and is available for download. Membership is free. This Web site can be accessed at *http://www.oracle.com/technology*.

International Oracle Users Group (IOUG)

The International Oracle Users Group (IOUG) is composed of more than 100 local and regional user groups who meet on a regular basis to share information about Oracle products. Members of IOUG can access the repository of knowledge accumulated from individuals who work with various Oracle products on a daily basis. In addition, members can receive publications, discounts, and special offers from various vendors, and they can have access to discussion forums. Contact information for user groups, conference information, etc. is provided on the IOUG Web site at *www.ioug.org*.

APPENDIX **G**

SQL*LOADER

INTRODUCTION

SQL*Loader is a utility packaged with the Oracle database to load data from external files into your tables.

The utility provides a great deal of flexibility for reading data in various formats, as well as for filtering and

manipulating data during the load operation. This appendix demonstrates two examples of file loads to get

you acquainted with the tool. Explore further options in the SQL*Loader documentation at the Oracle

Technology Network (OTN) Web site or *otn.oracle.com*.

READ A FIXED FORMAT FILE

SQL*Loader typically operates with two components: a control file with instructions for the data load and a data file. Both of these files are plaintext or ASCII files. Figure G-1 depicts the contents of these two components along with the command to execute SQL*Loader.

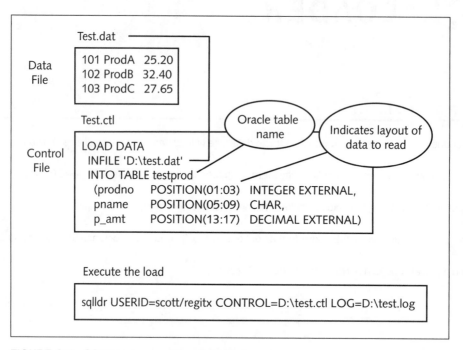

FIGURE G-1   SQL*Loader example—fixed file format

Note that the data is formatted in fixed placement, and, therefore, the instructions in the control file identify the specific positions to read for data that feeds into each column. For example, the value in character spaces 1 through 3 on each line in the data file is loaded or inserted into the Prodno column.

SQL*Loader is a line-command utility that can be invoked with a number of arguments. The example command, also shown executed in Figure G-2, contains arguments for the user login, identifying the control file, and identifying the log file.

The log file contains information regarding the results of the loader execution, as shown in Figure G-3. Notice that the list at the bottom indicates rows rejected based on errors, filtering conditions, or NULL values.

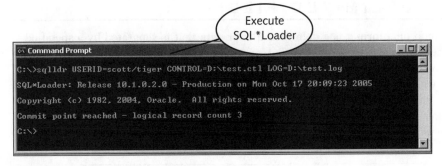

FIGURE G-2　SQL*Loader command execution

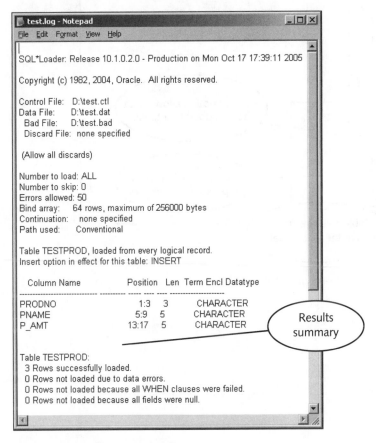

FIGURE G-3　SQL*Loader log file

READ A DELIMITED FILE

Another common data format is a delimited file, or data that is separated by a specified character. The example shown in Figure G-4 loads a comma-delimited data file. Notice that the field termination character is identified in the control file, as well as the order of the columns into which the data should be read. In addition, if data already exists in a table and you need the loader action to add the data to the existing rows, you must also specify the APPEND option in the control file, as shown in this example. If the APPEND option is not specified, the table must be empty for SQL*Loader to operate properly.

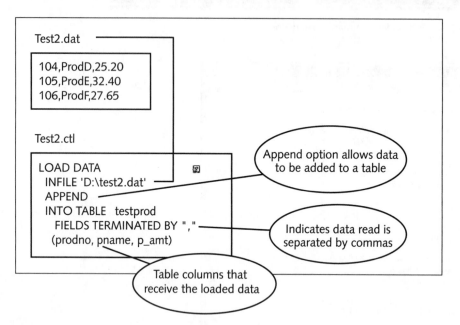

FIGURE G-4   SQL*Loader example—comma-delimited file format

aggregate functions *See* **group functions**

American National Standards Institute (ANSI) One of two industry-accepted committees that sets standards for SQL.

application cluster environment A high-volume work environment in which multiple users simultaneously request data from a database.

argument Values listed within parentheses in a function.

authentication The process of validating the identity of computer users.

authorization The granting of object privileges to a user based on the user's identity.

bridging entity An entity created to eliminate a many-to-many relationship between entities by adding two one-to-many relationships.

buffer pool The shared cache memory area of the database server.

Cartesian join Links table data so each record in the first table is matched with each individual record in the second table. Also called a *Cartesian product* or *cross join*.

Cartesian product *See* **Cartesian join**.

case conversion functions Allow a user to temporarily alter the case of data stored in a column or character string.

character The basic unit of data. It can be a letter, number, or special symbol.

character field A field composed of non-numeric data. This field will not display a heading longer than the width of the data stored in the field.

clause Each section of a statement that begins with a keyword (SELECT clause, FROM clause, WHERE clause, etc.).

column In a relational database, fields are commonly represented as columns and may be referred to as "columns."

column alias A name substituted for a column name. A column alias is created in a query and displayed in the results.

column qualifier Indicates the table containing a referenced column.

common column A column that exists in two or more tables and contains equivalent data.

common field A column that exists in two tables and is used to "join" two tables.

comparison operator A search condition that indicates how data should relate to a given search value (equal to, greater than, less than, etc.). Common comparison operators include >, <, >=, and <= .

composite primary key A combination of columns that uniquely identifies a record in a database table.

concatenation The combining of the contents of two or more columns or character strings. Two vertical bars, or pipes (||), instruct Oracle9*i* to concatenate the output of a query.

condition A portion of a SQL statement that identifies what must exist, or a requirement that must be met. When a query is executed, any record meeting the given condition will be returned in query results.

constraints Rules that ensure the accuracy and integrity of data. Constraints prevent data that violate these rules from being added to tables.

Constraints include PRIMARY KEY, FOREIGN KEY, UNIQUE, CHECK, and NOT NULL.

correlated subquery   A subquery that references a column in the outer query. The outer query executes the subquery once for every row in the outer query.

cross join   *See* **Cartesian join**.

data anomaly   An inconsistency in data.

data definition language (DDL)   Commands, basically SQL commands, that create or modify database tables or other objects.

data dictionary   Where Oracle9*i* stores all information about database objects. Stored information includes an object's name, type, structure, owner, and the identity of users who have access to the object.

data manipulation language (DML) Commands used to modify data. Changes to data made by DML commands are not accessible to other users until the data are committed.

data mining   Analyzing historical sales data and other information stored in an organization's database.

data redundancy   Refers to having the same data in different places within a database, which wastes space and complicates updates and changes.

database   A collection of interrelated files.

database management system (DBMS)   A generic term that applies to a software product that allows users to interact with a database to create and maintain the structure of the database, and then to enter, manipulate, and retrieve the data it stores.

database object   A defined, self-contained structure in Oracle9*i*. Database objects include tables, sequences, indexes, and synonyms.

datatype   Identifies the type of data Oracle9*i* will be expected to store in a column.

dimension   Any category used in analyzing data such as time, geography and product line.

encryption   The scrambling of data to make it unreadable to anyone other than the sender and receiver.

entity   Any person, place, or thing with characteristics or attributes that will be included in a database. In the E-R Model, an entity is usually represented as a square or rectangle.

Entity-Relationship (E-R) Model   A diagram that identifies the entities and data relationships in a database. The model is a logical representation of the physical system to be built.

equality join   Links table data in two (or more) tables having equivalent data stored in a common column. Also called an *equijoin* or *simple join*.

equality operator   A search condition that evaluates data for exact, or equal, values. The equality operator symbol is the equal sign (=).

equijoins   *See* **equality join**.

exclusive lock   When DDL operations are performed, Oracle 10*g* places this lock on a table so no other user can alter the table or place a lock on it. *See* **table lock**.

field   One attribute or characteristic of a

file   A group of records about the same type

first-normal form (1NF)   The first step in the normalization process in which repeating groups of data are removed from database records.

foreign key   When a common column exists in two tables, it will usually be a primary key in one table and will be called a foreign key in the second table.

full table scan   The type of scan in which each row of the table is read and the zip value is checked to determine whether it satisfies the condition.

function   A named PL/SQL block, or predefined block of code, that accepts zero or more input parameters and returns one value.

group functions Process groups of rows, returning only one result per group of rows processed. Also called *multiple-row functions* and *aggregate functions*.

heap-organized table An unordered collection of data.

index A separate database object that stores frequently referenced values so they can be quickly located. An index can either be created implicitly by Oracle9*i* or explicitly by a user.

inline view A temporary view of underlying database tables that exists only while a command is being executed. It is not a permanent database object and cannot be referenced again by a subsequent query.

inner join Joins that display data if there was a corresponding record in each table queried. *Equality joins*, *non-equality joins*, and *self-joins* are all classified as inner joins.

International Standards Organization (ISO) One of two industry-accepted committees that sets standards for SQL.

Julian date Represents the number of days that have passed between a specified date and January 1, 4712, b.c.

key-preserved table A table that contains the primary key that a view uses to uniquely identify each record displayed by the view.

keywords Words used in a SQL query that have a predefined meaning to Oracle9*i*. Common keywords include SELECT, FROM, and WHERE.

logical operators Used to combine two or more search conditions. The logical operators include AND and OR. The NOT operator reverses the meaning of search conditions.

manipulation functions Allow the user to control data (e.g., determine the length of a string, extract portions of a string) to yield a desired query output.

materialized view A view that allows you to store the data retrieved by the view query, and lets you reuse this data without executing the view query again.

multiple-column subquery A nested query that returns more than one column of results to the outer query. It can be listed in the FROM, WHERE, or HAVING clause.

multiple-row functions *See* **group functions**.

multiple-row subquery A nested query that returns more than one row of results to the parent query. It is most commonly used in a WHERE and HAVING clause and requires a multiple-row operator.

nesting One function is used as an argument inside another function.

non key-preserved table Does not uniquely identify the records in a view.

non-equality join Links data in two tables that do not have equivalent rows of data.

normal distribution When a large number of data values are obtained for statistical analysis, they tend to cluster around some "average" value This dispersion of values is called normal distribution.

normalization A multistep process that allows designers to take the raw data about an entity and evolve the data into a form that will reduce a database's data redundancy.

NULL value Means no value has been stored in that particular field. Indicates the absence of data, not a blank space.

object privileges Allow users to perform DML or retrieval operations on the data contained within database objects.

object relational database management system (ORDBMS) Oracle9*i* is an ORDBMS because it can reference individual data elements and objects composed of individual data elements.

optional keyword In a SQL query, the keyword AS is optional when creating a column alias.

outer join   Links data in tables that do not have equivalent rows. An outer join can be created in either the WHERE clause with an outer join operator (+) or by using the OUTER JOIN keywords.

outer join operator   The plus (+) symbol enclosed in parentheses, used in an outer join operation.

outer query   The main query in a SQL statement; a subquery passes its results back to the parent query, also known as the outer query. An outer query incorporates the value obtained from a subquery into its processing to determine the final output.

parent query   *See* **outer query**.

partial dependency   When the fields contained within a record are dependent on only one portion of the primary key.

primary key   A field that serves to uniquely identify a record in a database table.

private synonym   An alias used by an individual to reference objects owned by that individual. *See* **synonym**.

privileges   Allow database access to users. Oracle9*i* has *system privileges* and *object privileges*.

projection   Choosing specific column(s) in a SELECT statement.

public synonym   An alias that can be used by others to access an individual's database objects. *See* **synonym**.

query   A question posed to the database.

record   A collection of fields describing the attributes of one database element. In PL/SQL, a *record* is a composite datatype that can assume the same structure as the row being retrieved.

referential integrity   When a user refers to something that exists in another table, the REFERENCES keyword is used to identify the table and column that must already contain the data being entered.

regular expression   Allows the description of complex patterns in textual data.

relational database management system (RDBMS)   A software program used to create a relational database. It has functions that allow users to enter, manipulate, and retrieve data.

role   A group, or collection, of privileges. In most organizations, roles correlate to users' job duties.

row   A group of column values for a specific occurrence of an entity. In a database, records are commonly represented as rows.

schema   A collection of database objects owned by one user. By grouping objects according to the owner, multiple objects that have the same object name can exist in the same database.

secondary sort   When two or more columns are specified in the ORDER BY clause, data in the second column (or additional columns) provide an alternative field on which to order data if an exact match occurs between two or more rows in the first, or primary, sort.

second-normal form (2NF)   The second step in the normalization process in which partial dependencies are removed from database records.

selection   A process that allows you to see only the records that meet a certain condition or conditions.

self-join   Links data within a table to other data within the same table. A self-join can be created with a WHERE clause or by using the JOIN keyword with the ON clause.

sequence   A database object that generates sequential integers that can be used for an organization's internal controls. A sequence can also serve as a primary key for a table.

set operators   Combine the results of two (or more) SELECT statements. Valid set operators in Oracle9*i* are UNION, UNION ALL, INTERSECT, and MINUS.

shared lock   A table lock that lets other users access portions of a table but not alter the structure of the table. *See* **table lock**.

simple join   *See* **equality join**.

single value   The output of a single-row subquery.

single-row functions   Return one row of results for each record processed.

single-row subquery   A nested subquery that can return to the outer query only one row of results and consist of only one column. The output of a single-row subquery is a single value.

SQL*Plus®   The interface that allows users to issue SQL statements that create, maintain, and search stored data.

standard deviation   A calculation used to determine how closely individual values are to the mean, or average, of a group of numbers.

statistical group functions   Perform basic statistical calculations for data analysis. Oracle9*i*'s functions include standard deviation and variance.

string literal   Alphanumeric data, enclosed within single quotation marks, that instructs the software to interpret "literally" what has been entered and to show it in the resulting display.

structured query language (SQL)   The industry standard for interacting with a relational database. It is a data sublanguage, and unlike a programming language, it processes sets of data as groups and can navigate data stored within various tables.

subquery   A is a nested query—one complete query inside another query.

substitution variable   Instructs Oracle9*i* to use a substituted value in place of a specific variable at the time a command is executed. Used to make SQL statements or PL/SQL blocks interactive.

substring   A portion of a string of data.

synonym   An alternative name given to a database object with a complex name. Synonyms can be either private or public.

syntax   The basic structure, pattern, or rules for a SQL statement. For a SQL statement to execute properly, the correct syntax must be used.

system privileges   Allow access to the Oracle9*i* database and let users perform DDL operations such as CREATE, ALTER, and DROP database objects. An object privilege combined with the keyword ANY is also considered a system privilege.

system variable   *See* **environment variable**.

Systems Development Life Cycle (SDLC)   A series of steps for the design and development of a system.

table alias   A temporary name for a table, given in the FROM clause. Table aliases are used to reduce memory requirements or the number of keystrokes needed when specifying a table throughout a SQL statement.

third-normal form (3NF)   The third step in the normalization process in which transitive dependencies are removed from database records.

"TOP-N" analysis   When an inline view and a pseudocolumn ROWNUM are merged together to create a temporary list of records in a sorted order, then the top "n," or number of records, are retrieved.

transaction   A series of DML statements is considered to be one transaction. In Oracle9*i*, a transaction is simply a series of statements that have been issued and not committed. The duration of a transaction is defined by when a COMMIT implicitly or explicitly occurs.

transaction control   Data control statements that either save modified data or undo uncommitted changes made in error.

transitive dependency A means that at least one of the values in the record is not dependent upon the primary key, but upon another field in the record.

uncorrelated subquery A subquery that follows this method of processing: The subquery is executed, then the results of the subquery are passed to the outer query, and finally the outer query is executed.

unnormalized Refers to database records that contain repeating groups of data (multiple entries for a single column).

view Displays data in the underlying base tables. Views are used to provide a shortcut for users not having SQL training or to restrict users' access to sensitive data. Views are database objects, but they do not store data.

wildcard characters Symbols used to represent one or more alphanumeric characters. The wildcard characters in Oracle9*i* are the percent sign (%) and the underscore symbol (_). The percent sign is used to represent any number of characters; the underscore symbol represents one character.

INDEX

O

P